Organic Spectroscopic Structure Determination

Organic Spectroscopic Structure Determination

A Problem-Based Learning Approach

DOUGLASS F. TABER

OXFORD
UNIVERSITY PRESS
2007

Oxford University Press, Inc., publishes works that further
Oxford University's objective of excellence
in research, scholarship, and education.

Oxford New York
Auckland Cape Town Dar es Salaam Hong Kong Karachi
Kuala Lumpur Madrid Melbourne Mexico City Nairobi
New Delhi Shanghai Taipei Toronto

With offices in
Argentina Austria Brazil Chile Czech Republic France Greece
Guatemala Hungary Italy Japan Poland Portugal Singapore
South Korea Switzerland Thailand Turkey Ukraine Vietnam

Published by Oxford University Press, Inc.
198 Madison Avenue, New York, New York 10016

www.oup.com

Library of Congress Cataloging-in-Publication Data
Taber, D. F. (Douglass F.), 1948–
Organic spectroscopic structure determination : a problem-based learning approach / D. F. Taber.
p. cm.
ISBN 978-0-19-531470-0
1. Organic compounds—Structure. 2. Nuclear magnetic resonance spectroscopy. 3. Chemistry, Organic—Problems, exercises, etc. 4. Spectrum analysis—Problems, exercises, etc. I. Title.
QD272.S6T33 2007
547'.122—dc22 2006035525

9 8 7 6 5 4 3 2

Printed in the United States of America
on acid-free paper

Preface

The purpose of these exercises is to engage the imagination of the student. Too often, especially for those taking introductory classes, science is a set of facts to be memorized. I have found that even beginning organic students are able to deduce organic structures from simple data sets—and they enjoy doing it! This is designed to be a beginning book, a first introduction to the elucidation of molecular structures.

Whether working from a natural product extract or the results of an organic reaction, the first task is to isolate a pure organic substance. Once that substance is in hand, a variety of spectroscopic techniques will be applied, including advanced (including 2D) ^{1}H NMR methods, to develop a data set for the molecule. The third task is using that data set to assemble the final structure. This third task is introduced in this text.

At the University of Delaware, the sophomore organic students start working spectroscopic problems in the first week of class in September. I have found it convenient to teach this as part of the organic lab, covering organic spectroscopy in lab lecture, but the basic material outlined here could as easily be presented in the lecture part of the course. I have arranged the material in such a way that students can work the problems and learn the procedures on their own to minimize the time taken in lecture.

The first problems in this text can be solved from just the ^{13}C NMR data. I then add ^{1}H NMR and IR. Information is also included about mass spectrometry and ultraviolet for more advanced problems. Though the average sophomore organic class might not get to the most challenging problems included here, this text would also be appropriate for a more advanced organic spectrosopy course, where those problems may be needed. I have also found that the able sophomores enjoy the challenge of the advanced problems.

This text would be useful in a full-semester spectroscopy course, if used in conjunction with a more advanced text on acquisition of spectroscopic data, such as Crews, Rodriguez, and Jaspars, *Organic Structure Analysis* (Oxford University Press, 1998). For much more detailed tables of spectroscopic data, see Pretsch, Bühlmann, and Affolter, *Structure Determination of Organic Compounds: Tables of Spectral Data* (Springer-Verlag; Berlin, 2000).

As a service to instructors adopting this text, we will supply, on request, ten new spectroscopy problems each year to use for exams.

Contents

Section I
Strategies for Spectroscopic Structure Determination

1 ¹³C NMR

INTRODUCTION

Why are there wiggles?
And why are there squiggles?
It's all in the quantum mechanics . . .

Assume that you have a pure organic compound. How can you figure out the structure? You can't tell just by looking at it—it usually looks like a blob of vegetable oil, or it might be white and crystalline, like sugar. So you apply several spectroscopic techniques and deduce from the data—the images that the instrument made on a computer screen or a piece of chart paper—what the structure is. This is immediately enjoyable; within a week you will learn to decipher simple structures—and it can become a lifelong quest, figuring out more and more complicated systems.

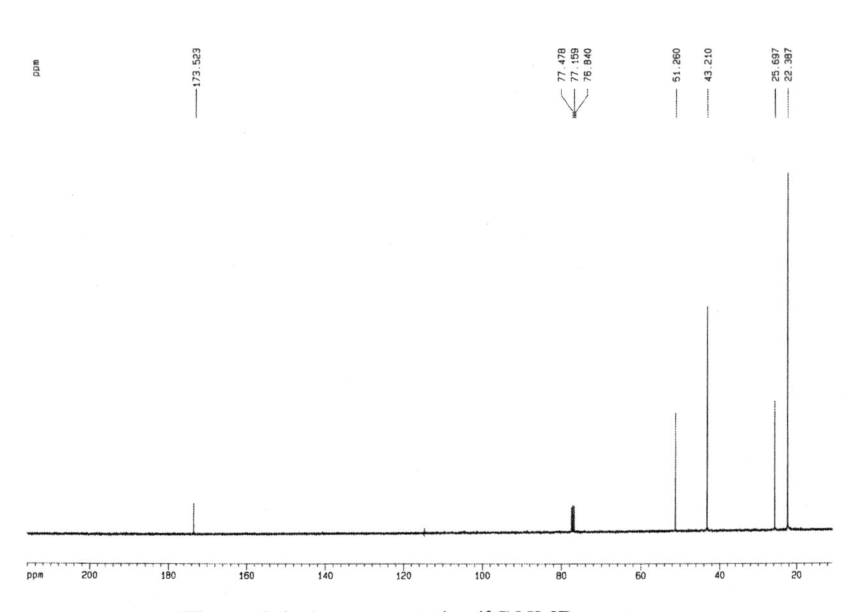

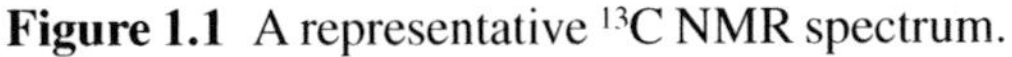

Figure 1.1 A representative ^{13}C NMR spectrum.

THE QUEST, LEVEL ONE: FINDING THE PIECES

This I know, that whereas I was blind, now I see.

The pieces that make up an organic molecule are the organic functional groups and the hydrocarbon framework. You will be able to find both the organic functional groups and the pieces of the hydrocarbon framework from ^{13}C nuclear magnetic resonance (NMR).

A ^{13}C NMR spectrum as it would come from the spectrometer is illustrated in Fig. 1.1.

Altogether, there are eight lines in this spectrum. The three centered around 77 are due to the solvent used, $CDCl_3$. The other five lines are due to the ester, as illustrated in Fig. 1.2.

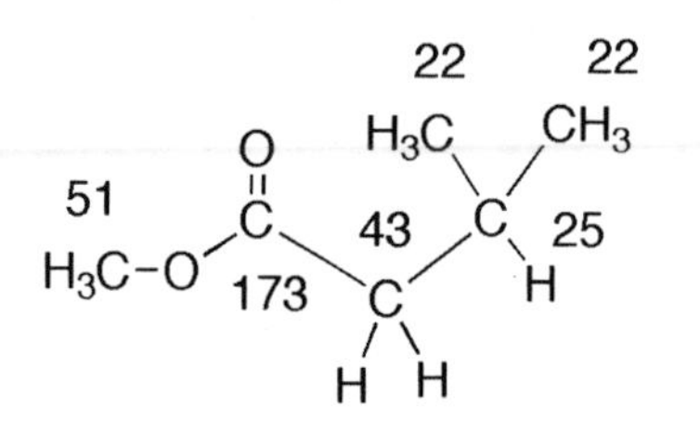

Figure 1.2 Methyl 3-methylbutanoate with ^{13}C chemical shifts.

These numbers (22, 25, 43, 51, and 173) are called the *chemical shift*, explained in more detail in the notes at the end of this chapter. In addition to the chemical shift, we would like to know how many H's are attached to each carbon. This information is summarized by single letters, *s* for zero H's attached, *d* for one, *t* for two, and *q* for three. The derivation of these letters, termed the *multiplicity*, is also explained in the notes at the end of this chapter. We would also like to know how many of each kind of carbon are present. Putting all of this information together, the ^{13}C NMR of the ester above is summarized in Fig. 1.3.

So, how do we use this information to find the organic functional groups in the molecule? Actually, there are two kinds of organic functional groups, those that have sp^2 or sp-hybridized carbons, and those that have only sp^3 hybridized carbons. We call the former unsaturated functional groups, and the latter saturated functional groups.

22, q (2)	51, q
25, d	173, s
43, t	

Figure 1.3 The ^{13}C NMR spectrum of methyl 3-methylbutanoate.

Representative unsaturated functional groups are illustrated in Fig. 1.4 and representative saturated functional groups in Fig. 1.5. For details beyond the examples here, see Tables C.1 through C.11 in section IV.

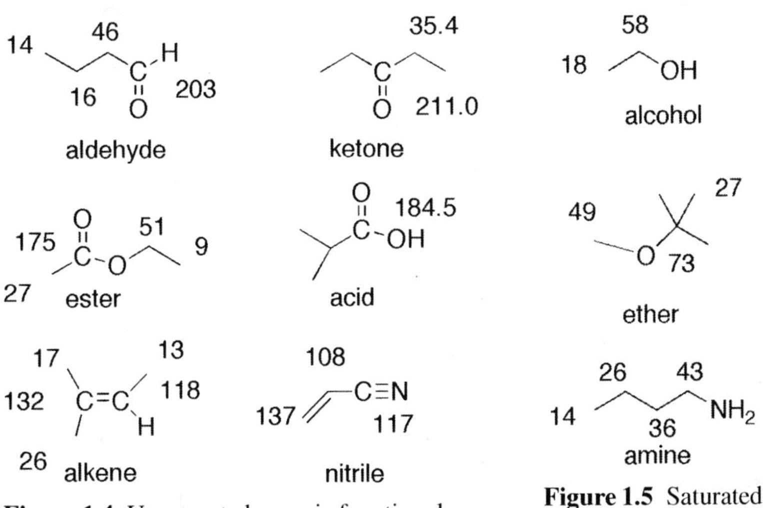

Figure 1.4 Unsaturated organic functional groups with ^{13}C chemical shifts.

Figure 1.5 Saturated organic functional groups with ^{13}C chemical shifts.

It is easy to tell if a molecule has any unsaturated functional groups by considering the molecular formula. The example at the beginning of this chapter has a molecular formula of $C_6H_{12}O_2$. A saturated hydrocarbon will have the formula C_nH_{2n+2}. If the ends of the hydrocarbon meet to form a ring, or if two H's are removed to make an alkene, the formula becomes C_nH_{2n}. This count is not affected by oxygen. The IHD (index of hydrogen deficiency) is the number of pairs of H's that are missing. This is the sum of double bonds plus rings in the molecule. For $C_{13}H_{16}O_2$, for instance, the IHD = 6. There are other considerations. Although oxygen does not affect the IHD, nitrogen does. Because N is trivalent, each nitrogen brings with it an extra hydrogen. For instance, the IHD of $C_9H_{14}N_2O_2$ is four. The H_{14} includes two extra H's, one for each of the N's. Halogens (F, Cl, Br, I) in the formula count as H's. In this example, one pair of H's is missing, so there is one ring or double bond in the molecule.

A double bond can be either a C=C bond or a C=O bond. There is O in the formula of the example, so there could be a C=O bond. From the examples above and the tables in section IV, we know that the C=O carbon of a ketone or aldehyde will come around 210, while the C=O carbon of an ester or acid will come around 170. We observe a singlet at 173, and we have two O's in the formula, so we know that we have either an ester or an acid, as illustrated in Fig. 1.6—but which is it?

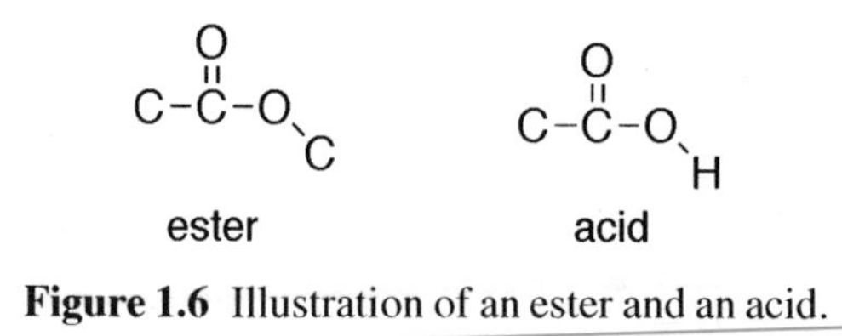

Figure 1.6 Illustration of an ester and an acid.

A key difference between an ester and an acid is that with an ester, all of the H's are attached to C. With an acid, one of the H's is attached to an O. We can count in the ^{13}C NMR spectrum how many H's are attached to each C by the multiplicity (s = 0 H, d = 1 H, t = 2 H, q = 3 H). In this example, if we add up all the H's attached to C, the total comes to 12. That is the same number of H's as in the formula, so there are no H's attached to O. Therefore, the unsaturated functional group in this example must be an ester.

Next, we explore around the unsaturated functional group. An ester has two carbons attached to it, one connected to the sp^2-hybridized C, and one connected to the sp^3-hybridized O. We would like to know how many H's are attached to each of these two C's. From Table C.9, we find that the C that is attached to the sp^3-hybridized O comes around 50 or 60. That must be the signal in our data at 51. That signal is a quartet (q), so there must be three H's attached to the C that is attached to the sp^3-hybridized O. Again from Table C.9, we observe that the C connected to the sp^2-hybridized C comes in the range 20–45. We conclude that this must be the signal from our data at 43. That signal is a triplet (t), so there must be two H's attached to the C that is connected to the sp^2-hybridized C of the carbonyl. Putting all of this together gives us the partial structure shown in Fig. 1.7.

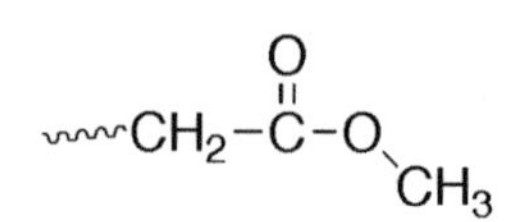

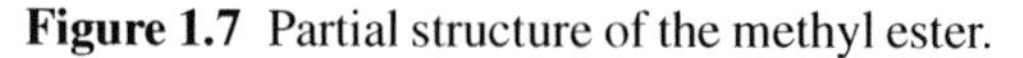

Figure 1.7 Partial structure of the methyl ester.

Saturated Functional Groups

After you are finished with all of the unsaturated functional groups, are there heteroatoms (O, N, S) in the formula that you have not yet dealt with? If so, now would be the time to find the saturated functional groups. In this example, we had two O's in the formula, and we have already accounted for both of them, so in this case there are no saturated functional groups.

Hydrocarbon Framework

Which signals in the carbon spectrum have you not yet accounted for? In this case, there are two carbons 22, q, so they are symmetrical. There is also 25, d, a C-H.

Putting It All Together

In this case, there is only one way to connect the pieces, so the structure must be that illustrated in Fig. 1.8.

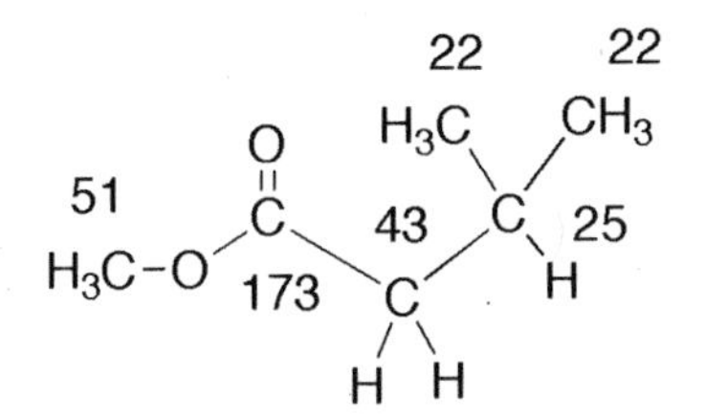

Figure 1.8 Full structure of the methyl ester.

To solve a structure, it is necessary to assemble a significant amount of data in an orderly way. This effort can often be confusing. To aid in this process, we use a worksheet. A sample of the worksheet is reproduced on p. xx. Especially in the early going, you may find it useful to make several copies of the worksheet and write directly on them.

Note that there is a pattern to ^{13}C chemical shifts—the fewer the H atoms on a carbon attached to the same functional group, the larger the chemical shift. Consider methyl ethyl ether (Table C.7). The methyl carbon directly attached to the oxygen (Fig. 1.9) comes at 57.6, whereas the methylene carbon directly attached to the oxygen comes at 67.9. If you study the examples in Table C.7 and the surrounding tables, you will see that this is a general pattern.

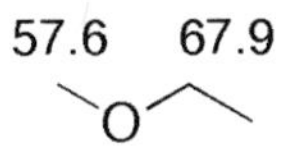

Figure 1.9 The ^{13}C NMR spectrum of methyl ethyl ether.

You should now be ready to work through Examples A–C.

Example A	Example B	Example C
$C_7H_{14}O$	$C_6H_{14}O$	$C_5H_6N_2$
23, q (2)	11, q (2)	16, t (2)
28, d	23, t (2)	22, t
29, q	44, d	119, s (2)
33, t	65, t	
42, t		
206, s		

As you follow the worksheets for each of these problems, you will find it instructive to draw structural isomers and see why they could not fit the data. After you complete these examples, you will be ready to use your worksheets to solve problems 1–15 in section II.

NOTES

1. Chemical shift: The NMR spectrum is measured by suspending the sample in a strong magnetic field. The nuclei then resonate at a particular radio frequency that depends on the field perceived by the nucleus. That magnetic field perceived is the sum of the external field, and the magnetic field due to the local structure around the nucleus being measured. Different carbon nuclei in the structure therefore resonate at different frequencies. The chemical shift is the difference between the resonance observed for a given carbon and the carbon of a molecule whose chemical shift is defined to be zero, $(CH_3)_4Si$, tetramethylsilane. This difference is expressed in parts per million of the external field.

2. Multiplicity: The H's attached to a C act as small magnets. If there is only one, it can be either aligned with the external field, or opposed to it. This creates two different magnetic environments for the C, so it appears as a doublet, d. If there are two H's on the C, they can both be in the same direction or in opposite directions, a total of three different magnetic environments for the C. It would appear as a triplet, t. The same reasoning makes a C with three H's attached to it four lines, a quartet, q. A C with no attached H's would appear as a singlet, s. The ^{13}C spectrum in Fig. 1.1 was recorded under conditions such that the C–H coupling was suppressed.

WORKSHEET

1. Molecular Formula

   IR

   UV

2. Index of Hydrogen Deficiency =

   Exploring around the unsaturated functional groups:

   \# >=O

   \# >=<

   \# sp (R-C≡N 105-120,s / —≡— 65-90)

   \# rings:

   ____

   IHD =

3. Other heteroatoms, exploring around the saturated functional groups:

4. Other pieces:

   a. Methyl groups

   b. ^{1}H NMR > 2.0

   c. Other carbons

5. Putting it all together:

WORKSHEET A

1. Molecular Formula $C_7H_{14}O$

   IR

   UV

2. Index of Hydrogen Deficiency = 1

   Exploring around the unsaturated functional groups:

   # C=O 1

   # C=C

   # sp (R-C≡N 105-120,s ; C≡C 65-90)

   # rings:

   IHD = 1

From the chemical shift (206) and multiplicity s), we know that we have a ketone, a carbonyl flanked by two carbons. By consulting Table C.8, it is apparent that there are two H's on one of the flanking carbons (42, t) and three on the other (29, q)

-CH2-C(=O)-CH3

3. Other heteroatoms, exploring around the saturated functional groups:

   no other heteroatoms

4. Other pieces:

   a. Methyl groups There are two additional methyl groups, and they are symmetrical. There is no other symmetry in the data.

   b. ^{1}H NMR > 2.0

   c. Other carbons 28, d 33, t

5. Putting it all together:

For detailed instructions for solving this problem, see http://valhalla.chem.udel.edu/SpecBook.pdf.

WORKSHEET B

1. Molecular Formula $C_6H_{14}O$

   IR there are 13 H's on carbon, so there must be one -OH

   UV

2. Index of Hydrogen Deficiency = 0

   Exploring around the unsaturated functional groups:

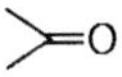

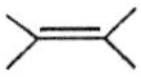

   # C=O

   # C=C

   # sp (R-C≡N 105-120,s ; C≡C 65-90)

   # rings:

   IHD =

3. Other heteroatoms, exploring around the saturated functional groups:

   From Table C.6, there are two H's on the carbon that has the alcohol $-CH_2-OH$

4. Other pieces:

   a. Methyl groups There are two methyl groups, that are symmetrical. There are also two symmetrical $-CH_2$-'s.

   b. ^1H NMR > 2.0

   c. Other carbons 44, d C-H

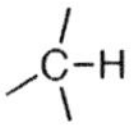

5. Putting it all together: OH

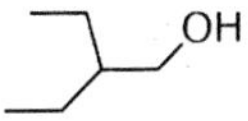

For detailed instructions for solving this problem, see http://valhalla.chem.udel.edu/SpecBook.pdf.

WORKSHEET C

1. Molecular Formula $C_5H_6N_2$

   IR

   UV

2. Index of Hydrogen Deficiency = 4

   Exploring around the unsaturated functional groups:

   # C=O

   # C=C

   # sp (R-C≡N 105-120,s / C≡C 65-90) 2 119, s (2)

   # rings:

   IHD = 2

3. Other heteroatoms, exploring around the saturated functional groups:

   There are no other heteratoms

4. Other pieces:

   a. Methyl groups

   b. ^1H NMR > 2.0

   c. Other carbons There are three CH_2-'s

5. Putting it all together: NC–CH2–CH2–CH2–CN

For detailed instructions for solving this problem, see http://valhalla.chem.udel.edu/SpecBook.pdf.

2 ^{1}H NMR and IR

THE QUEST, LEVEL TWO: CONNECTING THE PIECES TOGETHER

Hip bone connected to the thigh bone...

So far, you have learned how to find the functional groups in an unknown molecule. You have also learned to find the pieces of the molecule, and in some cases you have been able to find the carbons directly attached to the functional groups. For the compounds you examined so far, once you found the pieces, it was easy to put them together to get the final structure. With more advanced compounds, you will also need information from the ^{1}H NMR spectrum to settle on the structure. With ^{1}H NMR, you will be able to see how one piece of the unknown structure is connected to the other pieces. The theory of ^{1}H NMR (proton magnetic resonance) is discussed in detail in your organic text.

A typical ^{1}H NMR spectrum can be seen in Fig. 2.1.

In this text, a ^{1}H NMR spectrum will be summarized as shown in Fig. 2.2.

ANALYZING THE ^{1}H SPECTRUM

Taking the first entry of this spectrum as an example, with each entry you will see the chemical shift (0.87), the multiplicity (d), the number of H's at that chemical shift (6), and the coupling constants (J values), in this case = 6.5 Hz.

The Chemical Shift

The chemical shift d (parts per million downfield from the internal standard, tetramethylsilane) of the H's attached to a carbon is a function of the environment of those H's. These effects are summarized in Table H.1. In this case, 0.87 is typical for a methyl group that is not shifted by some other functional group.

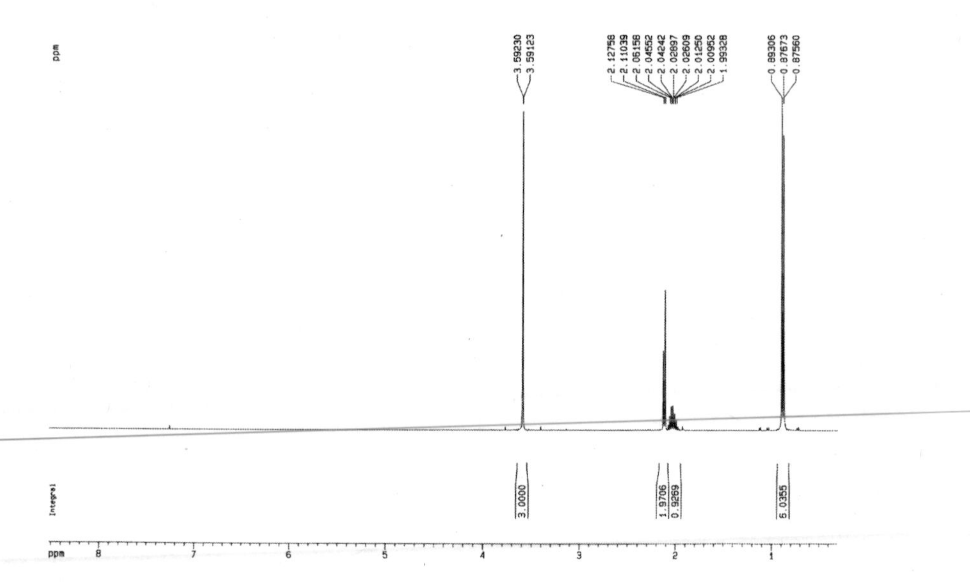

Figure 2.1 A representative ^{1}H NMR spectrum.

^{1}H NMR

0.87, d, 6H, J = 6.5 Hz
2.03, m, 1H
2.11, d, 2H, J = 6.9 Hz
3.59, s, 3H

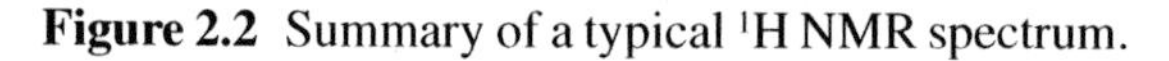

Figure 2.2 Summary of a typical ^{1}H NMR spectrum.

The Multiplicity

The multiplicity (d, in this case) is the number of lines in the signal. As with chapter 1, s = singlet, one line; d = doublet, two lines; t = triplet, three lines, q = quartet, four lines. In addition, you will see examples such as dd = doublet of doublets, and m = multiplet. In the ^{1}H NMR spectrum, multiple lines tell you the number of ***neighboring H's***. In this case, there would be one neighboring H, which would suggest the partial structure shown in Fig. 2.3. **Multiplicity** is explained in more detail under *coupling constant (J)* below.

The Number of H's—Integration

In the ^{1}H spectrum, it is typical to have two, three, or even more H's with the same chemical shift. The vertical distance on the integral of the signal at a given chemical shift is proportional to the peak area, and thus also to the number of H's having that chemical shift. If the total number of hydrogens is known, one can divide the total vertical integration by that number to get the vertical distance per hydrogen. In the summaries, you will be given the actual number of H's in the signal. In this first entry, there are six H's, so you might guess (corrrectly!) that this represents two methyl groups (see Fig. 2.3).

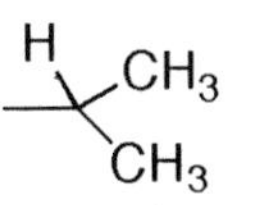

Figure 2.3 A fragment of the unknown structure.

Coupling Constant (J)

The effective magnetic field at a given nucleus is the sum of the imposed external magnetic field, H_0, added to all the smaller magnetic fields from surrounding nuclei. Consider the case (Fig. 2.4) of H_a, attached to carbon A, with one proton, H_b, on the adjacent carbon. The actual magnetic field experienced by H_a will be the sum of H_0 plus the field due to spin of H_b. On average, half the H_a's will see an H_b having a spin aligned with the external field, and half the H_a's will see an H_b with the spin opposed to the magnetic field. Thus, there will be two populations of H_a, and so two resonances. We say that the signal due to H_a is ***split***, because of coupling to H_b. The magnitude of the coupling, the *coupling constant (J)*, is measured in *Hertz* (=one cycle per second). For values of J in different situations, see Table H.2.

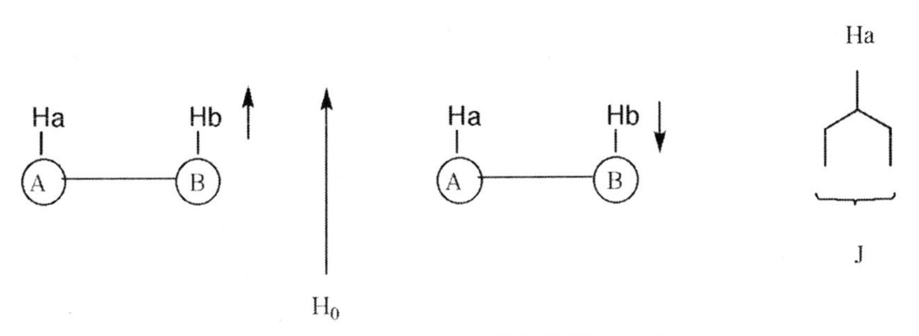

Figure 2.4 Illustration of ^{1}H NMR coupling.

H's with the same chemical shift will not show coupling. H's on the same carbon ("geminal") will show coupling if they have different chemical shifts, as, for instance, if the carbon is part of a ring (Fig. 2.5).

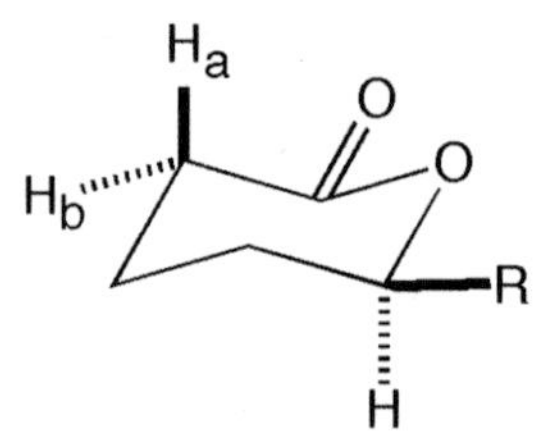

Figure 2.5 H_a and H_b are not equivalent. They will have different chemical shifts and will couple to each other.

If H's are coupled to each other, they will show the same numerical value for J ($J_{AB} = J_{BA}$). This is very useful for following connectivity in an unknown structure.

Multiplicity—The Number of Neighbors

If a proton is coupled to more than one other proton, the J values may or may not be equal. If they are equal, then the multiplicity of the signal for that first proton will be the number of hydrogens to which it is coupled with that J value plus one (Fig. 2.6). If there are no H's attached to the neighboring carbons, the ^1H signal will appear as a singlet.

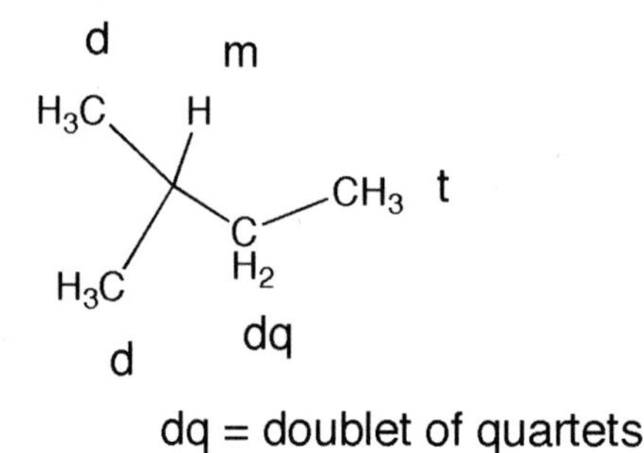

dq = doublet of quartets

Figure 2.6 The multiplicity of a ^1H NMR signal is the number of neighbors plus one.

Other Information in the ^1H Spectrum

In general, you should look for signals that stand out from the others (step 4 on the worksheet). Methyl groups often stand out, as do H's shifted downfield by adjacent functional groups. It is often possible to make a reasonable guess as to what that functional group is, first based on the ^{13}C data, and then using the ^1H chemical shift data in Table H.1. It should be remembered that a given proton could be adjacent to two or even three functional groups. As you will see, the proton shifts due to the nearness of two or more functional groups are very nearly additive.

Summary

The proton spectrum can be used in two ways: (a) confirming deductions from the ^{13}C spectrum about the presence and substitution pattern of organic functional groups in the molecule, and (b) establishing carbon–carbon connectivity.

Example A

$C_7H_{12}O$

^{13}C NMR	^{1}H NMR
206.7, d	1.08, s, 6H
133.1, d	2.21, d, J = 7.2 Hz, 2H
118.4, t	5.08, d, J = 11.8 Hz, 1H
45.7, s	5.11, d, J = 15.5 Hz, 1H
41.5, t	5.75, ddt, J = 11.8, 15.5, 7.2 Hz, 1H
21.2, q (2)	9.49, s, 1H

Example B

$C_{10}H_{20}O_2$

^{13}C NMR	^{1}H NMR
177.0, s	0.89, t, J = 7.3 Hz, 3H
64.4, t	1.26, d, J = 6.5 Hz, 6H
34.1, d	1.4, m, 6H
31.6, t	1.64, m, 2H
28.8, t	2.52, m, 1H
25.7, t	4.05, t, J = 7.1 Hz, 2H
22.6, t	
19.1, q (2)	
14.0, q	

Work these problems, then compare your answer to the completed Worksheet A and Worksheet B.

WORKSHEET A

1. Molecular Formula $C_7H_{12}O$

   IR

   from the ^{13}C spectrum, all H's are attached to carbon

   UV

2. Index of Hydrogen Deficiency = 2

   Exploring around the unsaturated functional groups:

   No H's on adjacent carbon

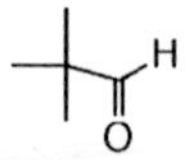

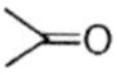

   \# C=O 1 d, 206.5, so must be aldehyde

   confirmed by aldehyde H at 9.49, s

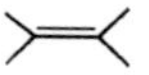

   \# C=C 1 133, d 118, t

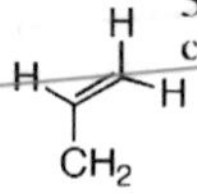

   5.08, d, J=11.8 Hz cis coupling constant Table H.2

   \# sp (R-C≡N 105-120,s ; C≡C 65-90)

   5.75, ddt

   so there must be a neighboring CH_2

   5.11, d, J=15.5 Hz trans coupling constant Table H.2

   \# rings:

   IHD = 2

3. Other heteroatoms - exploring around the saturated functional groups :

   There are no other heteroatoms

4. Other pieces:

   a. Methyl groups Two methyls, symmetrical

   b. ^1H NMR > 2.0 2.21, d, J = 7.2 Hz. This is the CH_2 next to the alkene, since the coupling constant is the same as that observed at 5.75

   c. Other carbons None

5. Putting it all together:

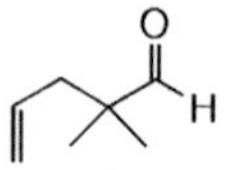

   Why is it not 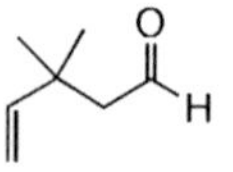 ?

For detailed instructions for solving this problem, see http://valhalla.chem.udel.edu/SpecBook.pdf.

WORKSHEET B

1. Molecular Formula $C_{10}H_{20}O_2$

   IR

   from the ^{13}C spectrum, all H's are attached to carbon

   UV

2. Index of Hydrogen Deficiency = 1

   Exploring around the unsaturated functional groups:

   # C=O 1 From the chemical shift, this must be a carboxylic acid derivative (Table C.9). Since the only other heteroatom is an O, and since all H's are attached to carbon, this must be an ester.

   # C=C

   # sp (R-C≡N 105-120,s; C≡C 65-90)

   # rings:

   64.4, t The C next to the sp^3 hybridized O has two H's. This comes at 4.05 in the ^1H NMR (Table H.1) as a t, so there are two H's on the adjacent C also

   The H next to the carbonyl comes (Table H.1) at 2.52

   IHD = 1

3. Other heteroatoms - exploring around the saturated functional groups:

   There are no other heteroatoms

4. Other pieces:

   a. Methyl groups — Three methyls, two of them symmetrical. There is no other symmetry in the molecule. The symmetrical methyls appear at 1.26 as a d, so they are adjacent to one H. There is only one methine in the molecule. The other methyl is adjacent to a CH_2.

   b. ^1H NMR > 2.0 — None

   c. Other carbons — There are a total of five CH_2's.

5. Putting it all together:

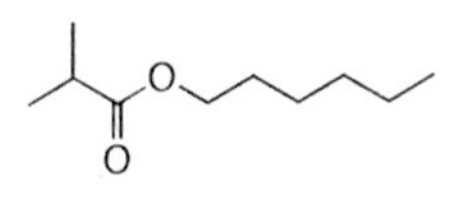

For detailed instructions for solving this problem, see http://valhalla.chem.udel.edu/SpecBook.pdf.

THE QUEST, LEVEL THREE: RINGS

Around and around and around we go . . .

You have assigned all of the unsaturated functional groups, but you still have IHD for which you have not accounted—you must have a ring! How do we approach rings?

Branch Points and End Groups

Imagine three cartoon rings. These cartoons are drawn as eight-membered rings, but are meant to be generic, any rings. The rings (Fig. 2.7) could be unbranched (A), singly branched (B), or multiply branched (C). How are you going to be able to tell?

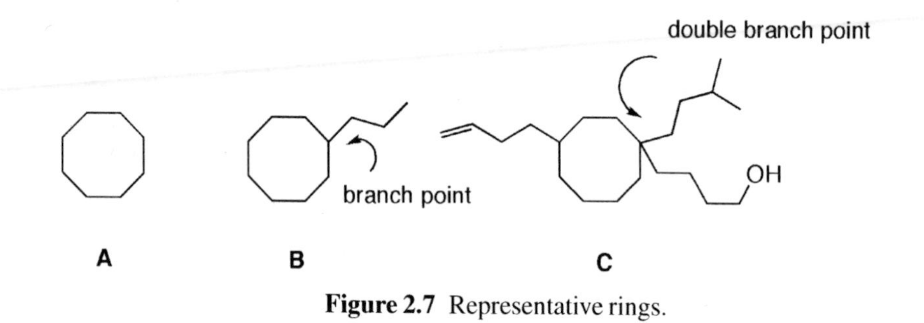

Figure 2.7 Representative rings.

We will use branch points and end groups to narrow the possibilities for our unknown ring. An end group is a group that ends a chain, such as a methyl group or a primary alcohol. A branch point is a methine, such as that in B. Example C also includes a methine, as well as an sp^3-hybridized quaternary carbon, a double branch point. Ring A has no methine or quaternary carbons, and it has no end groups, so there could not be any branches off the ring. Ring B has one branch point and one end group, so it must have one branch off the ring. Ring C has multiple branch points and end groups, including a branch point on a chain away from the ring. By considering how many branch points and end groups a molecule has, it is possible to more quickly come to a beginning idea of how to assemble the structure.

Infrared Spectroscopy (IR)

There are two uses for IR at this level of spectroscopic structure determination. The first is to confirm the presence of organic functional groups. The C–H stretch comes between 2800 and 3000 cm^{-1}, so everything will have that. Any IR signal >1600 cm^{-1}, except 2800–3000 cm^{-1}, should be accounted for. The alcohol OH stretch comes 3300–3400 cm^{-1}.

The alkyne C-H stretch comes at 2100–2150 cm^{-1}, and the aldehyde C–H stretch comes as a doublet at 2700–2800 cm^{-1}.

Although IR is diagnostic for a variety of organic functional groups, it is especially valuable for determing the ring size of cyclic esters (known as lactones) and cyclic ketones. Six-membered and larger lactones and ketones have the same IR as acyclic, 1705–1725 cm^{-1} for ketones and 1735–1755 cm^{-1} for esters and lactones. Cyclopentanones come at 1740–1750 cm^{-1}, and cyclobutanones come at 1760–1780 cm^{-1}. Five-membered lactones (γ-lactones) come at 1760–1780 cm^{-1}, and four-membered lactones (β-lactones) come at about 1820 cm^{-1}. Note that α,β-unsaturated carbonyl compounds in general absorb at 20 to 40 cm^{-1} lower frequency than the saturated analogs, so an α,β-unsaturated γ-lactone would come at about 1740 cm^{-1}.

Many of the problems from here on will have IR data. In step 5 of the worksheet, check prospective solutions against the IR. It may be possible that several otherwise acceptable alternative ways of assembling the unknown molecule will not be compatible with the IR spectrum.

After you have completed Example C and compared your answer to Worksheet C, you will be prepared to do problems 16–30 in section II. You will also be prepared to do many of the problems in section III.

Example C

$C_7H_{12}O_2$

IR: 2980, 2890, 1775, 1470, 1370, 1350, 1190, 1020, 980, 925 cm^{-1}

^{13}C NMR	^{1}H NMR
13.1, q	0.96, t, J = 6.8, 3H
17.9, t	1.4-1.9, m, 6H
27.3, t	2.4-2.6, m, 2H
28.1, t	4.5, m, 1H
36.9, t	
80.1, d	
176.6, s	

WORKSHEET C

1. Molecular Formula $C_7H_{12}O_2$ from the ^{13}C spectrum, all H's are attached to carbon

   IR 1775 cm^{-1}

   UV

2. Index of Hydrogen Deficiency = 2

   Exploring around the unsaturated functional groups:

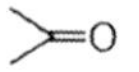

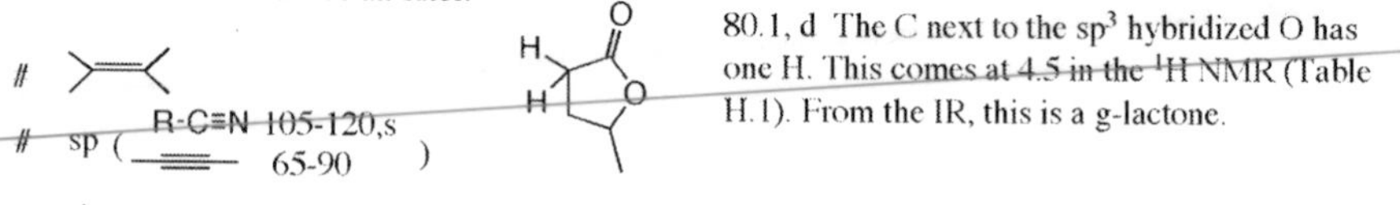

From the chemical shift, this must be a carboxylic acid derivative (Table C.9). Since the only other heteratom is an O, and since all H's are attached to carbon, this must be an ester.

80.1, d The C next to the sp^3 hybridized O has one H. This comes at 4.5 in the ^1H NMR (Table H.1). From the IR, this is a g-lactone.

rings:

The CH_2 next to the carbonyl comes (Table H.1) at 2.4-2.6.

IHD = 2

3. Other heteroatoms - exploring around the saturated functional groups:

   There are no other heteroatoms

4. Other pieces:

   The only methyl is adjacent to a CH_2.

   a. Methyl groups

   b. ^1H NMR > 2.0 None

   c. Other carbons There is only one branch point, the d at 80.1

5. Putting it all together:

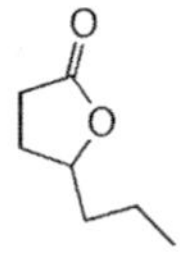

For detailed instructions for solving this problem, see http://valhalla.chem.udel.edu/SpecBook.pdf.

3 Calculating ^{13}C NMR, Mass Spectrometry, and UV

Down the rabbit hole, into Wonderland...

In this chapter, you will learn about more advanced concepts in structural elucidation, including the use of spectroscopic tables to predict ^{13}C and ^{1}H chemical shifts. These concepts, however, are just the beginning. Every day, the detailed three-dimensional structures of complex natural products are elucidated by professionals in the field. Leading references to this work, including detailed instruction on the several useful variations of NMR pulse sequences, are listed at the end of this chapter.

THE QUEST, LEVEL FOUR: CALCULATING ^{13}C CHEMICAL SHIFTS

Hear the calculators clicking...

Consider the unknown listed in Fig. 3.1. Using the approach outlined in chapter 1, take the time to work through this problem. You should be able to narrow it down to two structures, A and B. How could you tell these apart, just from the ^{13}C data? The fragments of the structure are listed in Fig. 3.2. The question is how to assemble those fragments. Two structures are possible, A and B. These are illustrated in Fig. 3.3. How can we tell which it is?

Unknown 3.1

$C_6H_{14}O$

^{13}C NMR
72.3, d
39.4, t
30.3, t
19.4, t
14.0, q
9.9, q

Figure 3.1 The data for unknown 3.1.

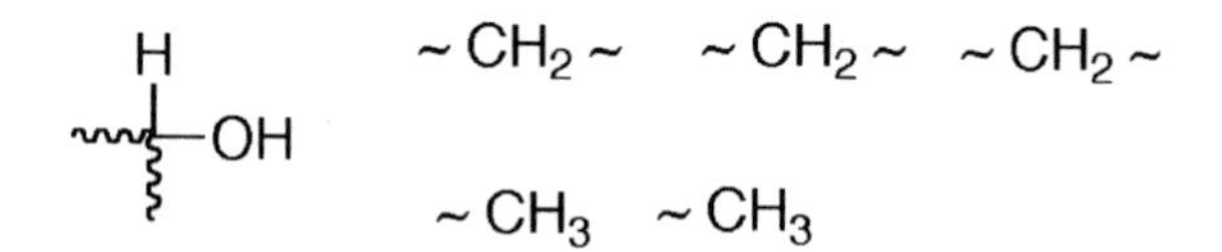

Figure 3.2 Structural fragments for unknown 3.1.

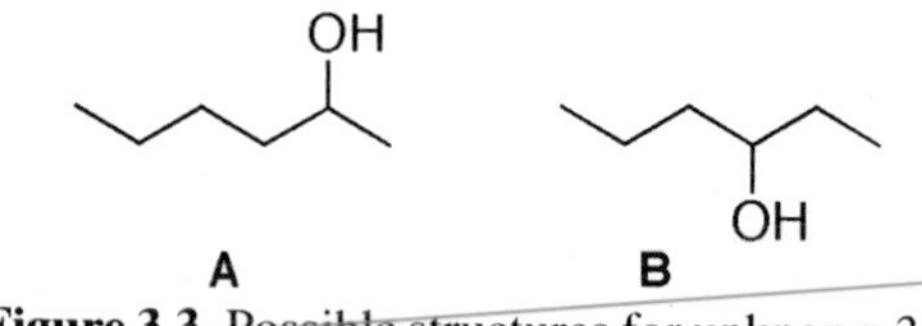

Figure 3.3 Possible structures for unknown 3.1.

To calculate ^{13}C chemical shifts, you will use the data in Tables C.10 and C.11. Table C.10 lists chemical shifts of representative hydrocarbons. Table C.11 lists the changes in chemical shifts induced by the attachment of functional groups.

To illustrate, both structure **A** and structure **B** (Fig. 3.3) are derived from *n*-hexane. The chemical shifts for *n*-hexane are listed in Fig. 3.4.

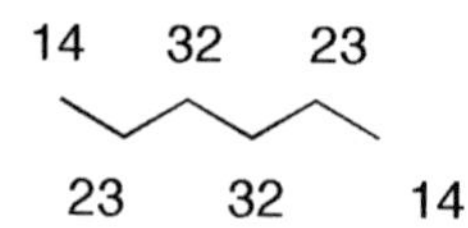

Figure 3.4 The chemical shifts for *n*-hexane.

Using the data in Table C. 11, we can calculate the chemical shifts expected for structure A (Fig. 3.5). Note that we use the internal shifts for the alcohol functional group, because we know that it is attached in the middle of the chain. The α shift refers to the carbon where the functional group is attached. The β carbon is next to the α carbon, and the γ carbon is next to the β carbon.

From Table C.11, we estimate the following chemical shifts (Fig. 3.5).

32 - 5 = 27 OH 23 + 41 = 64
14 14 + 8 = 22
23 32 + 8 = 40

Figure 3.5 Calculated ^{13}C chemical shifts for 2-hexanol (A).

Using this same approach, we can calculate the chemical shifts expected for structure B (Fig. 3.6).

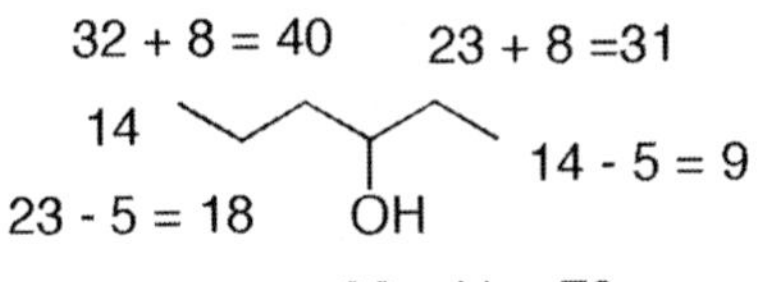

Figure 3.6 Calculated 13C chemical shifts for 3-hexanol (B).

A	B	3.1
14, q	9, q	9.9, q
22, q	14, q	14.0, q
23, t	18, t	19.4, t
27, t	31, t	30.3, t
40, t	40, t	39.4, t
64, d	73, d	72.3, d

Figure 3.7 Comparison of calculated ^{13}C chemical shifts for A and B with unknown 3.1.

With both sets of data calculated, it is easy to list each set in order and compare each list to the unknown (Fig. 3.6). It is apparent that B fits much more closely, so unknown 3.1 is 3-hexanol.

It would be a useful exercise at this point to practice by calculating the ^{13}C chemical shifts of the unknowns in chapters 1 and 2. In addition to the data in Tables C.10 and C.11, the structures in Tables C.1–C.9 can be used as starting hydrocarbons.

THE QUEST, LEVEL FIVE: ALKENES

How to find the number of alkenes? Count the number of alkene carbons and divide by two!

The ^{13}C chemical shifts of representative alkenes are summarized in Tables C.3 and C.4. A careful perusal of Table C.3 reveals a consistent pattern: alkene carbons bearing two hydrogens resonate in the range 105 to 120, those bearing one hydrogen resonate in the range 120 to 140, and those bearing no hydrogen resonate in the range 130 to 150.

In Table C.4, one observes some carbons that fall in the ranges specified, but many that do not. This is because these alkenes have either heteroatoms or electron-withdrawing groups directly attached to ("conjugated with") them. The chemical shift of a carbon is a function of the electron density observed at that carbon. An increase in electron density will cause a carbon to resonate at higher field (smaller δ = smaller chemical shift), whereas a decrease in electron density will cause the carbon to resonate at lower field (larger δ = larger chemical shift).

A carbonyl conjugated with an alkene shifts the β-carbon downfield. It does not affect the α-carbon. This can be rationalized by considering the polarization of the bonding electrons. The β-carbon has lost electron density, so it will resonate at lower field (larger numbers). The α-carbon is unchanged (Fig. 3.8).

A heteroatom conjugated with an alkene shifts *both* carbons. The α-carbon is shifted downfield (larger chemical shift) and the β-carbon is shifted upfield (smaller chemical shift). This can be rationalized by considering that the alkene electron cloud is repelled by the nonbonding electrons of the heteroatom, while at the same time the electronegative heteroatom withdraws electron density from the α-carbon. Take a moment to draw the resonance forms that illustrate these electron shifts.

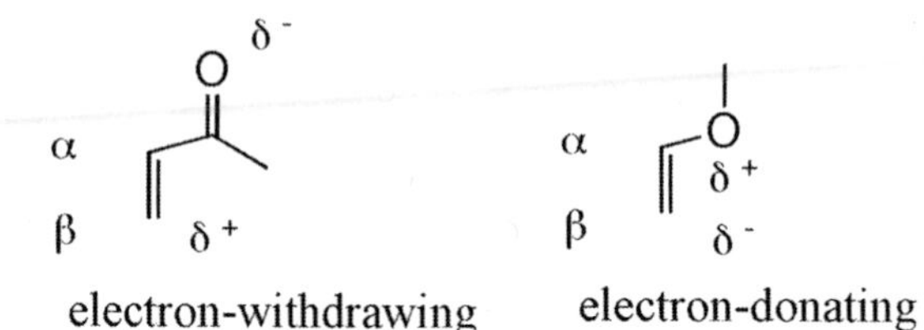

Figure 3.8 Polarization of alkenes by electron-withdrawing and electron-donating groups.

These effects are summarized by Fig. 3.9. Remember that "normal" chemical shift for an alkene carbon depends on how many alkyl groups are attached. From Tables C.3 and C.4, we can see that a "normal" alkene carbon with two H's attached would come at 105–120, a "normal" alkene with one H attached would come at 120–140, and a "normal" alkene with no H's attached would come at 130–150.

Note that each alkene has two carbons. If you think a signal upfield (e.g., 100) might be half of a heteroatom-polarized alkene, look for the other alkene carbon shifted downfield. If it is not there, then that signal at 100 is not an alkene carbon.

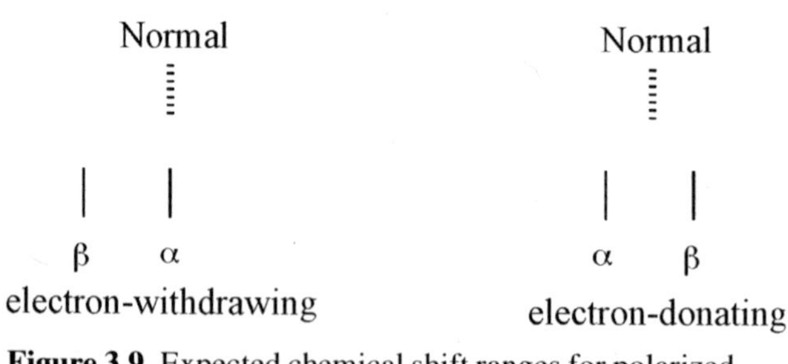

Figure 3.9 Expected chemical shift ranges for polarized and nonpolarized alkenes.

Calculating the 1H NMR Chemical Shifts of Alkenes

^{1}H NMR shifts of alkenes are easily calculated using Table H.6. Note that for polarized alkenes, the H's are shifted the same way as the carbons. For carbonyl-polarized alkenes, the β-H's are shifted downfield, and the α-H's are normal. For heteroatom-polarized alkenes, the α-H's are shifted downfield, and the β-H's are shifted upfield. Note (Table H.2) that H's *trans* on an alkene share a J value of about 17 Hz, whereas H's *cis* on an alkene share a J value of about 10 Hz. Fig. 3.10 is the diagram on which Table H.6 is based, and Fig. 3.11 is an example that illustrates the use of the data in Table H.6.

Starting value = 5.25, then add incremental shift for Gem, Cis, Trans

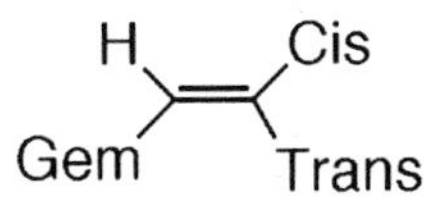

Figure 3.10 Diagram on which Table H.6 is based.

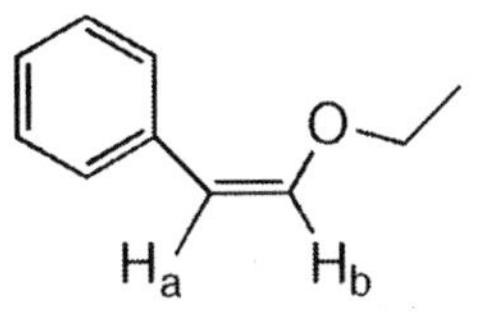

Figure 3.11 Using Table H.6 to calculate the ^{1}H NMR chemical shifts of an alkene.

Practice using Table H.6 with the alkene illustrated (Fig. 3.11). The starting value is always 5.25. H substituents neither add to nor subtract from this value. For H_a and H_b, then, we need only to consider the shifts due to the geminal substituent and to the trans substituent. The predicted chemical shift for H_a is then $5.25 + 1.34 - 1.28 = 5.32$, and for H_b the predicted chemical shift is $5.25 + 1.18 - 0.10 = 6.33$.

THE QUEST, LEVEL SIX: BENZENE DERIVATIVES

^{13}C of Benzene Derivatives

The ^{13}C chemical shift of a carbon in a benzene derivative can be calculated using the data in Table C.12. This can best be illustrated with an example (Fig. 3.12).

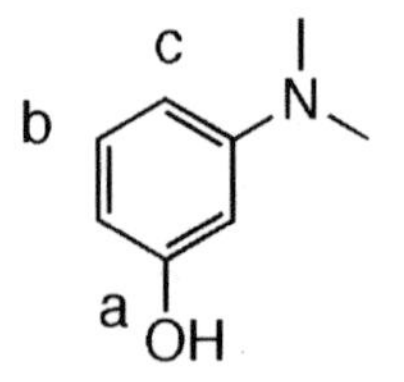

Figure 3.12 Using Table C.12 to calculate the ^{13}C NMR chemical shifts of a benzene derivative.

Carbon a: The starting value for any benzene carbon is 128.5, from the data in Table C.12. We then add the incremental contribution from each substituent. The increment shift from H is zero, so we only need to account for effects due to substituents on the ring. For carbon a there are two substituents, a C-1 OH and a C-3 -NMe_2. The C-1 OH contributes +26.6. The C-3 -NMe_2 contributes +0.8. The calculated chemical shift for carbon a is therefore $128.5 + 26.6 + 0.8 = 155.9$.
Carbon b: The starting value for carbon b is 128.5. With regard to b, the –OH is at C-3 and contributes +1.6. The -NMe_2 substituent is at the other C-3, and so contributes an increment of +0.8. The calculated chemical shift for carbon b is therefore $128.5 + 1.6 + 0.8 = 130.9$.
Carbon c: The starting value for carbon c is 128.5. With regard to c, the –OH is at C-4 and contributes–7.3. The -NMe_2 substituent is at C-2, and so contributes an increment of–15.7. The calculated chemical shift for carbon c is therefore $128.5 - 7.3 - 15.7 = 105.5$.

Often, the only way to decipher the substitution pattern on a highly substituted benzene derivative is to calculate the chemical shifts of the aromatic carbons for each possible substitution pattern, and compare the calculated values to the data given. Remember to consider multiplicity as well as chemical shift in making these comparisons.

^1H NMR of Benzene Derivatives

The ^1H NMR chemical shifts of a benzene derivative can be calculated using the data in Table H.4. These are approximately additive. Again, the use of this table can best be illustrated with an example (Fig. 3.13).

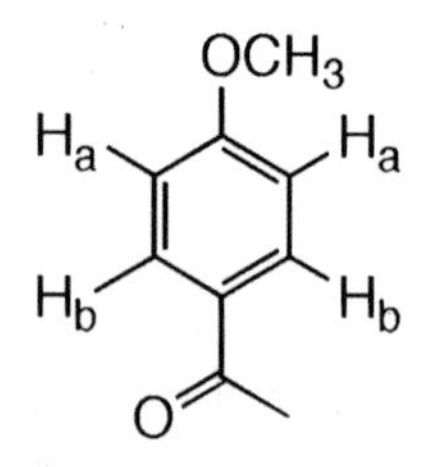

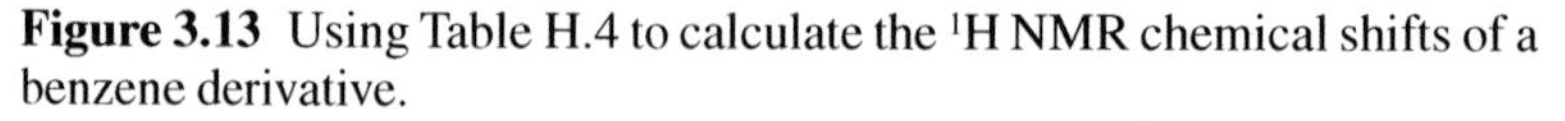

Figure 3.13 Using Table H.4 to calculate the ^1H NMR chemical shifts of a benzene derivative.

The H's on an unsubstituted or alkyl-substituted benzene (first entries in Table H.4) come at about 7.26. A proton *ortho* ("1,2-") to an $-OCH_3$ would, according to this table, come at about 7.0, a shift upfield of 0.26. A proton *meta* ("1,3") to a ketone carbonyl would come at 7.5, a downfield shift of 0.24. It follows that H_a should resonate at about $7.26 - 0.26 + 0.24 = 7.24$. For H_b, a proton *meta* to an $-OCH_3$ would, according to Table H.4, come at about 7.35, a shift downfield of 0.09. A proton *ortho* to a ketone carbonyl would come at 7.9, a downfield shift from 7.26 of 0.64. It follows that H_b should resonate at about $7.26 - 0.09 - 0.64 = 7.99$.

H's *ortho* to each other (Table H.2) have a 9-Hz coupling constant, H's *meta* to each other have a 3-Hz coupling constant, and H's *para* to each other do not couple. Remember that H's with the same chemical shift (even if not chemically the same) do *not* couple to each other.

For the example illustrated, H_a and H_b would share a 9-Hz coupling constant. There would be no *meta* coupling, because H's with the same chemical shift do not couple to each other. The aromatic portion of the ^1H spectrum would then be summarized: 7.24, 2H, d, J = 9.0 Hz; 7.99, 2H, d, J = 9.0 Hz.

Solving Problems with Alkenes and Arenes

The key to solving problems with alkenes and arenes is in step 2 on the worksheet "Assigning and Exploring around the Unsaturated Functional Groups." First establish the alkene carbon count, and, where possible, try to pair the alkene carbons. The first clue that an unknown might contain a benzene ring will come when the unknown has at least one ring and at least three double bonds. Confirmation comes if there are also arene-type H's (6.5–8.0).

It is important at this point to figure out the substitution pattern of the arene, especially if there are more than three double bonds in the molecule. Count the arene-type hydrogens. There are six positions on the benzene ring. If there are only three aromatic H's, then there must also be three substituents on the ring.

Once you know how many substituents there are, look for symmetry. For instance, a disubstituted aromatic that has only two bands of H's (two doublets, each 2H, J = 9 Hz) must be *para*-disubstituted. Then decide what the substituents might be. This can often be deduced from the chemical shifts of the arene H's, and other information about functional groups in the molecule. Once the substituents on the benzene ring are known and you have deduced the pattern of attachment, it should be possible to calculate the approximate chemical shifts (and multiplicity) of the arene carbons. Once these carbon signals are subtracted from the spectrum, assignment of the residual alkenes should be more straightforward.

Example A

$C_{12}H_{22}O_2$

IR: 2926, 2856, 1728, 1645, 1436, 1197, 1175, 819 cm^{-1}

^{13}C NMR		^{1}H NMR
166.5, s	29.0, t	6.14, dt, 1H, J = 7.5, 11.5 Hz
150.7, d	28.9, t	5.68, d, 1H, J = 11.5 Hz
119.0, d	28.8, t	3.61, s, 3H
50.6, q	22.5, t	2.57, dt, 2H, J = 7.5, 7.4 Hz
31.7, t	13.9, q	1.35, m, 2H
29.3, t		1.18, m, 10H
29.1, t		0.80, m, 3H

Example B

$C_{11}H_{16}O_2$

IR: 2980, 1615, 1585, 1470, 1370, 1305, 1188, 1040, 835 cm^{-1}

^{13}C NMR	^{1}H NMR
157.3, s	1.27, s, 9H
143.1, s	3.67, s, 3H
126.0, d (2)	6.80, d, J = 8.5 Hz, 2H
113.3, d (2)	7.25, d, J = 8.5 Hz, 2H
54.8, q	
33.9, s	
31.4, q	

WORKSHEET A

1. Molecular Formula $C_{12}H_{22}O_2$ from the ^{13}C spectrum, all H's are attached to carbon

   IR 1728 cm^{-1} 1645 cm^{-1}

   UV

2. Index of Hydrogen Deficiency = 2

   Exploring around the unsaturated functional groups:

   # C=O 1

   # C=C 1

   # sp (R-C≡N 105-120,s ; C≡C 65-90)

   # rings:

   IHD = 2

From the chemical shift, this must be a carboxylic acid derivative (Table C.9). Since the only other heteroatom is an O, and since all H's are attached to carbon, this must be an ester. From 50.6 q and 3.61, s, it is a methyl ester. The IR is only 1728 cm^{-1} however, suggesting α,β-unsaturation. The alkene stretch at 1645 cm^{-1} supports this, as does the ^{13}C NMR, with one normal alkene CH at 119 and one shifted downfield at 150.

H H CH2 OCH3 O

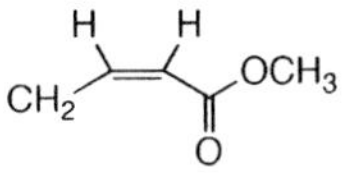

The alkene H's share J=11.5, so cis

The downfield alkene H is a triplet, so there is a flanking CH_2

3. Other heteroatoms - exploring around the saturated functional groups:

   There are no other heteroatoms

4. Other pieces:

   a. Methyl groups The only methyl is adjacent to a CH_2.

   b. ^1H NMR > 2.0 2,57 is the allylic CH_2

   c. Other carbons there are an additional six CH_2's

5. Putting it all together:

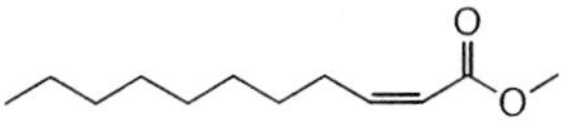

For detailed instructions for solving this problem, see http://valhalla.chem.udel.edu/SpecBook.pdf.

WORKSHEET B

1. Molecular Formula $C_{11}H_{16}O$ from the ^{13}C spectrum, all H's are attached to carbon

   IR 1615 cm^{-1} ?

   UV

2. Index of Hydrogen Deficiency = 4

   Exploring around the unsaturated functional groups:

   # C=O 0

   With three alkenes and a ring, a benzene derivative is likely. H's in the aromatic range (~ 7.0) confirm this. From the coupling pattern, this is a 1,4-disubstituted benzene.

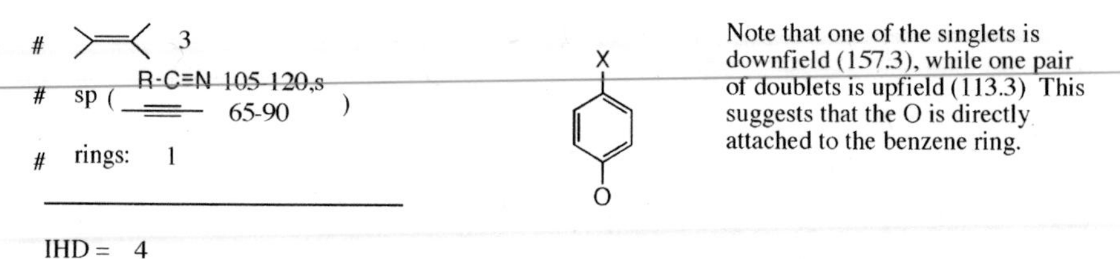

Note that one of the singlets is downfield (157.3), while one pair of doublets is upfield (113.3) This suggests that the O is directly attached to the benzene ring.

IHD = 4

3. Other heteroatoms - exploring around the saturated functional groups:

The O must be part of an ether, since all H's are attached to C. The other end of the ether is at 54.8, q, so this is - OCH_3

4. Other pieces:

   a. Methyl groups There are three symmetrical methyls, attached to a carbon with no H

   $-C(CH_3)_3$

   b. ^1H NMR > 2.0

   c. Other carbons

5. Putting it all together:

   $H_3C-C(CH_3)(CH_3)-C_6H_4-O-CH_3$

For detailed instructions for solving this problem, see http://valhalla.chem.udel.edu/SpecBook.pdf.

THE QUEST, LEVEL SEVEN: MASS SPECTROMETRY

A mass spectrometer is illustrated schematically in Fig. 3.14. A sample introduced into the source is vaporized and then ionized (an electron is removed). The resultant group of ions (some of which are falling apart!) is accelerated from the source. This group of ions then encounters a magnetic field. In that field, the moving ions are deflected by an amount inversely proportional to the mass/charge (m/z) ratio. Because most ions carry a single positive charge, they are thus separated, when they reach the detector, according to mass. An ion current is then measured at each value of m/z, and the result tabulated. That tabulated result is the ***mass spectrum***.

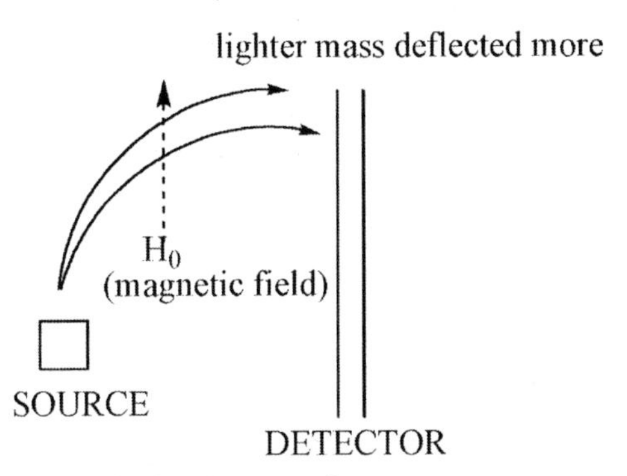

Figure 3.14 Diagram of a mass spectrometer.

Some portion of that ion current will be due to the molecular ion, the ion that is the entire molecule. Other "fragment" ions are also generated. In this section, we will learn how to use these fragment ions to assist in deducing the structure of an unknown. There are four processes that are important to understand: α-cleavage, β-cleavage, McLafferty rearrangement, and decarbonylation.

α-Cleavage

In the ionization process, an electron is removed from the molecule. If there is a heteroatom in the molecule, one of the nonbonding electrons on the heteroatom is likely to be the most easily removed. This gives a radical cation, the molecular ion. Radical cleavage (Fig. 3.15) of the adjacent carbon-carbon bond then gives a stable carbocation, a fragment ion.

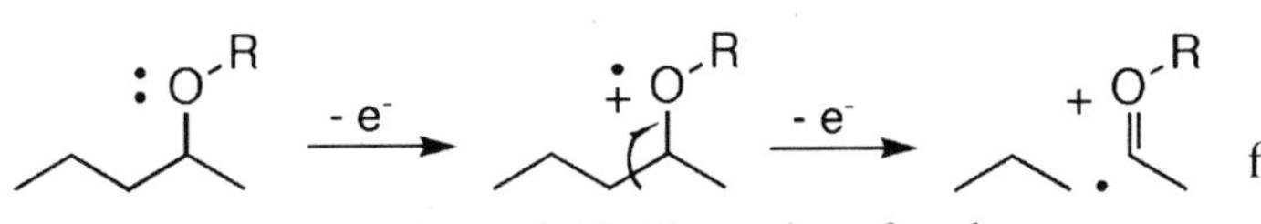

Figure 3.15 Illustration of α-cleavage.

β-Cleavage

Another process that is important, especially in cyclic systems, is β-cleavage. For example, α-cleavage of the molecular ion in a cyclic ketone, as shown in Fig. 3.16, gives a new species that has the same molecular weight as the molecular ion. Radical cleavage of the carbon–carbon bond β to the radical center then gives a fragment ion. β-Cleavage can proceed whenever there is a carbon–carbon bond β to a radical center.

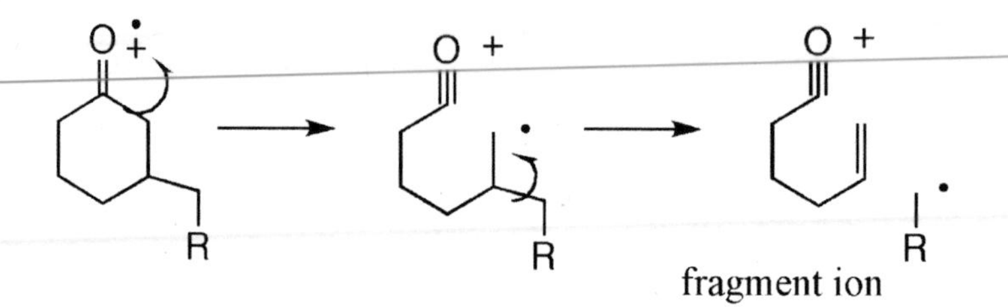

Figure 3.16 Illustration of β-cleavage.

McLafferty Rearrangement

When a carbonyl derivative has a proton on a carbon three carbons from the carbonyl, McLafferty rearrangement can proceed (Fig. 3.17). The product fragment ion from McLafferty rearrangement is itself the molecular ion of a new ketone, which can then proceed with α-cleavage, β-cleavage, or McLafferty rearrangement. The fragments resulting from McLafferty rearrangement stand out in the mass spectrum because (if the starting molecular weight is even) they are of *even mass*. Ions resulting from α-cleavage and β-cleavage, though more abundant, are all *odd mass*.

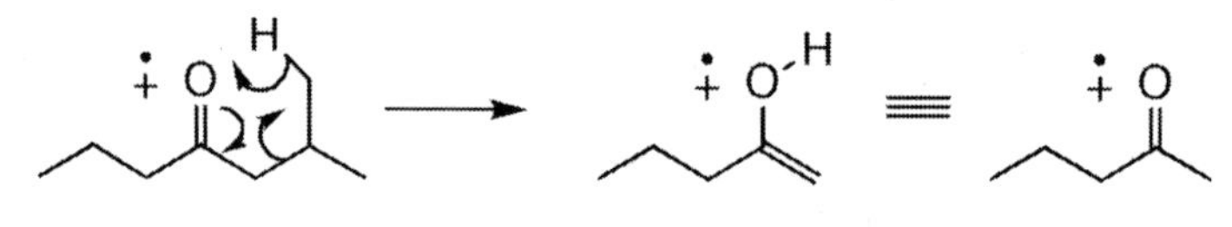

Figure 3.17 Illustration of McLafferty rearrangement.

Decarbonylation

The product of α-cleavage of a ketone, as shown in Fig. 3.18, is an acylium ion. Such an ion will often go on to lose carbon monoxide ("decarbonylation"), to give a new fragment ion. Such an ion will often fragment further to lose methylene units, as illustrated.

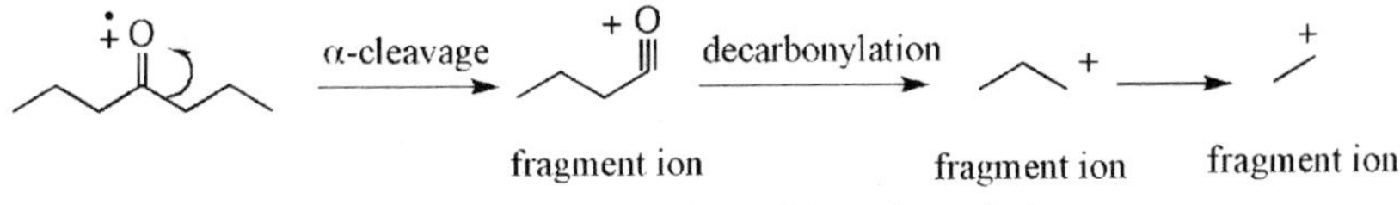

Figure 3.18 Illustration of decarbonylation.

Analysis of the Mass Spectrum

Although a variety of other fragmentation and rearrangement processes can and do proceed in the mass spectrometer, the four processes just outlined will often be sufficient to rationalize most of the prominent fragment ions. Use the mass spectrum in step 5 of the worksheet, focusing on fragmentations that differentiate between alternative structures. This is illustrated by the unknown whose data are summarized in Fig. 3.19.

$C_{19}H_{39}N$

MS: 280 (M+ - H, 4), 266 (10), 124 (3), 111 (5), 98 (100)

^{13}C NMR		^{1}H NMR
50.7, d	26.3, t	3.07, 1H, m
45.7, d	22.6, t	2.88, 1H, m
33.9, t	21.0, q	2.05, 1H, bs (exchanges)
32.8, t	19.4, t	1.6, 8H, m
31.8, t	14.0, q	1.3, 22H, m
30.6, t		1.07, 3H, d, J = 7.1 Hz
30.4, t (2)		0.88, 3H, t, J = 6.6 Hz
29.7, t (2)		
29.6, t (2)		
29.2, t (2)		

Figure 3.19 Data for unknown.

From the data, this is a secondary amine, with two methines (C–H) attached to the N. The methyl group that is a doublet must also be attached to one of those same methines. The unknown has a ring, and a side chain that ends in a methyl that is attached to a $-CH_2-$. This approximate structure is summarized in Fig. 3.20. The question is, how many methylenes are in the ring, and how many are in the side chain?

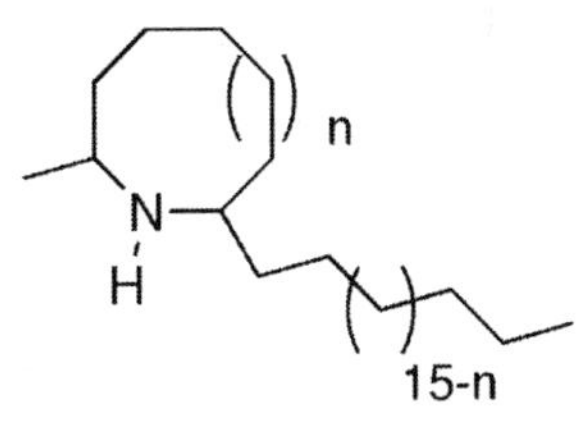

Figure 3.20 The approximate structure of the unknown.

The final structure can be found by analyzing the mass spectrum. α-Cleavage of the radical cation derived from the amine would proceed on either side. When the methyl group is lost, an ion of m/z = 266 will be generated. This is observed, but that does not tell us anything new—we already knew that the methyl group was attached to one of the methines. Loss of the other side chain would mean loss of the terminal methyl on that side chain, plus the accompanying methylenes. If the side chain were ethyl, for instance, α-cleavage would generate a fragment having m/z = 252. Loss of a propyl group would give m/z = 238. Working down, we eventually come to m/z = 98. This is loss of a C_{13} side chain, so the structure of unknown must be as depicted in Fig. 3.21.

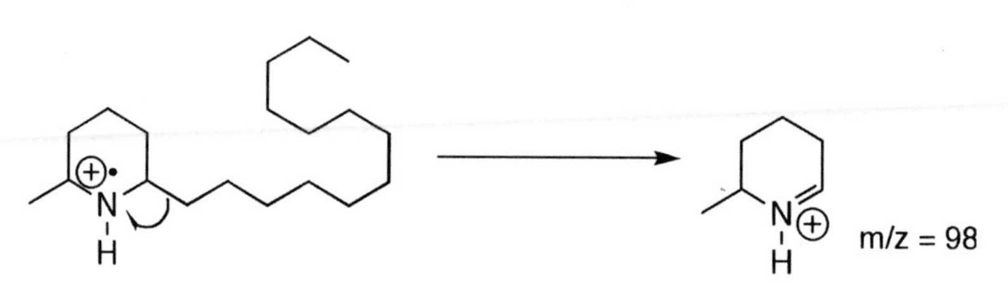

Figure 3.21 The structure of the unknown and its α-cleavage.

You are now ready to work on Example C.

Example C

$C_{16}H_{20}O_2$

IR: 2922, 1673, 1245, 755, 692 cm^{-1}

MS: 244 (M+, 43), 151 (42), 121 (16), 109 (100)

^{13}C NMR		^{1}H NMR
199.8, s	32.3, t	7.24 - 7.29, 2H, m
158.7, s	32.1, t	6.86 - 6.94, 3H, m
144.6, d	26.2, t	6.70, 1H, t, J = 4.5 Hz
135.4, s	15.6, q	3.93, 2H, t, J = 6.2 Hz
129.3, d (2)		2.55, 1H, dd, J = 11.4, 8.2 Hz
120.5, d		2.44, 1H, m
114.2, d (2)		2.10, 3H, m
67.4, t		1.80, 2H, m
44.4, t		1.77, 3H, s
35.3, d		1.52, 2H, m

WORKSHEET C

1. Molecular Formula $C_{16}H_{20}O_2$ from the ^{13}C spectrum, all H's are attached to carbon

   IR 1673 cm^{-1}

   UV

2. Index of Hydrogen Deficiency = 7

   Exploring around the unsaturated functional groups:

   # C=O 1

   # C=C 4

   # sp (R-C≡N 105-120,s; C≡C 65-90)

   # rings: 2

   IHD = 7

With three alkenes and a ring, a benzene derivative is likely. H's in the aromatic range (~ 7.0) confirm this. From the coupling pattern, this is a monosubstituted benzene. Note that one of the singlets is downfield (157.3), while one pair of doublets is upfield (113.3) This suggests that the O is directly attached to the benzene ring.

There is also an alkene, with a normal singlet and a downfield doublet, so the ketone is conjugated with the alkene.

6.70, t

3. Other heteroatoms - exploring around the saturated functional groups:

The O must be part of an ether, since all H's are attached to C. The other end of the ether is at 67.4, t, and 3.93, 2H, t, so this is - O-CH_2-CH_2

4. Other pieces:

   a. Methyl groups — There is one methyl, at 1.77, so this is attached to the alkene

   b. ^{1}H NMR > 2.0 — 2.55, dd, 1H and 2.44, m 1H and 44.4, t tell us that there are two H's adjacent to the ketone, and there is one H on the next carbon

   c. Other carbons — There is one more CH_2

5. Putting it all together:

The ketone must be in the ring. From the IR, it must be at least a six-membered ring. This gives us two possibilities, A and B. From the mass spec, it is A.

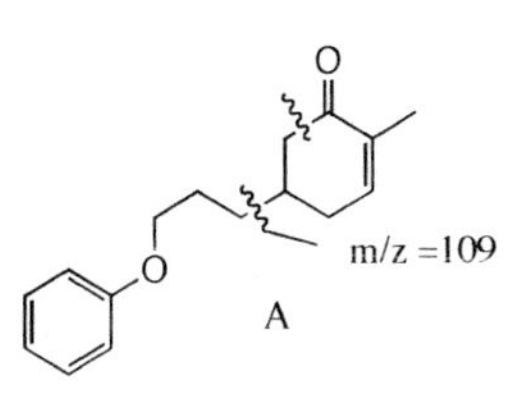

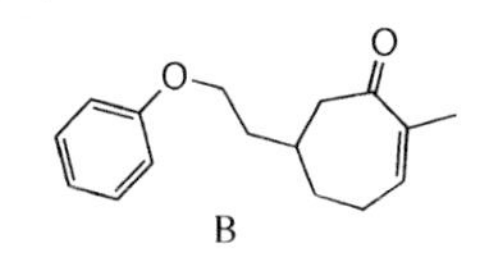

For detailed instructions for solving this problem, see
http://valhalla.chem.udel.edu/SpecBook.pdf.

THE QUEST, LEVEL EIGHT: ULTRAVIOLET (UV)

When two or more double bonds are conjugated, the bonding electrons are delocalized into *molecular orbitals*. Two or more double bonds in conjugation constitute a *UV chromophore*, illustrated in Fig. 3.22.

Figure 3.22 A representative UV chromophore.

The bonding electrons will be used to fill low-energy *bonding orbitals* (Fig. 3.23). When a photon of UV light is absorbed, an electron is promoted from a bonding orbital to an *antibonding* (high-energy) *orbital*.

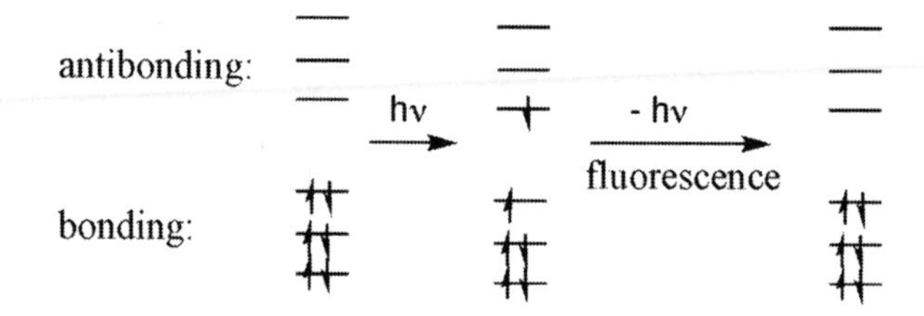

Figure 3.23 A molecular orbital diagram.

To get back to the ground state, a molecule must lose some energy. It can do this mechanically, by bumping into other molecules, or it can do it by emitting a photon of light. The latter process is called *fluorescence*.

Superimposed on both the ground-state and excited-state energy levels are a series of vibrational energy bands. As a result, UV absorption becomes a broad band, as illustrated in Fig. 3.24.

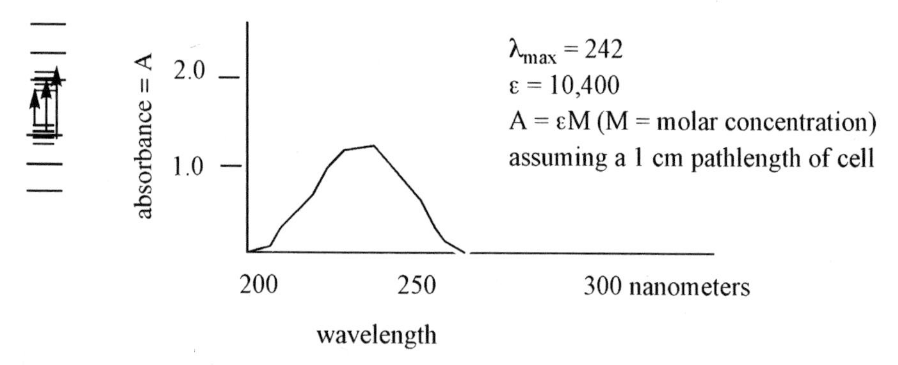

Figure 3.24 A typical UV spectrum.

UV chromophores are characterized by λ_{max}, the highest point on the UV band (maximum absorbance), and ε (molar absorbtivity), calculated by dividing the absorbance A by the molar concentration of the species.

We use λ_{max} to determine the substitution on a UV chromophore. We use ε to quantitate UV-absorbing species.

The three classes of UV chromophores that occur most commonly are illustrated in Fig. 3.25.

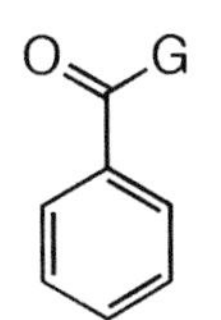

polyenes — Table UV.1

α,β-unsaturated carbonyl — Table UV.2

cabonyl-substituted benzene derivatives — Table UV.3

Figure 3.25 Three common UV chromophores.

Using the following tables, it is possible to approximate λ_{max} (within 5 nm) for a given chromophore. We will use UV in much the same way that we use IR. In step 1, UV absorption indicates the presence of a conjugated system. In step 5, any prospective solution must have a UV chromophore, *with a calculated* λ_{max} *consistent with the one given.*

Quantitative UV

Most of this text is devoted to qualitative organic analysis. In the future, you are most likely to encounter UV in a quantitative setting. For

Table UV.1.
Rules of Diene Absorption[a]

Base value for heteroannular diene	214
Base value for homoannular diene	253
Increments for	
Double bond extending conjugation	+ 30
Alkyl substituent or ring residue	+ 5
Exocyclic double bond	+ 5
Polar groupings	
OAc	+ 0
OAlk	+ 6
SAlk	+ 30
$N(Alk)_2$	+ 60

[a] For further information, see L. M. Fieser and M. Fieser, *Steroids*, (New York: Rheinhold, 1959), pp. 15–24; R. B. Woodward, *J. Am. Chem. Soc.* **1941**, *63*, 1126; **1942**, *64*, 72, 76; A. I. Scott, *Interpretation of the Ultraviolet Spectra of Natural Products* (New York: Pergamon Macmillan, 1964).

instance, chromatographic separations (e.g., HPLC) are often monitored using UV absorption (or fluorescence). It is also possible to assess the absolute amounts of a compound present in a mixture, using the formula A = eM.

Table UV.2.
α, β-Unsaturated Carbonyl Compounds and α, β, γ, δ-Unsaturated Carbonyl Compounds

Base Values		
Acyclic α, β-unsaturated ketone		215
Six-membered cyclic α, β-unsaturated ketone		215
Five-membered cyclic α, β-unsaturated ketone		202
α, β-Unsaturated aldehyde		210
α, β-Unsaturated acid or ester		195
Increments for		
Double bond extending conjugation		+ 30
Alkyl substituent or ring residue α		+ 10
Alkyl substituent or ring residue β		+ 12
Alkyl substituent or ring residue ≥γ		+ 18
Exocyclic double bond		+ 5
Homodiene component		+ 39
Polar groupings		
OH	α	+ 35
	β	+ 30
	γ	+ 50
OAc	α, β, γ	+ 6
OAlk	α	+ 35
	β	+ 30
	γ	+ 17
	δ	+ 31
SAlk	β	+ 85
$N(Alk)_2$	β	+ 95
Cl	α	+ 15
	β	+ 12
Br23	α	+ 25
	β	+ 30

These values are for EtOH. For other solvents, add the following increments: MeOH 0, dioxane 5, chloroform 1, ether 7, water –8, hexane 11, cyclohexane 11.

Table UV.3.
Substituted Benzene Derivatives

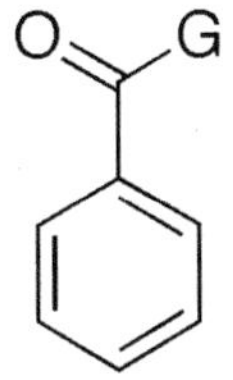

Parent Chromophore		
G = Alkyl or ring residue		246
G = H		250
G = OH, OAlk		230
Increment for Each Substituent		
Alkyl or ring residue	o, m	+ 3
	p	+ 10
OH, OR	o, m	+ 7
	p	+ 25
$-O^{(-)}$	o	+ 11
	m	+ 20
	p	+ 78
Cl	o, m	+ 0
	p	+ 10
Br	o, m	+ 2
	p	+ 13
NH_2	o, m	+ 13
	p	+ 58
NHAc	o, m	+ 20
	p	+ 45
NHMe, NMe_2	o, m	+ 20
	p	+ 85

The Quest, Level Nine: Solving Mechanism-Based Problems (Section III)

The problems in section III are different in that the starting material is given, as well as the reagent(s) with which the starting material has reacted. Rather than attempting to solve the spectroscopic data set de novo, it is more sensible to check to see which of the several components of the starting material are still there.

This text is only an introduction to spectroscopic structure determination. For additional information, see the following.

Bates, R. B., and Beavers, W. A. *Carbon-13 NMR Spectral Problems* (Clifton, NJ: Humana Press, 1981).

Crews, P., Rodriguez, J., and Jaspars, M. *Organic Structure Analysis* (New York: Oxford University Press, 1998).
Davis, R., Wells, C. H. J. *Spectral Problems in Organic Chemistry* (New York: Chapman and Hall, 1984).
Fuchs, P. L., and Bunnell, C. A. *Carbon-13 NMR Based Organic Spectral Problems* (New York: Wiley, 1979).
Graybeal, J. D. *Molecular Spectroscopy* (New York: McGraw-Hill, 1988).
Kemp, W. *Organic Spectroscopy*, 3rd ed. (New York: W. H. Freeman, 1991).
Lambert, J. B., Shurvell, H. F., Lightner, D. A., and Cooks, G. *Introduction to Organic Spectroscopy* (New York: Macmillan, 1987).
Macomber, R. S. *NMR Spectroscopy, Basic Principles and Applications* (New York: Harcourt Brace Jovanovich, 1987).
Pavia, D. L., Lampman, G. M., and Kriz, G. S. Jr. *Introduction to Spectroscopy, A Guide for Students of Organic Chemistry*, 2nd ed. (Philadelphia: Saunders, 1996).
Pretsch, E., Clerc, T., Seibl, J., and Simon, W. *Tables of Spectral Data for Structure Determination of Organic Compounds*, 2nd ed. (Berlin: Springer-Verlag, 1989).
Richards, S. A., *Laboratory Guide to Proton NMR Spectroscopy* (Oxford: Blackwell Scientific Publications, 1988).
Silverstein, R. M., Bassler, G. C., and Morrill, T. C. *Spectrometric Identification of Organic Compounds*, 5th ed. (New York: Wiley, 1991).
Sorrell, T. N. *Interpreting Spectra of Organic Molecules* (Mill Valley, CA: University Science Books, 1988).
Sternhall, S., and Kalman, J. R. *Organic Sructures from Spectra* (New York: Wiley, 1986).
Williams, D. H., and Fleming, I. *Spectroscopic Methods in Organic Chemistry*, 4th ed. (New York: McGraw-Hill, 1989).
Yoder, C. H., and Schaeffer, C. D. Jr. *Introduction to Multinuclear NMR* (Menlo Park, CA: Benjamin/Cummings, 1987).

Section II
Spectroscopic Data Sets and Solutions

Problems 1–15 can be done with just ^{13}C NMR. Problems 16–30 need both ^{13}C and ^{1}H NMR. Problems 31–50 need both ^{13}C and ^{1}H NMR, including alkenes and arenes (chapter 3).

2.1

$C_5H_{10}O$

^{13}C NMR:	**^{1}H NMR:**
23.5, t	1.7, m, 6H
26.7, t (2)	3.72, t, J = 7.4 Hz, 4H
68.9, t (2)	

WORKSHEET 2.1

1. Molecular Formula $C_5H_{10}O$ from the ^{13}C spectrum, all H's are attached to carbon

   IR

   UV

2. Index of Hydrogen Deficiency = 1

   Exploring around the unsaturated functional groups:

   # >=O

   # >=<

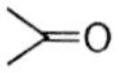

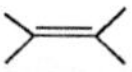

   # sp (R-C≡N 105-120,s ; ≡ 65-90)

   # rings: 1

   IHD = 1

3. Other heteroatoms, exploring around the saturated functional groups:

   There are two carbons at 68.9,t (2), so two CH_2-O, symmetrical. This must be an ether.

4. Other pieces:

   a. Methyl groups

   b. ^1H NMR > 2.0

   c. Other carbons $CH_2 \times 2$, symmetrical and one CH_2

5. Putting it all together: 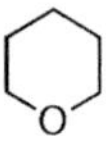 We know that there is one ring. This is the only way to have the symmetry that is observed.

For detailed instructions for solving this problem, see http://valhalla.chem.udel.edu/SpecBook.pdf.

2.2

$C_6H_{10}O$

^{13}C NMR:	^{1}H NMR:
17.9, t (2)	0.75, m, 2H
19.8, q	1.22, m, 2H
25.6, q	1.33, s, 3H
27.0, s	2.08, s, 3H
209.9, s	

WORKSHEET 2.2

1. Molecular Formula $C_6H_{10}O$ from the ^{13}C spectrum, all H's are attached to carbon

IR

UV

2. Index of Hydrogen Deficiency = 2

Exploring around the unsaturated functional groups:

>=O 1 209.9, s

C(=O)CH₃ — 27.0, s 25.6, q

The ketone carbon must have two more carbons flanking it. The CH_3 is apparent. The other carbon must have no H's, although the chemical shift is unusual for a quaternary carbon adjacent to a ketone carbonyl.

>=<

sp (R-C≡N 105-120,s ; ≡ 65-90)

rings: 1

IHD = 2

3. Other heteroatoms, exploring around the saturated functional groups:

None

4. Other pieces:

a. Methyl groups 25.6, q 19.8, q

b. ^1H NMR > 2.0

c. Other carbons CH_2 x 2 symmetrical, ring

5. Putting it all together:

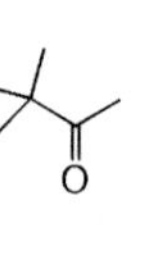

After we accommodate the ketone, the carbons flanking the ketone, and the other methyl group, we are left with only two more carbons to complete the ring.

Note that the carbons in the cyclopropane ring are shifted upfield (smaller numbers) compared to ordinary carbons. This is because of the unusual hybridization of the carbons in the three-membered ring.

For detailed instructions for solving this problem, see http://valhalla.chem.udel.edu/SpecBook.pdf.

2.3

C_4H_9I

^{13}C NMR:	^{1}H NMR:
29.8, s	1.93, s, 9H
40.5, q (3)	

WORKSHEET 2.3

1. Molecular Formula C_4H_9I from the ^{13}C spectrum, all H's are attached to carbon

   IR

   UV

2. Index of Hydrogen Deficiency = 0. Note that I counts as H when calculating the IHD.

   Exploring around the unsaturated functional groups:

   # >=O

   # >=<

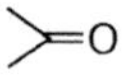

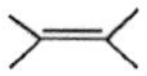

   # sp (R-C≡N 105-120,s; ≡ 65-90)

   # rings:

   IHD = 0

3. Other heteroatoms, exploring around the saturated functional groups:

   The I must be attached to a carbon. Since it cannot be one of the methyl groups, it must be the singlet, the carbon with no H's.

4. Other pieces:

   a. Methyl groups CH_3 x 3, symmetrical

   b. ^1H NMR > 2.0

   c. Other carbons

5. Putting it all together: (tert-butyl iodide structure, —I) Note the small alpha shift for I, and the large beta shift.

For detailed instructions for solving this problem, see http://valhalla.chem.udel.edu/SpecBook.pdf.

2.4

$C_6H_{14}O$

^{13}C NMR:	**^{1}H NMR:**
29.7, t	0.91, s, 9H
29.8, q (3)	1.53, t, J = 7.3 Hz, 2H
46.4, s	2.13, bs, 1H (exchanges)
60.0, t	3.70, t, J = 7.3 Hz, 2H

WORKSHEET 2.4

1. Molecular Formula $C_6H_{14}O$ from the ^{13}C spectrum, one H is not attached to carbon

   IR

   UV

2. Index of Hydrogen Deficiency = 0

   Exploring around the unsaturated functional groups:

   # >=O

   # >=<

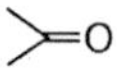

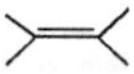

   # sp (R-C≡N 105-120,s ; —≡— 65-90)

   # rings:

   IHD = 0

3. Other heteroatoms, exploring around the saturated functional groups:

   There is an O in the molecular formula, and one H is not attached to carbon, so we have an alcohol functional group. The carbon to which the alcohol is attached appears at 60.0, t.

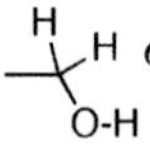

   H, H 60.0, t, O-H

   This is the H that exchanges

4. Other pieces:

   a. Methyl groups CH_3 x 3, symmetrical

   b. ^{1}H NMR > 2.0

   c. Other carbons CH_2

5. Putting it all together: —OH

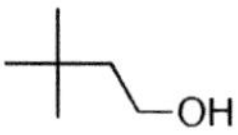

For detailed instructions for solving this problem, see http://valhalla.chem.udel.edu/SpecBook.pdf.

2.5

$C_7H_{12}O$

IR: 3370, 3295, 2925, 2110 cm^{-1}

^{13}C NMR:

18.2, t
24.8, t
28.1, t
32.0, t
62.3, t
68.2, d
84.3, s

^{1}H NMR:

1.5, m, 6H
1.94, t, J = 2.7 Hz, 1H
2.21, dt, J = 2.7, 6.6 Hz, 2H
3.12, bs, 1H (exchanges)
3.66, t, J = 6.0, 2H

WORKSHEET 2.5

1. Molecular Formula $C_7H_{12}O$ from the ^{13}C spectrum, one H is not attached to carbon

   IR

   UV

2. Index of Hydrogen Deficiency = 2

   Exploring around the unsaturated functional groups:

   # >=O

   # >=<

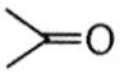

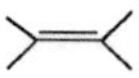

   # sp (R-C≡N 105-120,s / —≡— 65-90) one alkyne 68.2, d 84.3, s —C≡C-H

   # rings:

   IHD = 2

3. Other heteroatoms, exploring around the saturated functional groups:

   There is alcohol functional group. The carbon to which the alcohol is attached comes at 62.3, t.

4. Other pieces:

   a. Methyl groups

   b. ^1H NMR > 2.0

   c. Other carbons CH_2 x 4, no symmetry, no branching

5. Putting it all together:

   18.2, t, note the small alpha shift of an alkyne

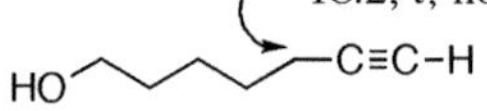

   HO —C≡C-H

For detailed instructions for solving this problem, see http://valhalla.chem.udel.edu/SpecBook.pdf.

2.6

$C_7H_{14}O_2$

^{13}C NMR:

22.2, q (2)
27.7, d
32.2, t
33.8, t
51.4, q
174.4, s

^{1}H NMR:

0.90, d, J = 7.5 Hz, 6H
1.55, m, dt, J = 7.4, 7.7 Hz, 2H
2.30, t, J = 7.7 Hz, 2H
3.67, s, 3H

WORKSHEET 2.6

1. Molecular Formula $C_7H_{14}O_2$ from the ^{13}C spectrum, all H's are attached to carbon

   IR

   UV

2. Index of Hydrogen Deficiency = 1

   Exploring around the unsaturated functional groups:

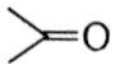

   # C=O one 174.4, s This is typical of a carboxylic acid derivative. All H's are attached to C, so it cannot be an acid—it must be an ester. This is borne out by 51.4, q, typical of a methyl ester.

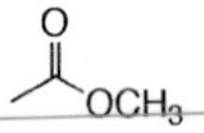

   # C=C

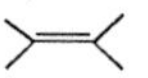

   # sp (R-C≡N 105-120,s ; C≡C 65-90)

   # rings:

   O ‖ C–OCH3 51.4, q

   IHD = 1

3. Other heteroatoms, exploring around the saturated functional groups:

   All O's accounted for

4. Other pieces:

   a. Methyl groups CH_3 x 2, symmetrical

   b. ^{1}H NMR > 2.0

   c. Other carbons CH_2 x 2 not symmetrical C-H

5. Putting it all together:

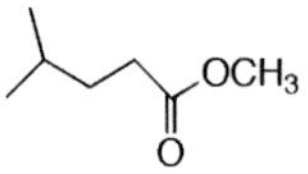

   OCH3 O

   Why would these alternative structures not fit the data?

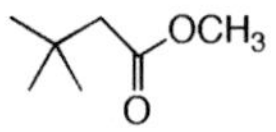

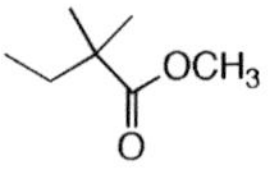

   OCH3 O OCH3 O

For detailed instructions for solving this problem, see http://valhalla.chem.udel.edu/SpecBook.pdf.

2.7

$C_{13}H_{26}O$

^{13}C NMR:

20.6, t
23.0, t (2)
23.1, t (2)
24.4, t (2)
24.9, t (2)
28.3, t (2)
56.1, q
78.7, d

^{1}H NMR:

1.3–1.6, m, 22 H
3.31, m, 1H
3.33, s, 3H

WORKSHEET 2.7

1. Molecular Formula $C_{13}H_{26}O$ from the ^{13}C spectrum, all H's are attached to carbon

   IR

   UV

2. Index of Hydrogen Deficiency = 1

   Exploring around the unsaturated functional groups:

   # >=O

   # >=<

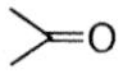

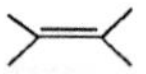

   # sp (R-C≡N 105-120,s ; ≡ 65-90)

   # rings: 1

   IHD = 1

3. Other heteroatoms, exploring around the saturated functional groups:

   There is an oxygen and all the H's are attached to carbon, so this must be an ether functional group. One carbon of the ether is 56.1, q, and the other is 78.7, d.

   H

   78.7, d >–OCH_3

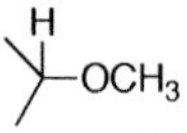

   56.1, q

4. Other pieces:

   a. Methyl groups

   b. ^1H NMR > 2.0

   c. Other carbons — symmetrical pairs of CH_2 x 5, single CH_2

5. Putting it all together:

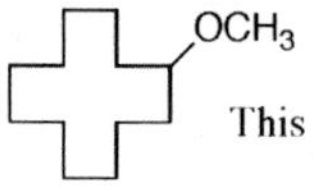

   OCH_3

   This is a convenient way to draw a 12-membered ring.

For detailed instructions for solving this problem, see http://valhalla.chem.udel.edu/SpecBook.pdf.

2.8

$C_5H_{11}Br$

^{13}C NMR:

11.2, q
18.4, q
27.6, t
36.8, d
41.0, t

^{1}H NMR:

0.93, t, J = 7.3 Hz, 3H
1.04, d, J = 7.6 Hz, 3H
1.27, m, 1H
1.48, m, 1H
1.71, m, 1H
3.31, dd, J = 6.8, 15.1 Hz, 1H
3.38, dd, J = 7.1, 15.1 Hz, 1H

WORKSHEET 2.8

1. Molecular Formula $C_5H_{11}Br$ from the ^{13}C spectrum, all H's are attached to carbon

   IR

   UV

2. Index of Hydrogen Deficiency = 0

   Exploring around the unsaturated functional groups:

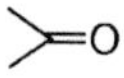

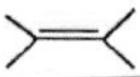

   # C=O

   # C=C

   # sp (R-C≡N 105-120,s; C≡C 65-90)

   # rings:

   IHD = 0

3. Other heteroatoms, exploring around the saturated functional groups:

   The Br is attached to a CH_2, 41.0, t $-CH_2-Br$

4. Other pieces:

   a. Methyl groups CH_3 x 2, no symmetry

   b. ^1H NMR > 2.0

   c. Other carbons CH, CH_2

5. Putting it all together: [structure] Br Why would it not be [structure] Br ?

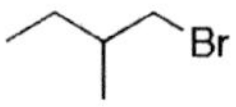

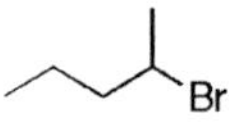

For detailed instructions for solving this problem, see http://valhalla.chem.udel.edu/SpecBook.pdf.

2.9

$C_7H_{14}O$

^{13}C NMR:	**^{1}H NMR:**
7.9, q	0.92, t, J = 7.3 Hz, 3H
13.8, q	1.05, t, J = 7.7 Hz, 3H
22.5, t	1.4, m, 2H
26.2, t	1.6, m, 2H
35.9, t	2.38, q, J = 7.7 Hz, 2H
42.1, t	2.44, t, J = 7.4 Hz, 2H
211.5, s	

WORKSHEET 2.9

1. Molecular Formula $C_7H_{14}O$ from the ^{13}C spectrum, all H's are attached to carbon

   IR

   UV

2. Index of Hydrogen Deficiency = 1

   Exploring around the unsaturated functional groups:

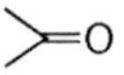

   # >=O 1 211.5, s is a ketone carbonyl; the carbons flanking the ketone are both CH_2's, 35.9, t and 42.1, t, but they are not symmetrical

   # >=<

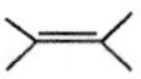

   # sp (R-C≡N 105-120,s; —≡— 65-90) $H_2C-C(=O)-CH_2$

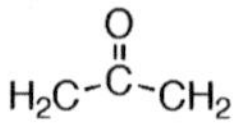

   # rings:

   IHD = 1

3. Other heteroatoms, exploring around the saturated functional groups:

   None

4. Other pieces:

   a. Methyl groups CH_3 x 2, not symmetrical

   b. ^{1}H NMR > 2.0

   c. Other carbons CH_2 x 2

5. Putting it all together:

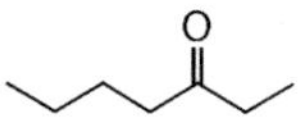

   In what ways would the ^{13}C spectrum have been different if it had been the isomeric ketone?

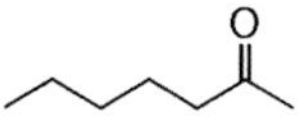

For detailed instructions for solving this problem, see http://valhalla.chem.udel.edu/SpecBook.pdf.

2.10

$C_6H_{11}N$

^{13}C NMR:

4.1, q
16.1, t
32.5, t
41.2, t
75.9, s
78.5, s

^{1}H NMR:

1.45, bs, 2H exchanges
1.63, m, 2H
1.78, s, 3H
2.20, t, J=7.1 Hz, 2H
2.80, t, J= 6.8 Hz, 2H

WORKSHEET 2.10

1. Molecular Formula $C_6H_{11}N$ — from the ^{13}C spectrum, two of the H's are not attached to carbon—they must be attached to the N, so this must be a primary amine

   IR

   UV

2. Index of Hydrogen Deficiency = 2

   Note that the N brings an extra H, so this is really C_6H_{10}.

   Exploring around the unsaturated functional groups:

   # >=O

   # >=<

   # sp (R-C≡N 105-120,s; ≡ 65-90) one alkyne 75.9, s and 78.5, s

   # rings:

   IHD =

3. Other heteroatoms, exploring around the saturated functional groups:

   The amine has an alpha shift of + 29, so we would look for the attached carbon at about 40. We find it at 41.2, t, so we have

   $-CH_2-NH_2$

4. Other pieces:

   a. Methyl groups CH_3 at 4.1, q—this very small alpha shift is typical of an alkyne.

   b. ^1H NMR > 2.0

   c. Other carbons CH_2 x 2

5. Putting it all together: $CH_3-C{\equiv}C-CH_2-CH_2-NH_2$

For detailed instructions for solving this problem, see http://valhalla.chem.udel.edu/SpecBook.pdf.

2.11

$C_7H_{12}O_2$

IR: 1746 cm^{-1}

^{13}C NMR:

23.8, t
33.9, t
35.4, t
51.1, q
115.9, t
137.3, d
174.8, s

^{1}H NMR:

1.8, m, 2H
2.12, dt, J=6.7, 7.2 Hz, 2H
2.41, t, J=7.6 Hz, 2H
3.76, s, 3H
4.9, m, 2H
5.64, ddd, J=6.7, 10.2, 15.1 Hz, 1H

WORKSHEET 2.11

1. Molecular Formula $C_7H_{12}O_2$ from the ^{13}C spectrum, all H's are attached to carbon

   IR

   UV

2. Index of Hydrogen Deficiency = 2

   Exploring around the unsaturated functional groups:

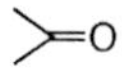

   # >=O 1

   174.8, s is the carbonyl of a carboxylic acid derivative. Since all H's are attached to carbon, and the only heteroatoms in the formula are O's, this must be an ester. Fom 51.1, q, it is a methyl ester.

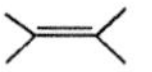

   # >=< 1

   From 137.3, d and 115.9, t, we know that we have a monosubstituted alkene.

   # sp (R-C≡N 105-120,s; —≡— 65-90)

   H H H

   O OCH3

   # rings:

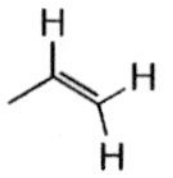

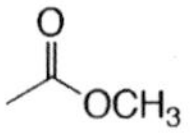

   IHD =

3. Other heteroatoms, exploring around the saturated functional groups:

   None

4. Other pieces:

   a. Methyl groups

   b. ^1H NMR > 2.0

   c. Other carbons CH_2 x 3

   O O

5. Putting it all together:

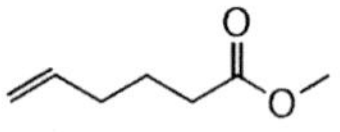

For detailed instructions for solving this problem, see http://valhalla.chem.udel.edu/SpecBook.pdf.

2.12

$C_6H_{10}O_3$

^{13}C NMR:

54.1, q (2)
107.1, d (2)
131.2, d (2)

^{1}H NMR:

3.38, s, 6H
5.58, d, J = 2.4 Hz, 2H
6.07, d, J = 2.4 Hz, 2H

WORKSHEET 2.12

1. Molecular Formula $C_6H_{10}O_3$ from the ^{13}C spectrum, all H's are attached to carbon

   IR

   UV

2. Index of Hydrogen Deficiency = 2

   Exploring around the unsaturated functional groups:

   # C=O

   # C=C 1 — Alkene 131.2, d (2) symmetrical

   # sp (R-C≡N 105-120,s / C≡C 65-90)

   # rings: 1

   IHD = 2

3. Other heteroatoms, exploring around the saturated functional groups:

   There are three O's, all ethers. Two of the ether carbons 54.1, q (2) are methyls—they must not be on the same O. The other two ether carbons are apparently each attached to two O's 107.1, d (2).

4. Other pieces:

   a. Methyl groups

   b. ^{1}H NMR > 2.0

   c. Other carbons. Don't forget the ring!

5. Putting it all together:

   CH$_3$O–(2,5-dihydrofuran ring, O)–OCH$_3$

For detailed instructions for solving this problem, see http://valhalla.chem.udel.edu/SpecBook.pdf.

2.13

C_7H_8

^{13}C NMR:

50.2, d (2)
75.3, t
143.3, d (4)

^{1}H NMR:

1.95, t, J = 7.2 Hz, 2H
3.56, tt, J = 7.2, 7.6 Hz, 2H
6.75, d, J = 7.6 Hz, 4H

WORKSHEET 2.13

1. Molecular Formula C_7H_8 from the ^{13}C spectrum, all H's are attached to carbon

   IR

   UV

2. Index of Hydrogen Deficiency = 4

   Exploring around the unsaturated functional groups:

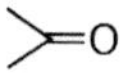

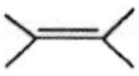

   # >=O

   # >=< 2   H–C=C–H (cis) x 2, all symmetrical

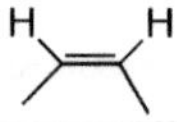

   # sp (R-C≡N 105-120,s; –C≡C– 65-90)

   # rings: 2

   IHD = 4

3. Other heteroatoms, exploring around the saturated functional groups:

   None

4. Other pieces:

   a. Methyl groups

   b. ^1H NMR > 2.0

   c. Other carbons   CH x 2 symmetrical CH_2 Remember the two rings!

5. Putting it all together:

For detailed instructions for solving this problem, see
http://valhalla.chem.udel.edu/SpecBook.pdf.

2.14

$C_8H_{16}O_2$

IR: 2969, 2877, 1700, 1479, 1457, 1465, 1049, 1070, 1029 cm^{-1}

^{13}C NMR:

14.8, q
25.8, q (3)
41.7, d
44.6, s
65.2, t
220.1, s

^{1}H NMR:

0.98, d, J = 7.1 Hz, 3H
1.09, s, 9H
2.56, m, 1H
3.1, bs, 1H (exchanges)
3.35, d, J = 10.6 Hz, 1H
3.57, d, J = 10.6 Hz, 1H

WORKSHEET 2.14

1. Molecular Formula $C_8H_{16}O_2$ from the ^{13}C spectrum, one H is not attached to carbon

   IR

   UV

2. Index of Hydrogen Deficiency = 1

   Exploring around the unsaturated functional groups:

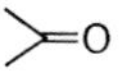

   # >=O 1 The carbonyl is a ketone, 220.1, s. The two carbons flanking the ketone carbonyl are 41.7, d and 44.6, s.

   # >=<

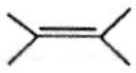

   # sp (R-C≡N 105-120,s ; ≡ 65-90)

   # rings:

   IHD = 1

3. Other heteroatoms, exploring around the saturated functional groups:

The OH is attached to a CH_2 65.2, t,

4. Other pieces:

   a. Methyl groups There is one set of CH_3 x 3, symmetrical, 25.8, q (3), and another CH_3, 14.8, q.

   b. ^{1}H NMR > 2.0

   c. Other carbons

5. Putting it all together: O OH

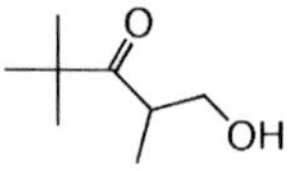

For detailed instructions for solving this problem, see http://valhalla.chem.udel.edu/SpecBook.pdf.

2.15

$C_9H_{16}O$

^{13}C NMR:

23.5, q (2)
26.7, t (2)
28.6, t (2)
79.8, d (2)
107.8, s

^{1}H NMR:

1.38, s, 6H
1.2–1.5, m, 4H
1.8–2.1, m, 4H
3.27, m, 2H

WORKSHEET 2.15

1. Molecular Formula $C_9H_{16}O$ from the ^{13}C spectrum, all H's are attached to carbon

   IR

   UV

2. Index of Hydrogen Deficiency = 2

   Exploring around the unsaturated functional groups:

   # >=O

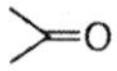

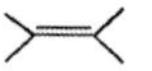

   # >=< The signal at 107.8, s could not be an alkene, since alkene carbons come two at a time.

   # sp (~~R-C≡N 105-120,s~~ ≡ 65-90)

   # rings: 2

   IHD = 2

3. Other heteroatoms, exploring around the saturated functional groups:

   There are two ether oxygens, so there must be four C-O bonds. There are only two carboms in the C-O range, 79.8, d (3), so one carbon must have two C-O bonds. It is doubly shifted, to 107.8, s.

   So: 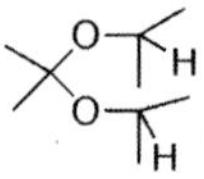 symmetrical

4. Other pieces:

   a. Methyl groups CH_3 x 2 symmetrical

   b. ^1H NMR > 2.0

   c. Other carbons CH_2 x 2 symmetrical x 2 Remember the two rings!

5. Putting it all together:

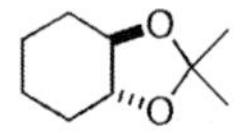

How do we know that it is not 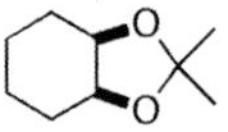 ?

For detailed instructions for solving this problem, see http://valhalla.chem.udel.edu/SpecBook.pdf.

2.16

$C_6H_{10}O$

^{13}C NMR:	**^{1}H NMR:**
8.8, q	1.05, t, J = 7.6 Hz, 3H
29.1, q	1.48, s, 3H
36.3, t	1.71, q, J = 7.6 Hz, 2H
68.5, s	2.35, s, 1H
71.2, d	1.44, bs, 1H (exchanges)
87.6, s	

WORKSHEET 2.16

1. Molecular Formula $C_6H_{10}O$ from the ^{13}C spectrum, one H is not attached to carbon

   IR

   UV

2. Index of Hydrogen Deficiency = 2

   Exploring around the unsaturated functional groups:

   # >=O

   # >=<

   # sp (R-C≡N 105-120,s; ≡ 65-90) one alkyne ≡-H

   # rings:

   IHD = 2

3. Other heteroatoms, exploring around the saturated functional groups:

   The alcohol is attached to a carbon with no H's, 68.5, s

   -OH

4. Other pieces:

   a. Methyl groups CH_3 x 2 From 1.05, t, J = 7.6 Hz, 3H, we know that one methyl is next to a CH_2. From 1.48, s, 3H, we know that one methyl is attached to the quaternary carbon.

   b. ^{1}H NMR > 2.0 2.35, s, 1H is the H on the alkyne.

   1.44, bs, 1H (exchanges) is the H on the alcohol.

   c. Other carbons

   From 1.71, q, J=7.6 Hz, 2H, we know that the CH_2 only sees neighboring H's from the methyl group.

5. Putting it all together:

H-≡-C-OH

For detailed instructions for solving this problem, see http://valhalla.chem.udel.edu/SpecBook.pdf.

2.17

C_5H_8O

^{13}C NMR:	**^{1}H NMR:**
10.6, t (2)	0.86, m, 2H
21.2, d	1.02, m, 2H
30.0, q	1.95,m, 1H
208.8, s	2.23, s, 3H

WORKSHEET 2.17

1. Molecular Formula C_5H_8O from the ^{13}C spectrum, all H's are attached to carbon

   IR

   UV

2. Index of Hydrogen Deficiency = 2

   Exploring around the unsaturated functional groups:

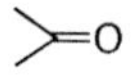

   # C=O 1 There is a ketone, 208.8, s. One of the carbons flanking the ketone is a methyl, 30.0, q and 2.23, s, 3H. From symmetry, the other carbon flanking the ketone must be 21.2, d.

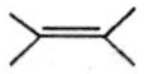

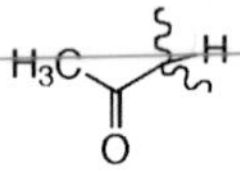

   # C=C

   # sp (R-C≡N 105-120,s; C≡C 65-90)

   # rings: 1

   H3C–C(=O)–CH

   IHD = 2

3. Other heteroatoms, exploring around the saturated functional groups:

   None

4. Other pieces:

   a. Methyl groups CH_3 next to ketone

   b. ^1H NMR > 2.0

   c. Other carbons CH_2 x 2 symmetrical ring

5. Putting it all together: cyclopropyl–C(=O)–CH3 Note (Table C.1)that the base value for cyclopropane is - 2.9.

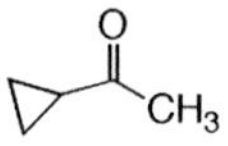

For detailed instructions for solving this problem, see http://valhalla.chem.udel.edu/SpecBook.pdf.

2.18

$C_9H_{16}O_2$

^{13}C NMR:

13.9, q
22.3, t
24.7, t
31.3, t
34.2, t
64.9, t
118.0, t
132.3, d
173.4, s

^{1}H NMR:

0.90, t, J = 7.6 Hz, 3H
1.3, m, 4H
1.65, m, 2H
2.32, t, J = 6.7 Hz, 2H
4.58, d, J = 7.8 Hz, 2H
5.21, d, J = 10.4 Hz, 1H
5.32, d, J = 15.9 Hz, 1H
5.92, ddt, J = 10.4, 15.9, 7.8 Hz, 1H

WORKSHEET 2.18

1. Molecular Formula $C_9H_{16}O_2$ from the ^{13}C spectrum, all H's are attached to carbon

IR

UV

2. Index of Hydrogen Deficiency = 2

Exploring around the unsaturated functional groups:

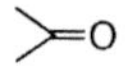

>=O 1

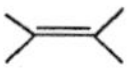

>=< 1

The carbonyl at 173.4, s is a carboxylic acid derivative. Since all the H's are attached to C, this must be an ester. The C attached to the O comes at 64.9, t, and 4.58, d, J=7.8 Hz, 2H. This further talks to the alkene at 5.92, ddt, J=10.4, 15.9, 7.8 Hz, 1H. The C attached to the carbonyl has two H's, 2.32, t, J=6.7 Hz, 2H. Putting it all together:

sp (R-C≡N 105-120,s ≡ 65-90)

rings:

O 4.58 O 2.32

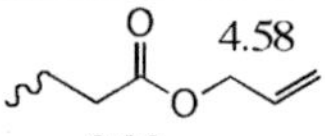

IHD = 2

3. Other heteroatoms, exploring around the saturated functional groups:

All the heteroatoms are all accounted for.

4. Other pieces:

a. Methyl groups CH_2-CH_3 0.90, t, J=7.6 Hz, 3H

b. ^1H NMR > 2.0

c. Other carbons CH_2

5. Putting it all together: O O

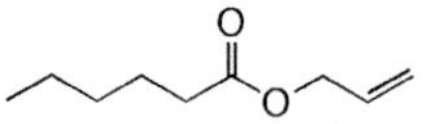

For detailed instructions for solving this problem, see http://valhalla.chem.udel.edu/SpecBook.pdf.

2.19

$C_6H_{11}BrO_2$

^{13}C NMR:

14.2, q
27.8, t
32.5, t
32.6, t
60.5, t
172.4, s

^{1}H NMR:

1.25, t, J = 7.6 Hz, 3H
2.18, m, 2H
2.58, t, J = 7.3 Hz, 2H
3.46, t, J = 6.7 Hz, 2H
4.15, q, J = 7.6 Hz, 2H

WORKSHEET 2.19

1. Molecular Formula $C_6H_{11}BrO_2$ from the ^{13}C spectrum, all H's are attached to carbon

   IR

   UV

2. Index of Hydrogen Deficiency = 1

   Exploring around the unsaturated functional groups:

   # C=O 1 Carboxylic acid derivative at 172.4, s. Since all H's are attached to C, this must be an ester. The C attached to the O of the ester appears at 60.5, t, and at 4.15, q, J=7.6 Hz, 2H, so it must be an ethyl ester. The carbon attached to the carbonyl carbon is also distinctive, 2.58, t, 2H. This shows us altogether five of the six carbons of the unknown.

   # C=C

   # sp (R-C≡N 105-120,s; C≡C 65-90)

   # rings:

   IHD = 1

   H H O 4.15 — O — 1.25; H H; 2.58, t, J=7.3 Hz, 2H

3. Other heteroatoms, exploring around the saturated functional groups:

   The C attached to the Br should come (Table H.1) at about 3.4. It does, 3.46, t, J = 6.7 Hz, 2H.

   so CH_2-CH_2-Br

4. Other pieces:

   a. Methyl groups

   b. ^1H NMR > 2.0

   c. Other carbons

5. Putting it all together:

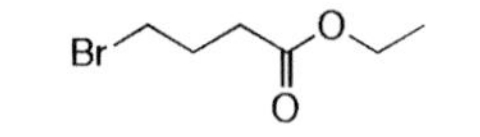

For detailed instructions for solving this problem, see http://valhalla.chem.udel.edu/SpecBook.pdf.

2.20

$C_8H_{16}O$

^{13}C NMR:

13.9, q
18.3, q (2)
22.5, t
25.6, t
40.0, t
40.8, d
215.0, s

^{1}H NMR:

0.90, t, J = 7.6 Hz, 3H
1.18, d, J = 7.1 Hz, 6H
1.4, m, 2H
1.6, m, 2H
2.44, t, J = 6.7 Hz, 2H
2.60, septet, J = 7.1 Hz, 1H

WORKSHEET 2.20

1. Molecular Formula $C_8H_{16}O$ from the ^{13}C spectrum, all H's are attached to carbon

   IR

   UV

2. Index of Hydrogen Deficiency = 1

   Exploring around the unsaturated functional groups:

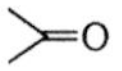

   # C=O 1

   # C=C

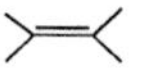

   # sp (R-C≡N 105-120,s ; C≡C 65-90)

   # rings:

   IHD = 1

There is a ketone carbonyl at 215.0, s. The carbons flanking the ketone are usually in the range 40–50, so probably 40.8, d and 40.0, t. The H's attached to these carbons are also distinctive, at 2.60, sept, J = 7.1 Hz, 1H and 2.44, t, J = 6.7 Hz, 2H, respectively. The H at 2.60 is clearly coupled (J = 7.6 Hz) to the two methyls at 1.18, and the two H's at 2.44 are clearly also coupled, to another CH_2. Putting all this together accounts for six of the eight carbons in the formula.

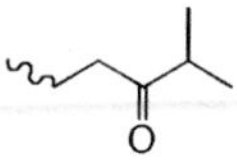

3. Other heteroatoms, exploring around the saturated functional groups:

   None

4. Other pieces:

   a. Methyl groups There is one more CH_3, at 0.90, t, J = 7.6 Hz, 3H, so it must be next to a CH_2.

   b. ^1H NMR > 2.0

   c. Other carbons

5. Putting it all together:

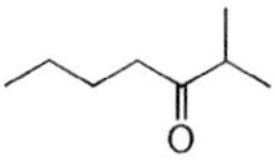

For detailed instructions for solving this problem, see http://valhalla.chem.udel.edu/SpecBook.pdf.

2.21

$C_8H_{16}O_2$

^{13}C NMR:

14.0, q
14.3, q
17.1, q
20.4, t
36.0, t
39.4, d
60.0, t
176.6, s

^{1}H NMR:

0.90, t, J = 7.6 Hz, 3H
1.14, d, J = 6.7 Hz, 3H
1.25, t, J = 7.1 Hz, 3H
1.3, m, 2H
1.7, m, 2H
2.63, m, 1H
4.13, q, J = 7.1 Hz, 2H

WORKSHEET 2.21

1. Molecular Formula $C_8H_{16}O_2$ from the ^{13}C spectrum, all H's are attached to carbon

   IR

   UV

2. Index of Hydrogen Deficiency = 1

   Exploring around the unsaturated functional groups:

   # C=O 1 — Carboxylic acid derivative at 176.6, s. Since all H's are attached to C, this must be an ester. The C attached to the O of the ester appears at 60.0, t, and at 4.13, q, J = 7.1 Hz, 2H, so it must be an ethyl ester. The carbon attached to the carbonyl carbon is also distinctive, 2.63, m, 1H. This shows us altogether four of the eight carbons of the unknown.

   # C=C

   # sp (R-C≡N 105-12, s; C≡C 65-90)

   # rings:

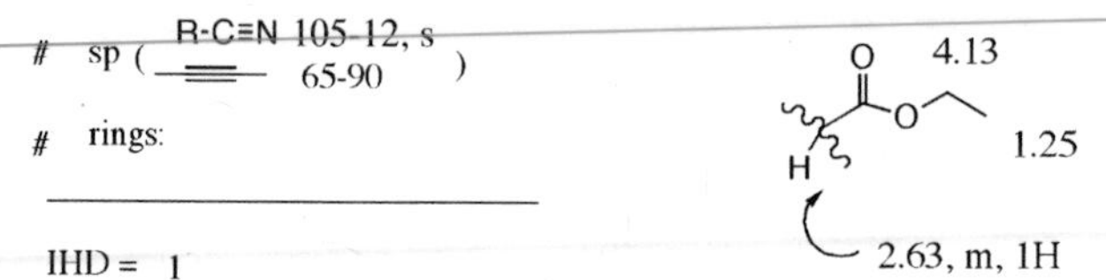

   IHD = 1

3. Other heteroatoms, exploring around the saturated functional groups:

   O's all accounted for

4. Other pieces:

   a. Methyl groups There are two more methyl groups:

   0.90, t, J = 7.6 Hz, 3H $-CH_2-CH_3$

   b. ^1H NMR > 2.0

   1.14, d, J = 6.7 Hz, 3H $-CH-CH_3$ There is only one CH in the molecule, so this must be the same CH that is next to the ester carbonyl.

   c. Other carbons CH_2

5. Putting it all together:

For detailed instructions for solving this problem, see http://valhalla.chem.udel.edu/SpecBook.pdf.

2.22

C_5H_7N

^{13}C NMR:	**^{1}H NMR:**
35.2, q	1.1, bs, 1H (exchanges)
40.0, t	2.24, t, J = 2.5 Hz, 1H
71.3, d	2.48, s, 3H
82.1, s	3.38, d, J = 2.5 Hz, 2H

WORKSHEET 2.22

1. Molecular Formula C_5H_7N from the ^{13}C spectrum, all one H is not attached to carbon

   IR

   UV

2. Index of Hydrogen Deficiency = 2

   Exploring around the unsaturated functional groups:

   # >=O

   # >=<

   # sp (R-C≡N 105-120,s ; ≡ 65-90) 1 — one alkyne 71.3, d ; 82.1, s — X—CH_2—C≡C—H

   Note (Table H.2) that there is long-range coupling between the alkyne H and the CH_2 on the other side of the alkyne. That CH_2 comes at 3.38 and has no additional coupling.

   # rings:

   ____

   IHD = 2

3. Other heteroatoms, exploring around the saturated functional groups:

   The N has one H attached, so there must be two alkyl groups. There is only one more carbon in the formula, 35.2, q, so this must be attached to the N, and the other group on the N must be the CH_2 at 3.38.

4. Other pieces:

   a. Methyl groups

   b. ^1H NMR > 2.0

   c. Other carbons

5. Putting it all together: H_3C–N(H)–CH_2–C≡C–H

For detailed instructions for solving this problem, see http://valhalla.chem.udel.edu/SpecBook.pdf.

2.23

$C_4H_6O_2$

IR: 1818 cm^{-1}

^{13}C NMR:

20.6, q
44.3, t
68.0, d
168.2, s

^{1}H NMR:

1.58, d, J = 7.2 Hz, 3H
3.06, dd, J = 7.6, 16.2 Hz, 1H
3.58, dd, J = 6.5, 16.2 Hz, 1H
4.7, m, 1H

WORKSHEET 2.23

1. Molecular Formula $C_4H_6O_2$ from the ^{13}C spectrum, all H's are attached to carbon

   IR

   UV

2. Index of Hydrogen Deficiency = 2

   Exploring around the unsaturated functional groups:

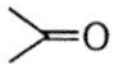

   # >=O 1

   # >=<

   # sp (R-C≡N 105-120,s / ≡ 65-90)

   # rings: 1

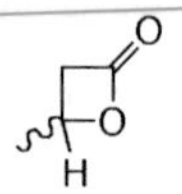

   IHD = 2

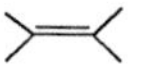

   The carbonyl at 168.2, s is a carboxylic acid derivative. Since all H's are attached to C, this must be an ester. Since there is a ring, there is the possibility that the ester is contained in that ring. The IR 1818 cm^{-1} confirms that it is, and that it is a four-membered ring (b-lactone).

   Further, the carbon attached to the ester oxygen appears at 68.0, d, so it has one H.

3. Other heteroatoms, exploring around the saturated functional groups:

   All the O's have been accounted for.

4. Other pieces:

   a. Methyl groups The CH_3 at 1.58, d, J=7.2 Hz, 3H is attached to a carbon with one H.

   b. ^1H NMR > 2.0

   c. Other carbons CH_2

5. Putting it all together:

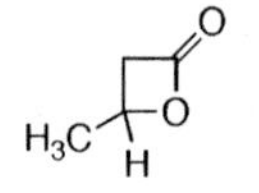

For detailed instructions for solving this problem, see http://valhalla.chem.udel.edu/SpecBook.pdf.

2.24

$C_6H_{10}O_3$

^{13}C NMR:

54.1, q (2)
107.1, d (2)
131.2, d (2)

^{1}H NMR:

3.38, s, 6H
5.58, d, J = 2.4 Hz, 2H
6.07, d, J = 2.4 Hz, 2H

WORKSHEET 2.24

1. Molecular Formula $C_6H_{10}O_3$ from the ^{13}C spectrum, all H's are attached to carbon

   IR

   UV

2. Index of Hydrogen Deficiency = 2

   Exploring around the unsaturated functional groups:

   # C=O

   # C=C 1

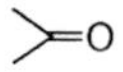

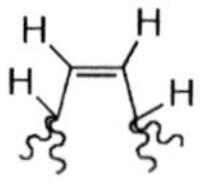

   Alkene 131.2, d (2) symmetrical, coupled 6.07, d, J = 2.4 Hz, 2H

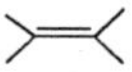

   # sp (R-C≡N 105-120,s; C≡C 65-90)

   # rings: 1

   IHD = 2

3. Other heteroatoms, exploring around the saturated functional groups:

   There are three O's, all ethers. Two of the ether carbons 54.1, q (2) are methyls—they must not be on the same O. The other two ether carbons are apparently each attached to two O's 107.1, d (2).

4. Other pieces:

   a. Methyl groups

   b. ^{1}H NMR > 2.0

   c. Other carbons Don't forget the ring!

5. Putting it all together:

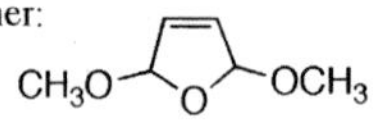

   CH_3O–(2,5-dihydrofuran ring, O)–OCH_3

For detailed instructions for solving this problem, see http://valhalla.chem.udel.edu/SpecBook.pdf.

2.25

$C_7H_{16}O$

^{13}C NMR:	**^{1}H NMR:**
14.4, q	0.88, t, J = 7.2 Hz, 3H
16.6, q	0.94, d, J = 6.7 Hz, 3H
20.1, t	1.3–1.7, m, 7H
35.5, t	2.13, bs, 1H (exchanges)
35.6, d	3.38, dd, J = 7.0, 14.2 Hz, 1H
68.3, t	3.45, dd, J = 7.4, 14.2 Hz, 1H

WORKSHEET 2.25

1. Molecular Formula $C_7H_{16}O$ from the ^{13}C spectrum, one H is not attached to carbon

   IR

   UV

2. Index of Hydrogen Deficiency = 0

   Exploring around the unsaturated functional groups:

   # >=O

   # >=<

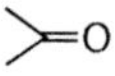

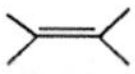

   # sp (R-C≡N 105-120,s; ≡ 65-90)

   # rings:

   IHD = 0

3. Other heteroatoms, exploring around the saturated functional groups:

   One H is attached to O, so this must be an alcohol. From 68.3, t and 3.38, dd, J = 7.0, 14.2 Hz, 1H and 3.45, dd, J = 7.4, 14.2 Hz, 1H we can deduce the partial structure:

   no symmetry

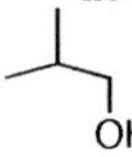

   OH

4. Other pieces:

   a. Methyl groups 0.88, t, J = 7.2 Hz, 3H CH_2-CH_3
   0.94, d, J = 6.7 Hz, 3H CH-CH3

   b. ^1H NMR > 2.0

   c. Other carbons CH_2 x 2

5. Putting it all together:

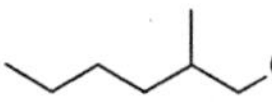

   OH

For detailed instructions for solving this problem, see http://valhalla.chem.udel.edu/SpecBook.pdf.

2.26

C_7H_9N

^{13}C NMR:	**^{1}H NMR:**
19.7, t	1.9, m, 2H
23.5, t	2.35, m, 4H
32.5, t	3.10, s, 2H
34.7, t	5.75, t, J=3.4 Hz, 1H
117.5, s	
129.5, d	
132.5, s	

WORKSHEET 2.26

1. Molecular Formula C_7H_9N from the ^{13}C spectrum, all H's are attached to carbon

   IR

   UV

2. Index of Hydrogen Deficiency = 4

   Exploring around the unsaturated functional groups:

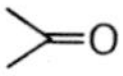

   # >=O 5.75, t, J=3.4 Hz, 1H

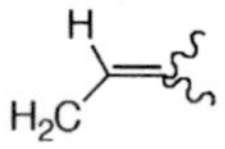

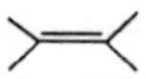

   # >=< 1 H, H_2C

   # sp (R-C≡N 105-120,s; ≡ 65-90) 1 117.5, s —C≡N

   # rings: 1

   IHD = 4

3. Other heteroatoms, exploring around the saturated functional groups:

   None

4. Other pieces:

   a. Methyl groups None

   b. ^1H NMR > 2.0 3.10, s, 2H is a CH_2 that is shifted by two groups, with no H's on adjacent carbons

   c. Other carbons CH_2 x 2 ring

5. Putting it all together: C≡N

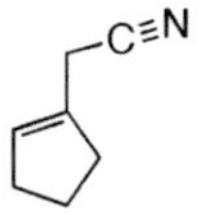

For detailed instructions for solving this problem, see http://valhalla.chem.udel.edu/SpecBook.pdf.

2.27

$C_6H_{10}O$

^{13}C NMR:

18.1, q
63.2, t
129.3, d
129.8, d
130.8, d
131.7, d

^{1}H NMR:

1.75, d, J = 7.2 Hz, 3H
2.4, bs, 1H (exchanges)
4.12, d, J = 7.6 Hz, 2H
5.7, m, 2H
6.12, dt, J = 15.5, 7.6 Hz, 1H
6.18, dq, J = 15.2, 7.2 Hz, 1H

WORKSHEET 2.27

1. Molecular Formula $C_6H_{10}O$ from the ^{13}C spectrum, one H is not attached to carbon

   IR

   UV

2. Index of Hydrogen Deficiency = 2

   Exploring around the unsaturated functional groups:

   # >=O

   # >=< 2

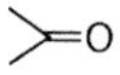

   There are four alkene carbons, so two alkenes. From the four signals 129.3–131.7, we know that each of the alkene carbons has one H. From 6.12, dt, J = 15.5, 7.6 Hz, 1H and 6.18, dq, J = 15.2, 7.2 Hz, 1H we know that both alkenes are trans.

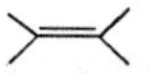

   # sp (R-C≡N 105-120,s ; ≡ 65-90)

   # rings:

   IHD = 2

3. Other heteroatoms, exploring around the saturated functional groups:

   Since one H is not attached to C, the one O is an alcohol. From 63.2, t, the carbon of the alcohol is a CH2. From 4.12, d, J=7.6 Hz, 2H, the H next to that C has one H, so it must be one of the alkene H's, and we find it at 6.12, dt, J=15.5, 7.6 Hz, 1H

4. Other pieces:

   a. Methyl groups

   From 1.75, d, J=7.2 Hz, 3H we know that the methyl is attached to a carbon having one H. This must be the alkene carbon, with H at 6.18, dq, J=15.2, 7.2 Hz, 1H.

   b. ^{1}H NMR > 2.0

   c. Other carbons

5. Putting it all together: OH

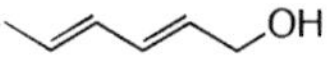

For detailed instructions for solving this problem, see http://valhalla.chem.udel.edu/SpecBook.pdf.

2.28

$C_7H_{12}O$

^{13}C NMR:	**^{1}H NMR:**
21.2, q (2)	1.08, s, 6H
41.5, t	2.21, d, J = 7.2 Hz, 2H
45.7, s	5.08, d, J = 11.8 Hz, 1H
118.4, t	5.11, d, J = 15.5 Hz, 1H
133.1, d	5.75, ddt, J = 11.8, 15.5, 7.2 Hz, 1H
206.7, d	9.49, s, 1H

WORKSHEET 2.28

1. Molecular Formula $C_7H_{12}O$ from the ^{13}C spectrum, all H's are attached to carbon

   IR

   UV

2. Index of Hydrogen Deficiency = 1

   Exploring around the unsaturated functional groups:

   # C=O 1 This is an aldehyde, 206.7, d. From the H spectrum, 9.49, s, 1H the adjacent carbon has no H.

   # C=C 1 H, H, H, CH_2 2.21, d, J = 7.2 Hz, 2H

   # sp (R-C≡N 105-120,s; C≡C 65-90)

   # rings:

   IHD = 2

3. Other heteroatoms, exploring around the saturated functional groups:

   All accounted for

4. Other pieces:

   a. Methyl groups CH_3 x 2 symmetrical 1.08, s, 3H attached to C with no H's

   b. ^1H NMR > 2.0

   c. Other carbons

5. Putting it all together:

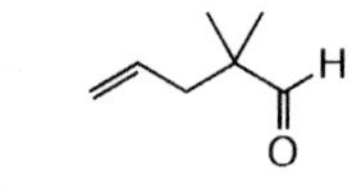

For detailed instructions for solving this problem, see http://valhalla.chem.udel.edu/SpecBook.pdf.

2.29

$C_9H_{16}O$

^{13}C NMR:

14.0, q
22.5, t
27.8, t
28.8, t
31.6, t
32.7, t
132.9, d
159.0, d
194.1, d

^{1}H NMR:

0.90, t, J = 7.6 Hz, 3H
1.3, m, 6H
1.5, m, 2H
2.55, dt, J = 7.3, 7.8 Hz, 2H
6.10, dd, J = 4.5, 15.7 Hz, 1H
6.72, dt, J = 15.7, 7.3 Hz, 1H
9.50, d, J = 4.5 Hz, 1H

WORKSHEET 2.29

1. Molecular Formula $C_9H_{16}O$ from the ^{13}C spectrum, all H's are attached to carbon

   IR

   UV

2. Index of Hydrogen Deficiency = 2

   Exploring around the unsaturated functional groups:

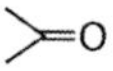

   # >=O 1 — From 194.1, d, we know that we have an aldehyde. From 9.50, d, J=4.5 Hz, 1H, we know that there is one H on the carbon next to the aldehyde carbon. From 6.10, dd, J = 4.5, 15.7 Hz, 1H, we know that that is an alkene carbon.

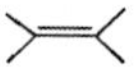

   # >=< 1 — There are two alkene carbons, so one alkene. Eachalkene carbon has one H. From 6.10, dd, J=4.5, 15.7 Hz, 1H and 6.72, dt, J=15.7, 7.3 Hz, 1H, we know that it is a trans alkene, with the aldehyde at one end and a CH_2 at the other.

   # sp (R-C≡N 105-120,s ; ≡ 65-90)

   # rings:

   IHD = 2

3. Other heteroatoms, exploring around the saturated functional groups:

   No more heteroatoms.

4. Other pieces:

   a. Methyl groups — From 0.90, t, J = 7.6 Hz, 3H, we know that there is a CH_3-CH_2-.

   b. ^1H NMR > 2.0

   c. Other carbons — CH_2 x 3

5. Putting it all together:

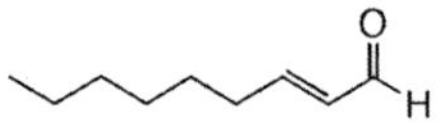

For detailed instructions for solving this problem, see http://valhalla.chem.udel.edu/SpecBook.pdf.

2.30

$C_{12}H_{18}O_4$

^{13}C NMR:

24.2, q
27.6, q (3)
41.8, t (2)
50.0, s
56.1, t
83.6, s
172.9, s
201.7, s (2)

^{1}H NMR:

1.31, s, 3H
1.37, s, 9H
2.51, d, J = 15.6 Hz, 2H
2.83, d, J = 15.6 Hz, 2H
3.28, d, J = 18.0 Hz, 1H
3.35, d, J = 18.0 Hz, 1H

WORKSHEET 2.30

1. Molecular Formula $C_{12}H_{18}O_4$ from the ^{13}C spectrum, all H's are attached to carbon

   IR

   UV

2. Index of Hydrogen Deficiency = 4

   Exploring around the unsaturated functional groups:

   # >=O 3 There are three carbonyls. Two of these are symmetrical ketones, 201.7, s (2). The other is a carboxylic acid derivative, 172.9, s. Since all H's are attached to C, this is an ester. The C attached to O appears at 83.6, s. Three of the methyls are attached to that single C, 1.37, s, 9H.

   # >=<

   # sp (R-C≡N 105-120,s; —≡— 65-90)

   # rings: 1

   IHD = 4

3. Other heteroatoms, exploring around the saturated functional groups:

   All accounted for.

4. Other pieces:

   a. Methyl groups 1.31, s, 3H CH3 attached to C with no H's

   b. ^{1}H NMR > 2.0 3.28, d, J=18.0 Hz, 1H; 3.35, d, J=18.0 Hz, 1H — H's shifted by *two* groups; not symmetrical

   c. Other carbons CH_2 x 2 symmetrical

5. Putting it all together:

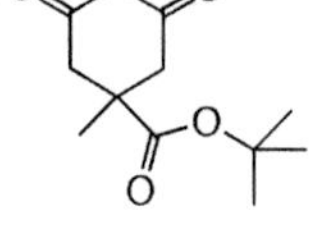

For detailed instructions for solving this problem, see http://valhalla.chem.udel.edu/SpecBook.pdf.

2.31

$C_5H_{12}O_2$

^{13}C NMR:	^1H NMR:
18.9, q	1.18, d, J = 7.7 Hz, 3H
38.7, t	1.72, m, 2H
55.9, q	3.10, bs, 1H (exchanges)
60.6, t	3.33, s, 3H
76.5, d	3.57, m, 1H
	3.74, m, 2H

WORKSHEET 2.31

1. Molecular Formula $C_5H_{12}O_2$ from the ^{13}C spectrum, all but one H is not attached to carbon

   IR

   UV

2. Index of Hydrogen Deficiency = 0

   Exploring around the unsaturated functional groups:

   # >=O

   # >=<

   # sp (R-C≡N 105-120,s ≡ 65-90)

   # rings:

   IHD = 0

3. Other heteroatoms, exploring around the saturated functional groups:

There are two O's. One H is not attached to C, so there must be one ether and one alcohol. There are three C's attached to O, 58.9, q, 66.2, d and 78.4, t. The carbon having three H's that is attached to an O must be part of the ether, so there are two possibilities:

$-CH_2-O-CH_3$ + $-CH-OH$ -or- $-CH-O-CH_3$ + $-CH_2-OH$

4. Other pieces:

   a. Methyl groups CH-CH_3 at 1.18, d, J = 7.7 Hz, 3H

   b. ^1H NMR > 2.0

   c. Other carbons CH_2

5. Putting it all together:

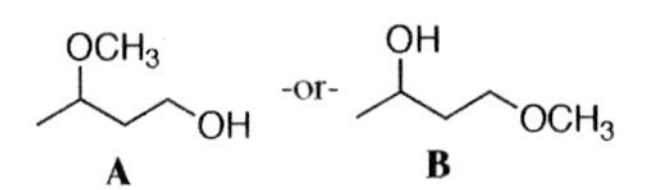

Calculate both using Table C.10 and Table C.11. You will see that **A** is the better fit.

For detailed instructions for solving this problem, see http://valhalla.chem.udel.edu/SpecBook.pdf.

2.32

$C_{10}H_{18}O_2$

MS: 170, 129, 111, 83

^{13}C NMR:

13.3, q
25.7, t (2)
27.9, t
28.5, t (2)
46.1, d
49.8, d
64.4, t
218.1, s

^{1}H NMR:

1.05, d, J = 6.5 Hz, 3H
1.1–1.4, m, 6H
1.6–1.9, m, 4H
2.46, m, 2H
2.9, m, 1H
3.53, dd, J = 4.4, 10.1 Hz, 1H
3.70, dd, J = 6.8, 10.1 Hz, 1H

WORKSHEET 2.32

1. Molecular Formula $C_{10}H_{18}O_2$ from the ^{13}C spectrum, one H is not attached to carbon

   IR

   UV

2. Index of Hydrogen Deficiency = 2

   Exploring around the unsaturated functional groups:

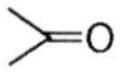

   # C=O 1 Ketone carbonyl at 218.1, s. The alpha shift of the ketone is + 30, so 46.1, d and 49.8, d are probably the carbons flanking the ketone carbonyl.

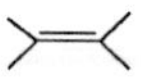

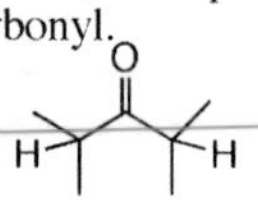

   # C=C

   O, H, H not symmetrical!

   # sp (R-C≡N 105-120,s; C≡C 65-90)

   # rings: 1

   IHD = 2

3. Other heteroatoms, exploring around the saturated functional groups:

   The carbon two, which the alcohol is attached is at 64.4, t $-CH_2-OH$

4. Other pieces:

   a. Methyl groups one CH_3

   b. ^1H NMR > 2.0

   c. Other carbons CH_2 x 2 symmetrical x 2 one CH_2

5. Putting it all together:

   Since the CH2's are symmetrical and the ketone is not, the ketone must be outside the ring.

   Correct: O, OH

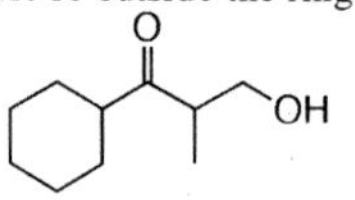

   Calculate the ^{13}C chemical shifts for the correct structure and for the two competing alternatives.

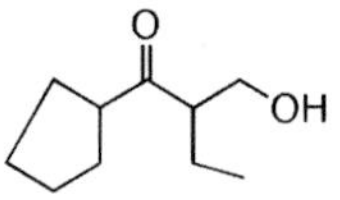

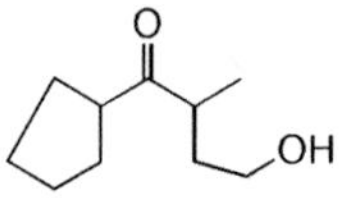

   O, OH O, OH

For detailed instructions for solving this problem, see http://valhalla.chem.udel.edu/SpecBook.pdf.

2.33

$C_6H_{10}O$

^{13}C NMR:

20.6, q
27.5, q
31.5, q
124.3, d
154.6, s
198.2, s

^{1}H NMR:

1.90, s, 3H
2.10, 3, 3H
2.15, s, 3H
6.09, s, 1H

WORKSHEET 2.33

1. Molecular Formula $C_6H_{10}O$ from the ^{13}C spectrum, all H's are attached to carbon

   IR

   UV

2. Index of Hydrogen Deficiency = 2

   Exploring around the unsaturated functional groups:

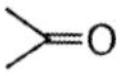

   # >=O 1 Ketone, 198.2, s. One flanking carbon is a CH_3,31.5, q, 2.15, s, 3H. 124.3, d.

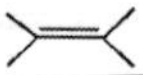

   # >=< 1 154.6, s This alkene is polarized by X, an electron-withdrawing group.

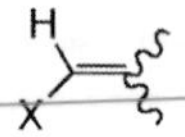

   # sp (R-C≡N 105-120,s; ≡ 65-90)

   # rings:

   IHD = 2

3. Other heteroatoms, exploring around the saturated functional groups:

   All are accounted for.

4. Other pieces:

   a. Methyl groups CH_3 x 2, both attached to a carbon with no H, 1.90, s, 3H, 2.10, 3, 3H.

   b. ^1H NMR > 2.0

   c. Other carbons

5. Putting it all together:

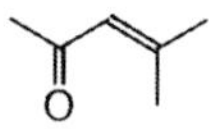

For detailed instructions for solving this problem, see http://valhalla.chem.udel.edu/SpecBook.pdf.

2.34

$C_7H_{14}O_3$

^{13}C NMR:	**^1H NMR:**
18.8, q	1.11, d, J = 6.2 Hz, 3H
30.0, t	1.77, m, 2H
31.3, t	2.37, t, J = 7.3 Hz, 2H
51.4, q	3.27, s, 3H
56.0, q	3.31, m, 1H
75.7, d	3.64, s, 3H
174.2, s	

WORKSHEET 2.34

1. Molecular Formula $C_7H_{14}O_3$ from the ^{13}C spectrum, all H's are attached to carbon

   IR

   UV

2. Index of Hydrogen Deficiency = 1

   Exploring around the unsaturated functional groups:

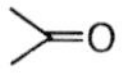

   # C=O 1

   This is a carboxylic acid derivative, 174.2, s, a methyl ester, 56.0, q, 3.64, s, 3H. The carbon next the ester carbonyl has two H's 2.37, t, J = 7.3 Hz, 2H, and there are two H's on the next carbon also.

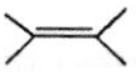

   # C=C

   # sp (R-C≡N 105-120,s; C≡C 65-90)

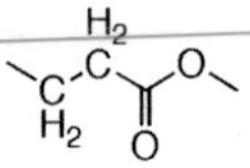

   # rings:

   IHD = 1

3. Other heteroatoms, exploring around the saturated functional groups:

   One other O, an ether. One of the attached C's has three H's, and the other has one: 75.7, d, 51.4, q, 3.27, s, 3H.

4. Other pieces:

   a. Methyl groups CH_3 attached to a C with one H 1.11, d, J = 6.2 Hz, 3H.

   b. ^1H NMR > 2.0

   c. Other carbons

5. Putting it all together:

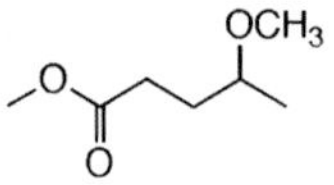

For detailed instructions for solving this problem, see http://valhalla.chem.udel.edu/SpecBook.pdf.

2.35

$C_8H_{12}O$

IR: 2921, 2853, 1678, 1657, 1615, 1356, 1260 cm^{-1}

^{13}C NMR:

16.7, q
21.3, t
30.2, q
34.2, t
41.0, t
135.6, s
154.1, s
198.3, s

^{1}H NMR:

1.82, m, 2H
2.09, s, 3H
2.25, s, 3H
2.50, t, J = 7.6 Hz, 2H
2.66, t, J = 7.2 Hz, 2H

WORKSHEET 2.35

1. Molecular Formula $C_8H_{12}O$ from the ^{13}C spectrum, all H's are attached to carbon

   IR

   UV

2. Index of Hydrogen Deficiency = 3

   Exploring around the unsaturated functional groups:

   # >=O 1 Ketone, 198.3, s. One flanking carbon is a CH_3,30.2, q, 2.25, s, 3H.

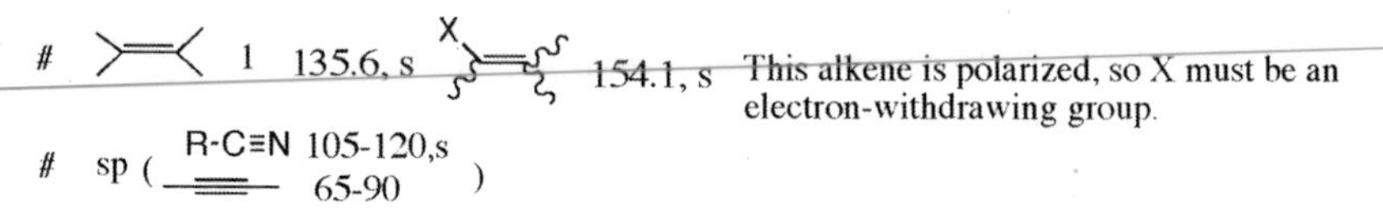

   # rings: 1

   IHD = 3

3. Other heteroatoms, exploring around the saturated functional groups:

   None

4. Other pieces:

   a. Methyl groups CH_3 attached to a C with no H's, 2.09, s, 3H. From the chemical shift, this CH_3 is probably attached to an alkene.

   b. ^1H NMR > 2.0

   c. Other carbons CH_2 x 3 ring

5. Putting it all together:

For detailed instructions for solving this problem, see http://valhalla.chem.udel.edu/SpecBook.pdf.

2.36

C_5H_8O

^{13}C NMR:

13.4, q
30.3, t
69.9, t
94.4, d
154.9, s

^{1}H NMR:

1.78, s, 3H
2.6, dt, J = 3.4, 7.6 Hz, 2H
4.30, t, J = 7.6 Hz, 2H
4.66, t, J = 3.4 Hz, 1H

WORKSHEET 2.36

1. Molecular Formula C_5H_8O from the ^{13}C spectrum, all H's are attached to carbon

   IR

   UV

2. Index of Hydrogen Deficiency = 2

   Exploring around the unsaturated functional groups:

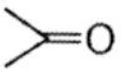

   # >=O

   154.9, s 94.4, d

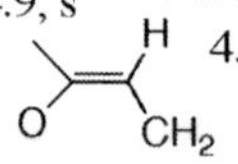

   H 4.66, t, J=3.4 Hz, 1H

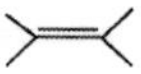

   # >=< 1

   O CH2

   From the chemical shifts of the alkene carbons, it is apparent that the alkene is polarized by a heteroatom. It would have to be an O.

   # sp (R-C≡N 105-120,s ≡ 65-90)

   # rings: 1

   IHD = 2

3. Other heteroatoms, exploring around the saturated functional groups:

   The other carbon on the ether oxygen is a CH_2, 69.9, t.

4. Other pieces:

   a. Methyl groups The CH_3 is attached to a C with no H's. That it is directly attached to the alkene is confirmed by the chemical shift, 1.78, s, 3H.

   b. ^1H NMR > 2.0

   c. Other carbons ring

5. Putting it all together:

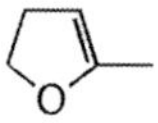

For detailed instructions for solving this problem, see http://valhalla.chem.udel.edu/SpecBook.pdf.

2.37

$C_7H_{10}O$

^{13}C NMR:	**^{1}H NMR:**
21.7, t	1.8, m, 4H
26.1, t	2.4, m, 2H
30.2, t	2.60, t, J = 6.4 Hz, 2H
43.5, t	6.00, d, J = 11.5 Hz, 1H
132.5, d	6.60, dt, J = 11.5, 5.8 Hz, 1H
146.4, d	
204.2, s	

WORKSHEET 2.37

1. Molecular Formula $C_7H_{10}O$ from the ^{13}C spectrum, all H's are attached to carbon

IR

UV

2. Index of Hydrogen Deficiency = 3

Exploring around the unsaturated functional groups:

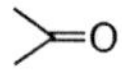

>=O 1 The carbonyl at 204.2, s is a ketone. One of the flanking carbons is a CH_2, 43.5, t. This comes at 2.60, t, J=6.4 Hz, 2H, so it is further connected to another CH_2.

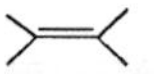

>=< 1

sp (R-C≡N 105-120,s / ≡ 65-90)

The alkene is polarized by an electron-withdrawing group, which must be the ketone. It is a Z alkene, from the 11.5 Hz coupling constant between the alkene H's, 6.00, d, J=11.5 Hz, 1H and 6.60, dt, J = 11.5, 5.8 Hz, 1H.

rings: 1

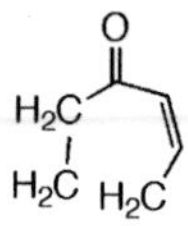

O
H_2C
H_2C H_2C
132.5, d
146.4, d

IHD = 3

3. Other heteroatoms, exploring around the saturated functional groups:

All accounted for.

4. Other pieces:

a. Methyl groups None

b. ^1H NMR > 2.0

c. Other carbons CH_2, ring

5. Putting it all together:

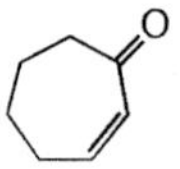

For detailed instructions for solving this problem, see http://valhalla.chem.udel.edu/SpecBook.pdf.

2.38

$C_9H_{14}O_2$

^{13}C NMR:	^{1}H NMR:
26.4, t	1.7, m, 2H
29.3. t	2.2, m, 2H
31.7, t	2.5, m, 2H
50.6, q	3.80, s, 3H
116.4, t	4.89, d, J=15.8 Hz, 1H
119.0, d	5.01, d, J=10.5, 1H
131.3, d	5.77, m, 1H
150.7, d	5.68, d, J=11.5 Hz, 1H
166.5, s	6.14, dt, J=11.5, 7.5 Hz, 1H

WORKSHEET 2.38

1. Molecular Formula $C_9H_{14}O_2$ from the ^{13}C spectrum, all H's are attached to carbon

   IR

   UV

2. Index of Hydrogen Deficiency = 3

   Exploring around the unsaturated functional groups:

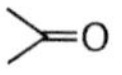

   # >=O 1

   # >=< 2

   The carbonyl is a carboxylic acid derivative, 166.5, s. Since all H's are attached to carbon, and we only have O in the formula, this must be an ester. The carbon attached to O comes at 50.6, q, so this is a methyl ester. This is confirmed by 3.80, s, 3H.

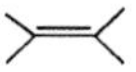

   # sp (R-C≡N 105-120,s ; —≡— 65-90)

   # rings :

   There are four alkene type carbons, so two alkenes. Three of the alkenes have one H attached, and one has two. One of the alkene carbons, 150.7, d, has a larger chemical shift than normal, so it is at the other end of the alkene from an electron-withdrawing group.

   IHD = 3

3. Other heteroatoms, exploring around the saturated functional groups:

   No other heteroatoms.

4. Other pieces:

   a. Methyl groups

   b. ^1H NMR > 2.0

   Note: There are no branch points, so this is a single straight chain. One end is the methyl ester. The other end must be the alkene carbon with two H's.

   c. Other carbons CH_2 x 3

5. Putting it all together:

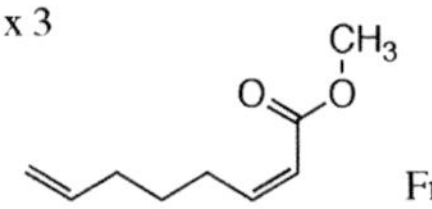

   CH_3 O O

   From 5.68, d, J = 11.5 Hz, 1H and 6.14, dt, J = 11.5, 7.5 Hz, 1H, the shared coupling constant between the alkenes is 11.5 Hz, so this must be a Z (cis) alkene.

For detailed instructions for solving this problem, see http://valhalla.chem.udel.edu/SpecBook.pdf.

2.39

$C_{10}H_{11}ClO_2$

^{13}C NMR:	**^{1}H NMR:**
27.0, t	2.05, m, 2H
34.4, t	3.08, t, J = 7.8 Hz, 2H
44.9, t	3.70, t, J = 7.3 Hz, 2H
115.1, d (2)	6.85, d, J = 8.4 Hz, 2H
128.0, s	7.76, d, J = 8.4 Hz, 2H
130.2, d (2)	10.2, bs, 1H (exchanges)
161.9, s	
196.8, s	

WORKSHEET 2.39

1. Molecular Formula $C_{10}H_{11}ClO_2$ from the ^{13}C spectrum, one H is attached to carbon

   IR

   UV

2. Index of Hydrogen Deficiency = 5

   Exploring around the unsaturated functional groups:

   # >=O 1 — From 196.8, s, there is a ketone. One of the carbons next to the ketone is probably 44.9, t. This corresponds to 3.08, t, J=7.8 Hz, 2H, so there is a CH_2 next to the CH_2 next to the ketone.

   # >=< 3

   # sp (R-C≡N 105-120,s / ≡ 65-90)

   # rings: 1

   With three alkenes (six alkene carbons) and a ring, we suspect a benzene derivative. This is confirmed by 6.85, d, J=8.4 Hz, 2H and 7.76, d, J=8.4 Hz, 2H. From the pattern of coupling constants, we can deduce that this is a symmetrical 1,4-disubstituted benzene.

   IHD = 5

3. Other heteroatoms, exploring around the saturated functional groups:

   The O is an OH. From 161.9, s and 10.2, bs, 1H (exchanges), we know that the OH is directly attached to the benzene ring.

4. Other pieces:

   a. Methyl groups None

   b. ^1H NMR > 2.0 From 3.70, t, J = 7.3 Hz, 2H, the Cl is attached to a CH_2.

   c. Other carbons CH_2 x 3 total

5. Putting it all together:

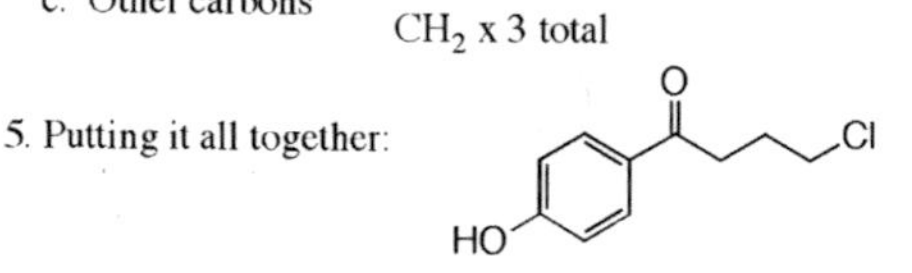

As an exercise, calculate the ^{13}C chemical shift of each of the four benzene carbons.

For detailed instructions for solving this problem, see http://valhalla.chem.udel.edu/SpecBook.pdf.

2.40

$C_{12}H_{15}N$

^{13}C NMR:	**^{1}H NMR:**
28.3, t	2.41, s, 3H
45.7, q	2.6, m, 2H
52.4, t	2.65, t, J = 5.5 Hz, 2H
55.1, t	3.12, d, J = 4.6 Hz, 2H
121.8, d (2)	6.05, t, J = 4.6 Hz, 1H
124.9, d	7.4, m, 5H
126.9, d	
128.2, d (2)	
134.8, s	
141.0, s	

WORKSHEET 2.40

1. Molecular Formula $C_{12}H_{15}N$ from the ^{13}C spectrum, all H's are attached to carbon

   IR

   UV

2. Index of Hydrogen Deficiency = 6

   Exploring around the unsaturated functional groups:

   # >=O

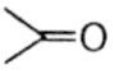

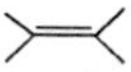

   # >=< 4

   # sp (R-C≡N 105-120,s ≡ 65-90)

   # rings: 2

   IHD = 6

With four alkenes (eight alkene carbons) and a ring, we suspect a benzene derivative. This is confirmed by 7.4, m, 5H. This is a monosubstituted benzene derivative. There is an additional alkene. One of those alkene carbons has one H, and one has none. The alkene H 6.05, t, J=4.6 Hz, 1H, there is a CH_2 next to the alkene H.

3. Other heteroatoms, exploring around the saturated functional groups:

   The nitrogen is attached to three carbons, 45.7, q; 52.4, t; 55.1, t. These correspond to 2.41, s, 3H; 2.65, t, J=5.5 Hz, 2H; 3.12, d, J=4.6 Hz, 2H.

4. Other pieces:

   a. Methyl groups

   Thinking about the other ring: There are two end groups, a phenyl and a methyl group, so there must be two branch points. One is the alkene carbon with no H's; the other is the N.

   b. ^{1}H NMR > 2.0

   c. Other carbons CH_2 x 1

5. Putting it all together:

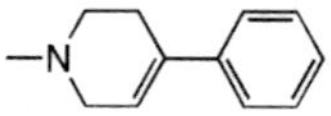

   —N

For detailed instructions for solving this problem, see http://valhalla.chem.udel.edu/SpecBook.pdf.

2.41

$C_{12}H_{15}NO_2$

^{13}C NMR:

21.5, t
28.9, t
45.8, t
59.2, d
66.3, t
128.4, d (2)
128.6, d
128.7, d (2)
135.5, s
168.6, s

^{1}H NMR:

2.0, m, 2H
2.1, m, 1H
2.4, m, 1H
2.48, dd, J = 7.2, 15.9 Hz, 1H
2.58, dd, J = 7.7, 15.9 Hz, 1H
4.50, dd, J = 6.8, 7.4 Hz, 1H
5.14, d, J = 16.2 Hz, 1H
5.23, d, J = 16.2 Hz, 1H
6.8, bs, 1H (exchanges)
7.25, bs, 5H

WORKSHEET 2.41

1. Molecular Formula $C_{12}H_{15}NO_2$ from the ^{13}C spectrum, one H is not attached to carbon

   IR

   UV

2. Index of Hydrogen Deficiency = 6

   Exploring around the unsaturated functional groups:

   # >=O 1 The carbonyl is a carboxylic acid derivative. It could be an acid, an amide or an ester. Since there are two carbons attached to N, 45.8, t; 59.2, d this is either an amide and an alcohol or a secondary amine and an ester.

   # >=< 3

   # sp (R-C≡N 105-120,s ; ≡ 65-90) With three alkenes (six alkene carbons) and a ring, we suspect a benzene derivative. This is confirmed by 7.25, bs, 5H. This is a monosubstituted benzene derivative.

   # rings: 2

   IHD = 6

3. Other heteroatoms, exploring around the saturated functional groups:

   The key to sorting out the heteroatoms is the observation of a carbon attached to oxygen, 66.3, t and 5.14, d, J = 16.2 Hz, 1H; 5.23, d, J = 16.2 Hz, 1H. This is a benzyl ester. There is also a secondary amine, 45.8, t and 59.2, d.

4. Other pieces:

   a. Methyl groups The second ring: There are two branch points, the N-H, and 59.2, d. This latter branch point is a carbon directly attached to the N.

   b. ^{1}H NMR > 2.0

   c. Other carbons CH_2 x 2

5. Putting it all together:

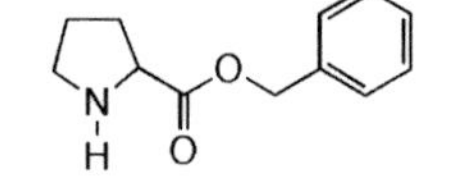

For detailed instructions for solving this problem, see http://valhalla.chem.udel.edu/SpecBook.pdf.

2.42

$C_{14}H_{19}NO_2$

^{13}C NMR:

14.2, q
22.8, t
29.3, t
53.2, t
58.8, t
60.4, t
65.3, d
127.0, d, (2)
128.1, d
129.1, d (2)
138.4, s
174.0, s

^{1}H NMR:

1.24, t, J=7.2 Hz, 3H
2.9, m, 4H
2.36, dd, J=9.3, 11.1 Hz, 1H
3.04, dd, J=3.4, 11.1 Hz, 1H
3.24, dd, J=4.4, 5.1 Hz, 1H
3.56, d, J=17.2 Hz, 1H
3.92, d, J=17.2 Hz, 1H
4.21, t, J=7.2 Hz, 2H
7.3, bs, 5H

WORKSHEET 2.42

1. Molecular Formula $C_{14}H_{19}NO_2$ from the ^{13}C spectrum, all H's are attached to carbon

   IR

   UV

2. Index of Hydrogen Deficiency = 6

   Exploring around the unsaturated functional groups:

   # C=O 1

   The carbonyl is a carboxylic acid derivative. It could be an amide or an ester. The key to this is the observation of a carbon attached to O, 60.4, t and 4.21, t, J=7.2 Hz, 2H. This is attached to a methyl group at 1.24, t, J=7.2 Hz, 3H, so it is an ethyl ester.

   # C=C 3

   # sp (R-C≡N 105-120,s ; C≡C 65-90)

   With three alkenes (six alkene carbons) and a ring, we suspect a benzene derivative. This is confirmed by 7.3, bs, 5H. This is a monosubstituted benzene derivative.

   # rings: 2

   IHD = 6

3. Other heteroatoms, exploring around the saturated functional groups:

   There are three carbons attached to N, 53.2, t; 58.8, t; 65.3, d. These are at 2.36–3.92. 3.56, d, J=17.2 Hz, 1H and 3.92, d, J=17.2 Hz, 1H particularly stand out. This is a CH_2 attached to N and to another electron-withdrawing group, with the ester carbonyl or the phenyl group. The chemical shift of this carbon at 58.8, t.

4. Other pieces:

   About the other ring: There are two branch points, the N and the methine carbon, and two end groups, the phenyl group and the ethyl ester.

   a. Methyl groups

   b. ^1H NMR > 2.0

   c. Other carbons CH_2 x 2

5. Putting it all together:

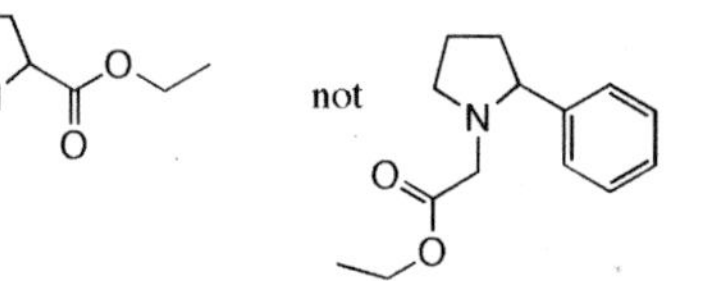

The solution to this is found in the C-1 of the benzene ring, 138.4, s, indicative of a secondary benzylic carbon.

For detailed instructions for solving this problem, see http://valhalla.chem.udel.edu/SpecBook.pdf.

2.43

$C_{12}H_{18}O$

IR: 1683 cm^{-1}

^{13}C NMR:

17.8, q
24.6, q
25.9, q
33.4, t
34.1, t
36.5, s
39.0, t
119.0, d
127.3, d
135.0, s
159.1, d
199.6, s

^{1}H NMR:

1.04, s, 3H
1.52. s, 3H
1.70, s, 3H
1.9, m, 1H
2.2, m, 3H
2.4, m, 2H
5.08, t, J = 7.2 Hz, 1H
5.80, d, J = 10.2 Hz, 1H
6.60, d, J = 10.2 Hz, 1H

WORKSHEET 2.43

1. Molecular Formula $C_{12}H_{18}O$ from the ^{13}C spectrum, all H's are attached to carbon

   IR

   UV

2. Index of Hydrogen Deficiency = 4

   Exploring around the unsaturated functional groups:

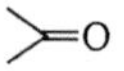

   # >=O 1 One ketone, 199.6, s. One of the carbons flanking the carbonyl carbon is a CH_2, 39.0, t.

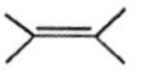

   # >=< 2 There are four alkene carboms, so two alkenes. Since one of the alkenes is shifted downfield, 159.1, d, that alkene must be conjuatged with the electron-withdrawing carbonyl. The H on this alkene carbon comes ar 6.60, d, J = 10.2 Hz, 1H and is adjacent to 5.80, d, J = 10.2 Hz, 1H. This neighboring carbon to the carbon bearing this latter H must not have any H attached to it.

   # sp (R-C≡N 105-120,s ; ≡ 65-90)

   # rings: 1

   IHD = 4

3. Other heteroatoms, exploring around the saturated functional groups:

   All accounted for

4. Other pieces:

   a. Methyl groups One CH_3 1.04, s, 3H is attached to a carbon with no other H's attached.

   CH_3 x 2 1.52. s, 3H, 1.70, s, 3H are attached to a fully substituted alkene carbon.

   b. ^1H NMR > 2.0

   c. Other carbons *Considering the ring*: The fully substituted carbon is a double branch point, so there must be two end groups. These are the methyl group at 1.04, s, 3H, and the alkene carbon with the two methyls attached.

5. Putting it all together:

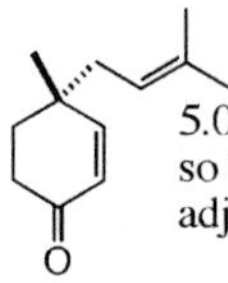

5.08, t, J=7.2 Hz, 1H, so there must be at least one adjacent CH_2

O

This ring must be at least a six-membered ring IR: 1683 cm^{-1}

For detailed instructions for solving this problem, see http://valhalla.chem.udel.edu/SpecBook.pdf.

2.44

$C_9H_{10}O_2$

^{13}C NMR:

29.8, q
46.0, t
116.6, d
120.6, d
121.0, s
128.8, d
130.9, d
154.7, s
210.3, s

^{1}H NMR:

2.30, s, 3H
3.75, s, 2H
6.8, m, 2H
6.99, bs, 1H (exchanges)
7.2, m, 1H
7.6, m, 1H

WORKSHEET 2.44

1. Molecular Formula $C_9H_{10}O_2$ from the ^{13}C spectrum, one H is not attached to carbon

   IR

   UV

2. Index of Hydrogen Deficiency = 5

   Exploring around the unsaturated functional groups:

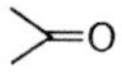

   # >=O 1 — The carbonyl is a ketone, 210.3, s. The two carbons flanking the carbonyl carbon are 29.8, q and 46.0, t. From 2.30, s, 3H, and 3.75, s, 2H the CH_2 is also attached to another electrron-withdrawing group, with no H's on that neighboring carbon.

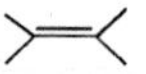

   # >=< 3

   # sp (R-C≡N 105-120,s / ≡ 65-90) — The three alkenes and the ring make up a benzene derivative. From 116.6, d through 154.7, s there is no symmetry. Further, those two signals indicate that the OH is directly attached to the benzene ring.

   # rings: 1

   IHD = 5

3. Other heteroatoms, exploring around the saturated functional groups:

   OH

4. Other pieces:

   a. Methyl groups

   b. ^1H NMR > 2.0

   c. Other carbons

5. Putting it all together:

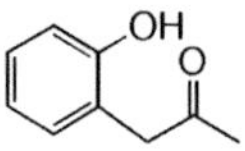

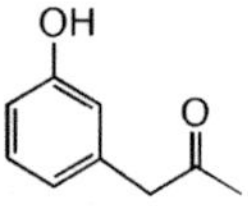

OH O OH O

Does not fit—as an exercise, calculate the ^{13}C chemical shifts for both structures.

For detailed instructions for solving this problem, see http://valhalla.chem.udel.edu/SpecBook.pdf.

2.45

$C_9H_8O_2$

^{13}C NMR:	**^{1}H NMR:**
37.7, t	2.87, t, J = 7.2 Hz, 2H
67.0, t	2.52, t, J = 7.2 Hz, 2H
117.8, d	6.95, d, J = 8.2 Hz, 1H
121.3, d	7.02, t, J = 8.2 Hz, 1H
127.0, d	7.44, t, J = 8.2 Hz, 1H
135.8, d	7.88, d, J = 8.2 Hz, 1H
142.1, s	
161.8, s	
191.7, s	

WORKSHEET 2.45

1. Molecular Formula $C_9H_8O_2$ from the ^{13}C spectrum, all H's are attached to carbon

   IR

   UV

2. Index of Hydrogen Deficiency = 6

   Exploring around the unsaturated functional groups:

   # >=O 1 This is a ketone carbonyl, 191.7, s. One of the carbons flanking the carbonyl is found at 37.7, t.

   # >=< 3

   # sp (R-C≡N 105-120,s ≡ 65-90)

   Three alkenes and a ring suggest a benzene derivative. It has two substituents, which from the coupling pattern (6.95-7.88) indicate cis-disubstitution with no symmetry. The 7.88, d, J=8.2 Hz, ^1H teaches that that position in ortho to the ketone carbonyl.

   # rings: 2

   IHD = 6

3. Other heteroatoms, exploring around the saturated functional groups:

   The other O is an ether, with a CH_2 67.0, t attached. The other carbon attached to the O is a carbon 161.8, s of the benzene ring.

4. Other pieces :

   a. Methyl groups

   b. ^1H NMR > 2.0

   c. Other carbons

5. Putting it all together:

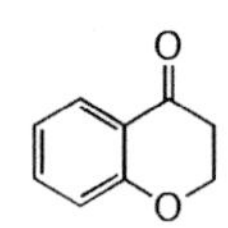

For detailed instructions for solving this problem, see http://valhalla.chem.udel.edu/SpecBook.pdf.

2.46

$C_{11}H_{12}ClN$

^{13}C NMR:	^{1}H NMR:
22.9, t	2.81, t, J = 4.2 Hz, 2H
39.7, t	3.2, bs, 1H (exchanges)
41.1, t	3.27, t, J = 4.2 Hz, 2H
117.6, d (2)	3.72, d, J = 4.6 Hz, 2H
126.4, d (2)	6.23, t, J = 4.6 Hz, 1H
128.3, d	7.42, d, J = 8.1 Hz, 2H
132.2, s	7.52, d, J = 8.1 Hz, 2H
133.1, s	
137.5, s	

WORKSHEET 2.46

1. Molecular Formula $C_{11}H_{12}ClN$ from the ^{13}C spectrum, one H is not attached to carbon

   IR

   UV

2. Index of Hydrogen Deficiency = 6

   Exploring around the unsaturated functional groups:

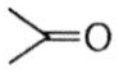

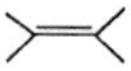

   # C=O

   # C=C 4

   Three of the alkenes and one of the rings make up a benzene derivative. From 7.42, d, J=8.1 Hz, 2H and 7.52, d, J=8.1 Hz, 2H, this is a 1,4-disubstituted benzene ring.

   # sp (~~R-C≡N 105-120,s~~ C≡C 65-90)

   # rings: 2

   IHD = 6

3. Other heteroatoms, exploring around the saturated functional groups:

   There are two carbons, 39.7, t and 41.1, t attached to the N-H.

4. Other pieces:

   The other ring: There are two branch points, the N and the alkene C with no H, and two end groups, the N-H and the phenyl ring. The alkene H is next to a CH_2 6.23, t, J=4.6 Hz, 1H. The other two CH_2's are next to each other 2.81, t, J=4.2 Hz, 2H, 3.27, t, J=4.2 Hz, 2H.

   a. Methyl groups

   b. ^1H NMR > 2.0

   c. Other carbons

5. Putting it all together:

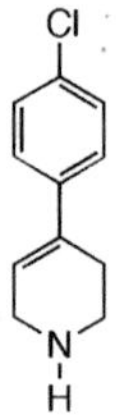

Cl

N

H

For detailed instructions for solving this problem, see http://valhalla.chem.udel.edu/SpecBook.pdf.

2.47

C_8H_9IO

^{13}C NMR:

21.9, q
56.6, q
82.3, s
112.6, d
123.8, d
139.4, d
140.3, s
158.3, s

^{1}H NMR:

2.37, s, 3H
3.91, s, 3H
6.59, d, J = 7.9 Hz, 1H
6.69, s, 1H
7.66, d, J = 7.9 Hz, 1H

WORKSHEET 2.47

1. Molecular Formula C_8H_9IO from the ^{13}C spectrum, all H's are attached to carbon

   IR

   UV

2. Index of Hydrogen Deficiency = 4

   Exploring around the unsaturated functional groups:

   # >=O

   # >=< 3

   Three alkenes and a ring make a benzene derivative. From 82.3, s through 158.3, s there are three subsituents on the benzene ring. The challenge is to identify those substituents and their positioning on the ring. 82.3, s is unusual, but is understandable as the carbon to which the iodine is directly attached.

   # sp (R-C≡N 105-120,s ; ≡ 65-90)

   # rings: 1

   IHD = 4

3. Other heteroatoms, exploring around the saturated functional groups:

   One of the carbons attached to the ether oxygen 56.6, q is a methyl ether. From the chemical shift, 3.91, s, 3H, the methoxy group is attached directly to the benzene ring.

4. Other pieces:

   a. Methyl groups From 2.37, s, 3H, the CH_3 is directly attached to the benzene ring.

   b. ^1H NMR > 2.0

   c. Other carbons

5. Putting it all together:

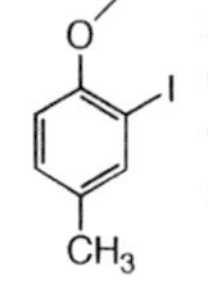

Note that 112.6, d teaches that one and only one carbon C-2 to the methoxy group has an H attached. Calculations of ^{13}C chemical shifts lead to the structure shown.

For detailed instructions for solving this problem, see http://valhalla.chem.udel.edu/SpecBook.pdf.

2.48

$C_{10}H_{17}NO$

^{13}C NMR:	**^{1}H NMR:**
28.8, q (2)	1.08, s, 6H
32.6, t	2.12, s, 2H
39.7, q (2)	2.39, s, 2H
40.8, s	3.00, s, 6H
49.3, t	5.13, s, 1H
97.4, d	
164.0, s	
196.0, s	

WORKSHEET 2.48

1. Molecular Formula $C_{10}H_{17}NO$ from the ^{13}C spectrum, all H's are attached to carbon

   IR

   UV

2. Index of Hydrogen Deficiency = 3

   Exploring around the unsaturated functional groups:

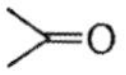

   # >=O 1

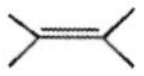

   # >=< 1

   # sp (R-C≡N 105-120,s ; —≡— 65-90)

   # rings: 1

   IHD = 3

3. Other heteroatoms, exploring around the saturated functional groups:

   This is a tertiary amine. Two of the carbons attached to the N are methyl groups, 39.7, q (2). The other carbon is probably the alkene carbon at 196.0, s.

4. Other pieces:

   a. Methyl groups CH_3 x 2 28.8, q (2) symmetrical, two methyl groups attached to a carbon with no H.

   b. ^{1}H NMR > 2.0

   c. Other carbons

5. Putting it all together:

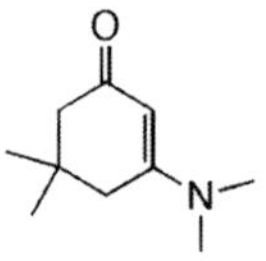

For detailed instructions for solving this problem, see http://valhalla.chem.udel.edu/SpecBook.pdf.

2.49

$C_{14}H_{20}$

^{13}C NMR:

20.8, q
22.7, t
26.4, t (2)
30.6, q
37.5, s
37.8, t (2)
125.7, d (2)
128.9, d (2)
134.5, s
147.0, s

^{1}H NMR:

1.16, s, 3H
1.4, m, 8H
1.98, dd, J=7.49, 13.00 Hz, 2H
2.31, s, 3H
7.12, d, J=8.12 Hz, 2H
7.23, d, J=8.12 Hz, 2H

WORKSHEET 2.49

1. Molecular Formula $C_{14}H_{20}$ from the ^{13}C spectrum, all H's are attached to carbon

   IR

   UV

2. Index of Hydrogen Deficiency = 5

   Exploring around the unsaturated functional groups:

   # >=O

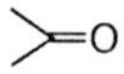

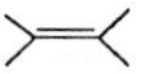

   # >=< 3 Three alkenes and a ring make a benzene derivative. From 7.12, d, J = 8.12 Hz, 2H and 7.23, d, J = 8.12 Hz, 2H, this is a 1,4-disubstituted benzene derivative.

   # sp (R-C≡N 105-120,s ; ≡ 65-90)

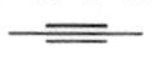

   # rings: 2

   IHD = 1

3. Other heteroatoms, exploring around the saturated functional groups:

   None

4. Other pieces:

   a. Methyl groups One CH_3 is aliphatic, attached to a C with no H's 1.16, s, 3H.

   b. ^1H NMR > 2.0 One CH_3 is attached to the benzene ring 2.31, s, 3H.

   c. Other carbons Two set of symmetrical CH_2's one other CH_2.

5. Putting it all together:

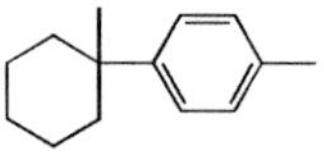

not

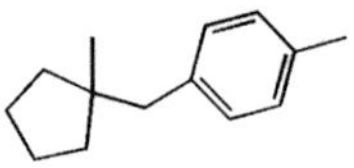

From 147.0, s, indicating a quaternary carbon directly attached to the benzene ring.

For detailed instructions for solving this problem, see http://valhalla.chem.udel.edu/SpecBook.pdf.

2.50

$C_{13}H_{13}NO$

^{13}C NMR:

55.2, q
103.4, d
110.3, d
116.9, d
117.8, d (2)
128.2, d
129.5, d
129.9, d (2)
139.4, s
142.5, s
161.1, s

^{1}H NMR:

2.76, bs, 1H (exchanges)
3.65, s, 3H
6.3, m, 3H
7.12, t, J = 7.5 Hz, 1H
7.33, t, J = 7.5 Hz, 1H
7.39, d, J = 7.5 Hz, 2H
7.56, t, J = 7.5 Hz, 2H

WORKSHEET 2.50

1. Molecular Formula $C_{13}H_{13}NO$ from the ^{13}C spectrum, one H is not attached to carbon

   IR

   UV

2. Index of Hydrogen Deficiency = 8

   Exploring around the unsaturated functional groups:

   # >=O

   # >=< 6

   # sp (R-C≡N 105-120,s / ≡ 65-90)

   # rings: 2

   IHD = 8

   Six alkenes and two rings make two benzenes. There are questions: what substituents are there, to which rings are they attached, and, how are the rings attached to each other. The problem is somewhat simplified, in that one of the benzene rings is monosubstituted, so the attachment to the other ring must be that one substituent.

3. Other heteroatoms, exploring around the saturated functional groups:

   We have to deal with two heteratoms, an O and an N. This is either an ether and a secondary amine, or an alcohol and a tertiary amine. From 55.2, q and 3.65, s, 3H, this is a methyl ether, directly attached to one of the benzene rings. It must also be a secondary amine.

4. Other pieces:

   a. Methyl groups

   b. ^{1}H NMR > 2.0

   c. Other carbons

   Notes about the benzene rings: From 117.8, d (2), the NH must be directly attached to the monosubstituted benzene ring. The other benzene ring has two substituents, the other on the second ring being the methoxy group. From 103.4, d, a CH on that second ring must be between the methoxy group and the amine.

5. Putting it all together:

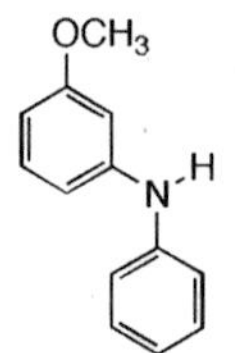

For detailed instructions for solving this problem, see http://valhalla.chem.udel.edu/SpecBook.pdf.

Section III

Mechanism-Based Problems

The problems in section III differ in that the starting material is given, as well as the reagent(s) with which the starting material has reacted. Rather than attempting to solve the spectroscopic data set de novo, it is more sensible to first decide which of the several components of the starting material are still there in the product.

Problem 3.1

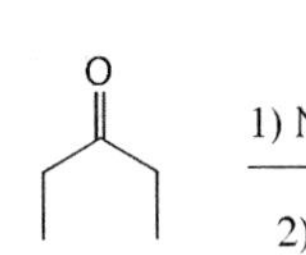

1) NaH / H–C(=O)–OEt → 2) H^+ / CH_3OH → A $C_8H_{16}O_3$

^{13}C NMR:	^1H NMR:
7.4, q	1.00, t, J = 7.6 Hz, 3H
12.6, q	1.07, d, J = 7.2 Hz, 3H
36.3, t	2.49, q, J = 7.6 Hz, 2H
48.7, d	2.89, m, 1H
52.7, q	3.34, s, 3H
55.6, q	3.45, s, 3H
106.3, d	4.39, d, J = 6.8 Hz, 1H
206.4, s	

WORKSHEET 3.1

From the molecular formula: $C_8H_{16}O_3$

$C_5H_{10}O$

$C_3H_6O_2$

Fragments of the starting materials that remain:

H O 206.4, s

2.49, q, J = 7.6 Hz, 2H

CH_3 CH_3

1.00, t, J = 7.6 Hz, 3H

1.07, d, J = 7.2 Hz, 3H

Fragments that have changed in the product:

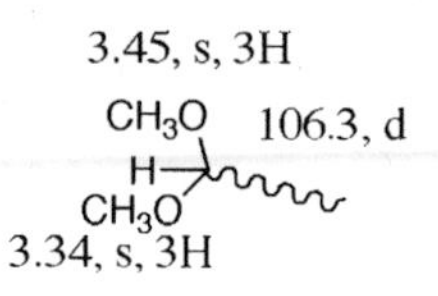

Mechanism and product:

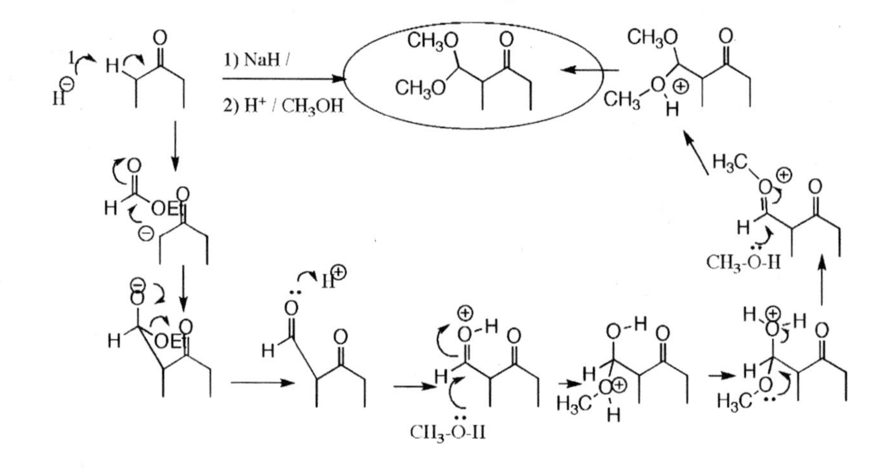

For detailed instructions for solving this problem, see
http://valhalla.chem.udel.edu/SpecBook.pdf.

Problem 3.2

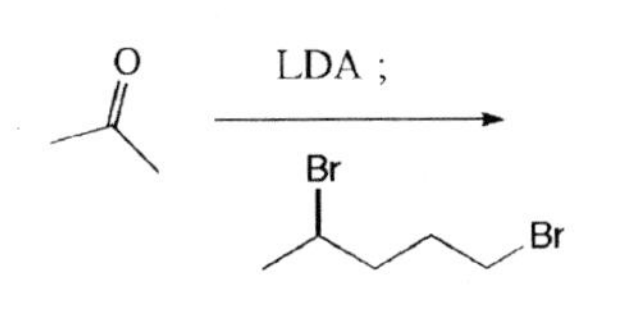

A $C_8H_{15}BrO$

^{13}C NMR:	^{1}H NMR:
208.2, s	4.13, tq, J = 1.5, 6.3 Hz, 1H
51.2, d	2.45, t, J = 7.3 Hz, 2H
43.2, t	2.14, s, 3H
40.7, t	1.70, d, J = 6.3 Hz, 3H
29.8, q	1.3–1.8, m, 6H
27.1, t	
26.3,q	
22.9, t	

WORKSHEET 3.2

From the molecular formula:

$C_5H_8Br_2$ + C_3H_6O = $C_8H_{14}Br_2O$ - $C_8H_{15}BrO$ = HBr

Fragments of the starting materials that remain:

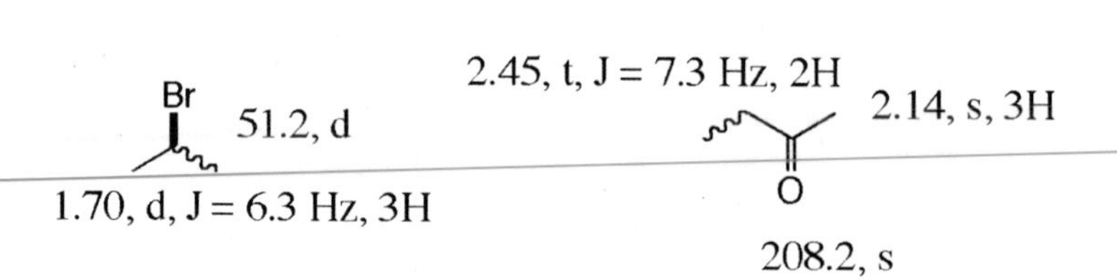

Fragments that have changed in the product:

None

Mechanism and product:

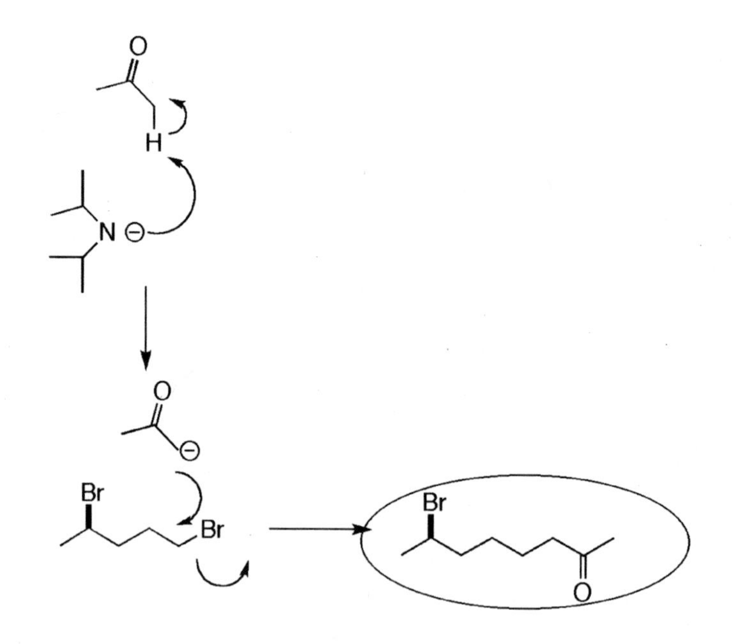

For detailed instructions for solving this problem, see http://valhalla.chem.udel.edu/SpecBook.pdf.

Problem 3.3

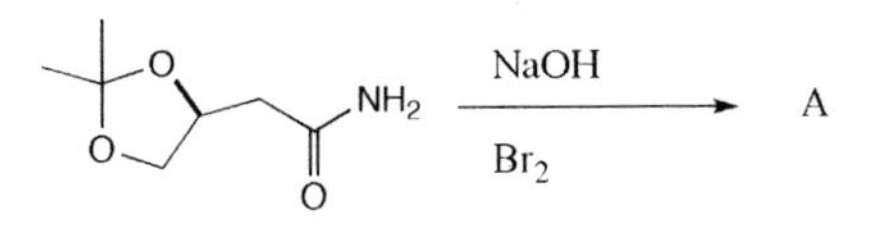

$C_6H_{13}NO_2$

^{13}C NMR:	^{1}H NMR:
25.3, q	1.31, bs, 2H (exchanges)
26.8, q	1.34, s, 3H
44.7, t	1.40, s, 3H
66.9, t	2.78, dd, J = 6.0, 13.2 Hz, 1H
77.4, d	2.85, dd, J = 4.2, 13.2 Hz, 1H
109.1, s	3.67, dd, J = 6.3, 8.1 Hz, 1H
	4.00, dd, J = 6.6, 8.1 Hz, 1H
	4.13, m, 1H

WORKSHEET 3.3

From the molecular formula:

$$C_7H_{13}NO_3 \; - \; C_6H_{13}NO_2 \; = \; - \; CO$$

Fragments of the starting materials that remain:

1.40, s, 3H

1.34, s, 3H

109.1, s

3.67, dd, J = 6.3, 8.1 Hz, 1H

4.00, dd, J = 6.6, 8.1 Hz, 1H

Fragments that have changed in the product:

CH_2–NH_2

2.78, dd, J = 6.0, 13.2 Hz, 1H

2.85, dd, J = 4.2, 13.2 Hz, 1H

Mechanism and product:

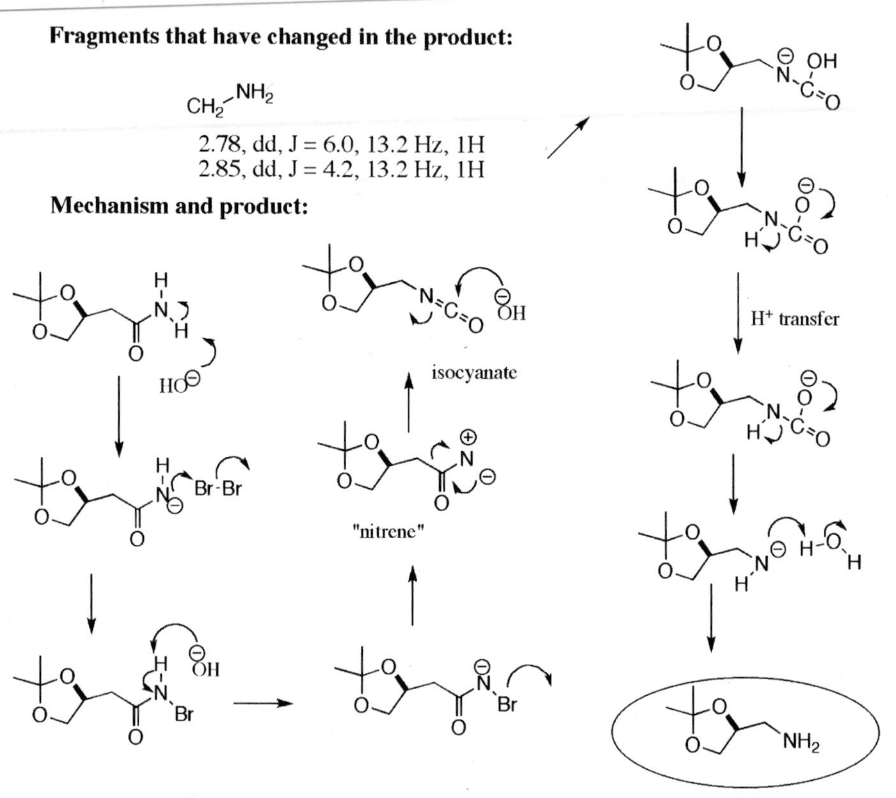

For detailed instructions for solving this problem, see http://valhalla.chem.udel.edu/SpecBook.pdf.

Problem 3.4

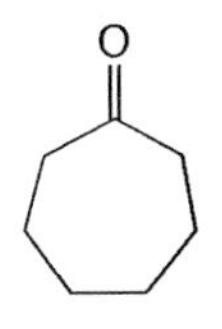

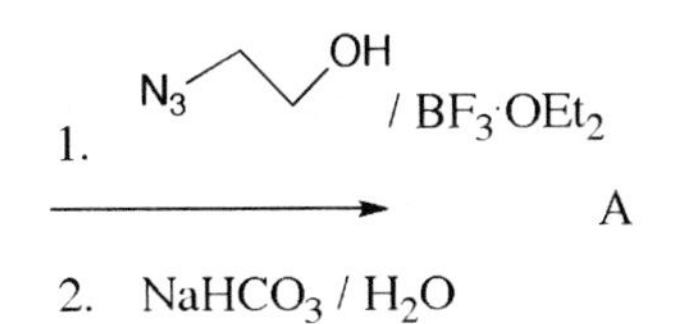

A $C_9H_{17}NO_2$

IR: 3340, 2920, 1715, 1345, 1240 cm^{-1}

^{1}H NMR:
1.42–1.59, m, 7H
1.82, m, 2H
2.39, t, 2H
2.68, t, J = 4.4 Hz, 2H
2.94, t, J = 4.8 Hz, 2H
4.22, t, J = 4.8 Hz, 2H

^{13}C NMR:
24.7, t
26.6, t
29.0, t
29.8, t
34.4, t
49.6, t
50.6, t
63.3, t
177.9, s

WORKSHEET 3.4

From the molecular formula:

$C_7H_{12}O \;+\; C_2H_5N_3O \;=\; C_9H_{17}N_3O_2 \;-\; C_9H_{17}NO_2 \;=\; N_2$

Fragments of the starting materials that remain:

2.39, t, 2H

4.22, t, J = 4.8 Hz, 2H

2.94, t, J = 4.8 Hz, 2H

2.68, t, J = 4.4 Hz, 2H

There are no branch points other than the N-H, so everything must in the one ring.

Fragments that have changed in the product:

N_3 is gone.

CH_2-OH is gone.

Mechanism and product:

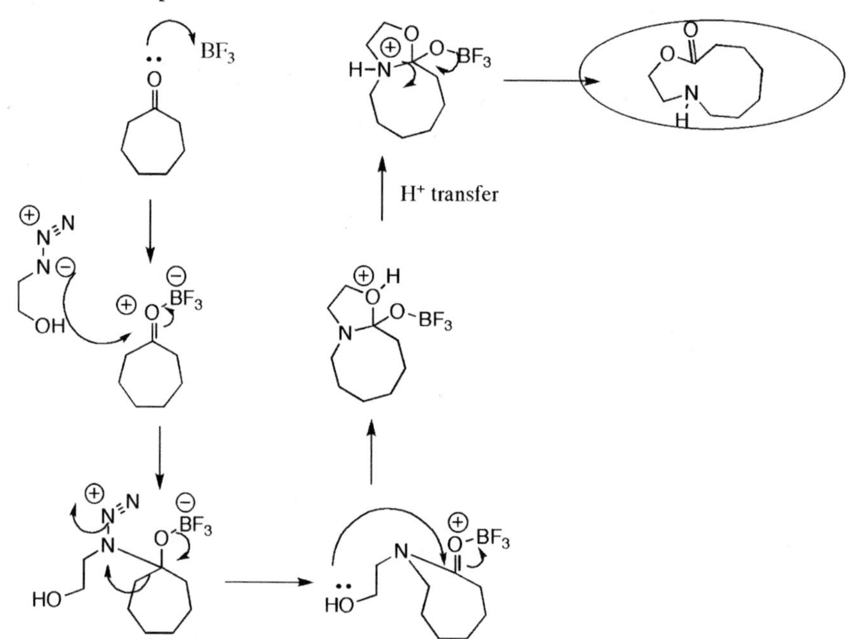

For detailed instructions for solving this problem, see http://valhalla.chem.udel.edu/SpecBook.pdf.

Problem 3.5

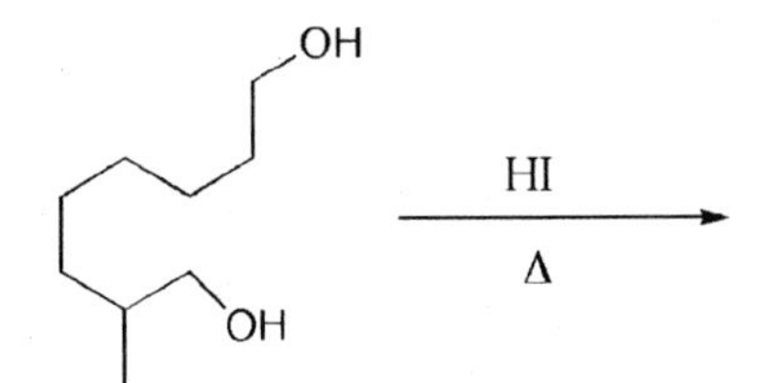

A $C_9H_{19}IO$

^{1}H NMR:

1.2–1.3, m, 4H
1.32, s, 6H
1.4–1.5, m, 4H
1.65, m, 2H
3.17, bs, 1H (exchanges)
3.60, t, J = 6.5 Hz, 2H

^{13}C NMR:

24.8, t
24.8, t
26.2, t
29.5, s
32.4, q (2)
33.2, t
40.7, t
42.4, t
63.0, t

WORKSHEET 3.5

From the molecular formula:

$C_9H_{20}O_2$ + HI = $C_9H_{21}IO_2$ - $C_9H_{19}IO$ = H_2O

Fragments of the starting materials that remain:

OH

3.17, bs, 1H (exchanges)
3.60, t, J = 6.5 Hz, 2H

Fragments that have changed in the product:

I

1.32, s, 6H 32.4, q (2)
29.5, s

Mechanism and product:

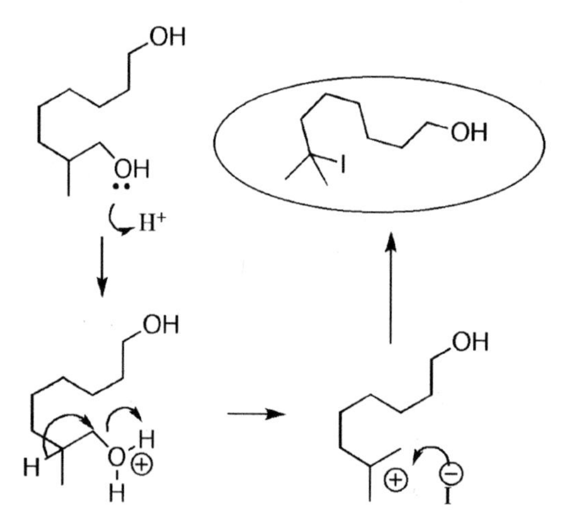

For detailed instructions for solving this problem, see
http://valhalla.chem.udel.edu/SpecBook.pdf.

Problem 3.6

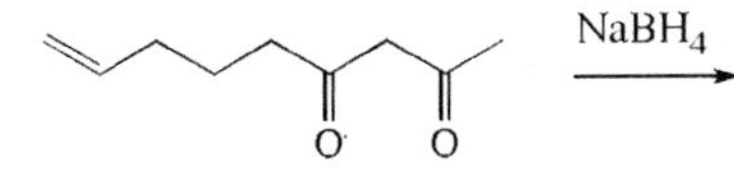

A $C_9H_{16}O_2$

IR: 3430, 3077, 1704, 1640 cm^{-1}

^{13}C NMR:	**^1H NMR:**
22.3, q	1.16, d, J = 6.4 Hz, 3H
22.5, t	1.6, m, 2H
33.0, t	2.1–2.5, m, 6H
42.6, t	3.08, bs, 1H (exchanges)
50.5, t	4.2, m, 1H
63.8, d	5.0, m, 2H
115.4, t	5.73, ddt, J = 17.1, 10.3, 6.8 Hz, 1H
137.8, d	
212.1, s	

WORKSHEET 3.6

From the molecular formula:

$C_9H_{14}O_2 + H = C_9H_{15}O_2 \quad - C_9H_{16}O_2 = H$

Fragments of the starting materials that remain:

5.0, m, 2H
5.73, ddt, J = 17.1, 10.3, 6.8 Hz, 1H

O
42.6, t
50.5, t

Note that the methyl ketone is gone

CH_3
O

Fragments that have changed in the product:

CH_3
OH

1.16, d, J = 6.4 Hz, 3H

Mechanism and product:

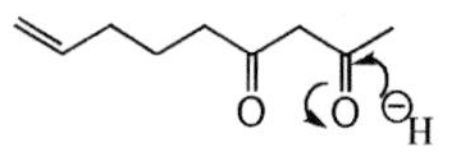

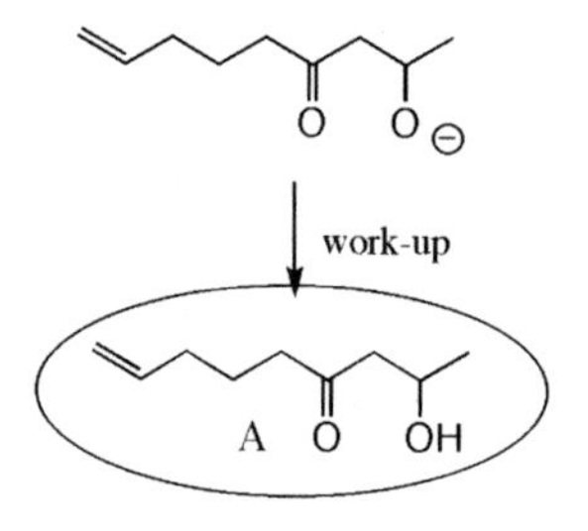

For detailed instructions for solving this problem, see http://valhalla.chem.udel.edu/SpecBook.pdf.

Problem 3.7

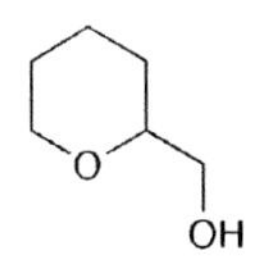

1. TsCl / pyridine
2. NaCN / DMSO

A $C_7H_{11}NO$

IR: 2238 cm^{-1}

^{13}C NMR:	^{1}H NMR:
22.8, t	1.4–1.8, m, 6H
24.9, t	2.49, d, J = 6.0 Hz, 2H
25.2, t	3.45, dt, J = 11.0, 3.0 Hz, 1H
31.0, t	3.55, m, 1H
68.7, t	4.00, m, 1H
72.8, d	
117.3, s	

WORKSHEET 3.7

From the molecular formula:

$C_6H_{12}O_2$ + NaCN = $C_7H_{12}NO_2$ - $C_7H_{11}NO$ = NaOH

Fragments of the starting materials that remain:

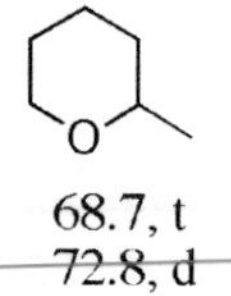

Fragments that have changed in the product:

CH_2-CN   IR: 2238 cm^{-1} CH_2OH is gone.

2.49, d, J = 6.0 Hz, 2H 117.3, s

Mechanism and product:

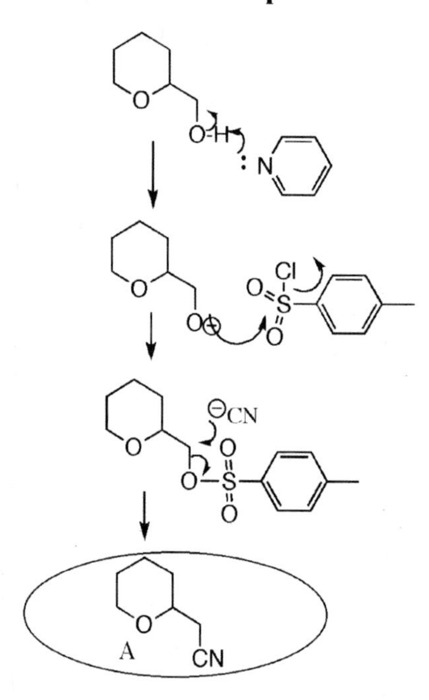

For detailed instructions for solving this problem, see http://valhalla.chem.udel.edu/SpecBook.pdf.

Problem 3.8

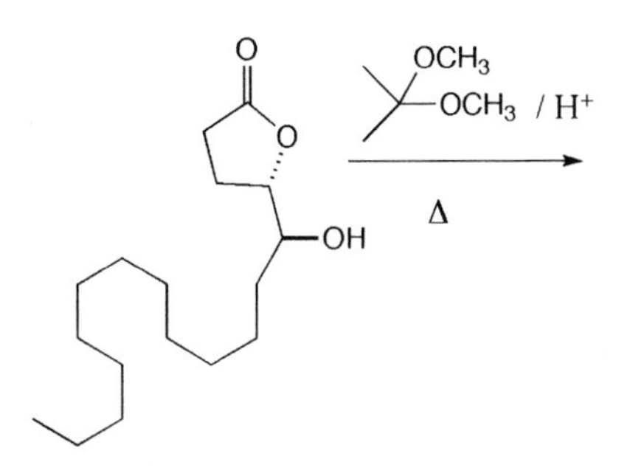

A $C_{21}H_{40}O_4$

¹³C NMR:	**¹H NMR:**
14.0, q	0.87, t, J = 7.0 Hz, 3H
22.5, t	1.25, bs, 18H
25.9, q	1.36, s, 6H
27.0, q	1.49, m, 3H
27.1, t	1.75, m, 2H
27.2, t	1.93, m, 1H
27.4, t	2.45, m, 1H
27.6, t	2.51, m, 1H
27.8, t	3.60, m, 2H
29.2, t	3.67, s, 3H
29.3, t	
29.4, t	
30.4, t	
31.8, t	
32.6, t	
32.7, t	
51.5, q	
79.7, d	
80.6, d	
108.0, s	
173.7, s	

WORKSHEET 3.8

From the molecular formula:

$C_{17}H_{32}O_3 + C_5H_{12}O_2 = C_{22}H_{44}O_5 \quad - C_{21}H_{40}O_4 = CH_3OH$

Fragments of the starting materials that remain:

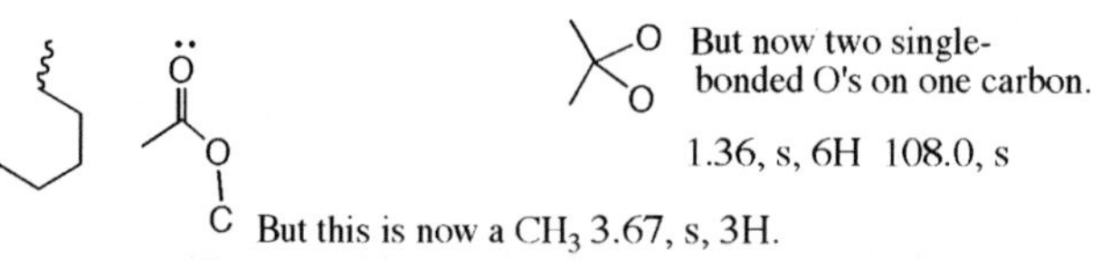

Fragments that have changed in the product:

-OH is gone.

Mechanism and product:

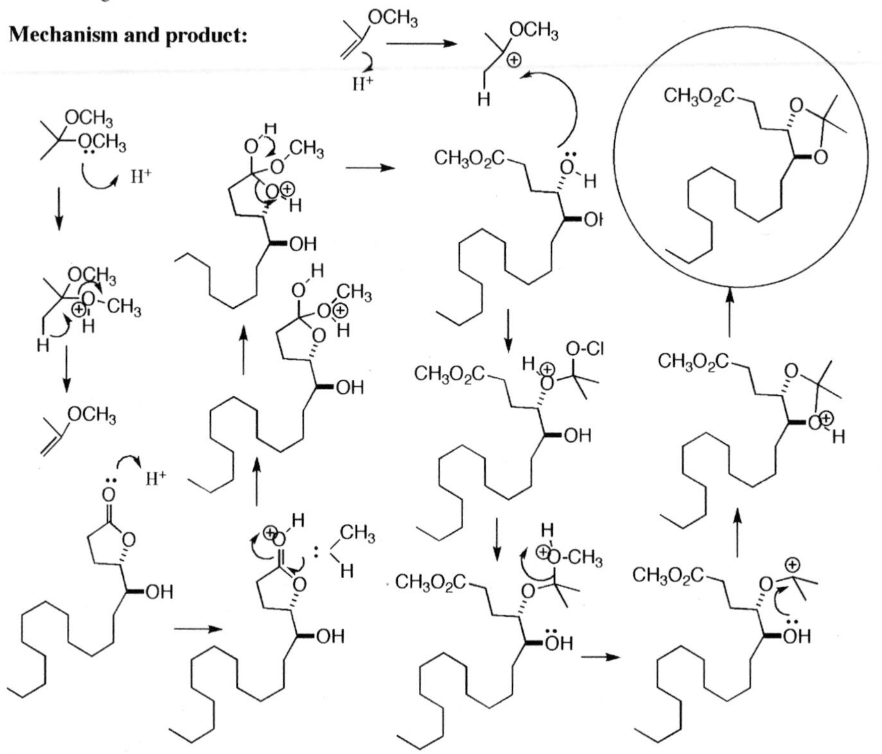

For detailed instructions for solving this problem, see http://valhalla.chem.udel.edu/SpecBook.pdf.

Problem 3.9

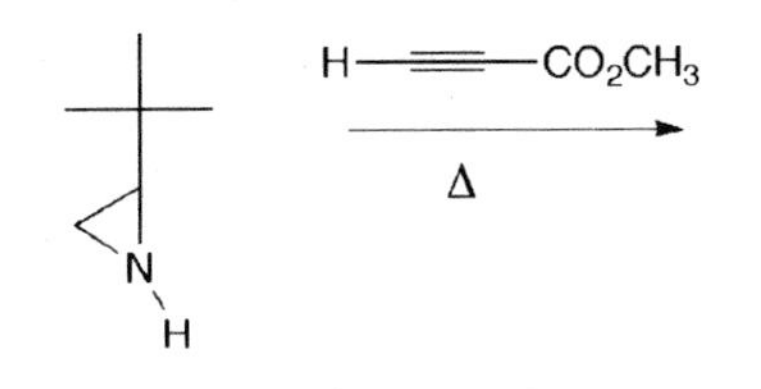

A $C_{10}H_{17}NO_2$

IR: 1154, 1255, 1624, 1714 cm^{-1}

^{13}C NMR:	**^{1}H NMR:**
26.1, q (3)	0.92, s, 9H
30.4, s	1.86, dd, J = 6.0, 3.7 Hz, 1H
30.6, t	1.91, d, J = 6.0 Hz, 1 H
49.6, d	2.53, d, J = 3.7 Hz, 1H
50.8, q	3.68, s, 3H
101.7, d	5.10, d, J = 8.8 Hz, 1H
156.6, d	6.64, d, J = 8.8 Hz, 1H
165.9, s	

WORKSHEET 3.9

From the molecular formula:

$C_6H_{13}N + C_4H_4O_2 = C_{10}H_{17}NO_2$

Fragments of the starting materials that remain:

$-CO_2CH_3$ 3.68, s, 3H

(tert-butyl) 0.92, s, 9H

(aziridine, N) 30.6, t
49.6, d

Fragments that have changed in the product:

(alkene, H, H) 5.10, d, J = 8.8 Hz, 1H
6.64, d, J = 8.8 Hz, 1H

N-H is gone.

Note: must be a conjugated alkene.

Mechanism and product:

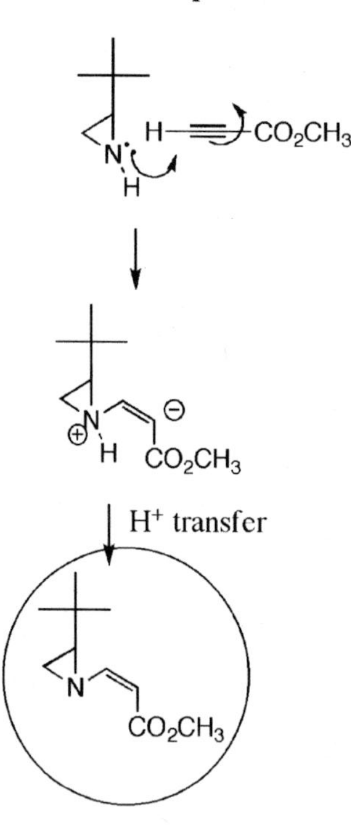

For detailed instructions for solving this problem, see http://valhalla.chem.udel.edu/SpecBook.pdf.

Problem 3.10

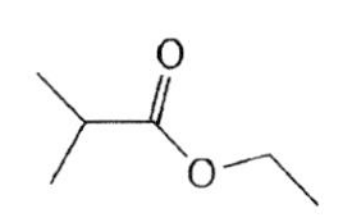

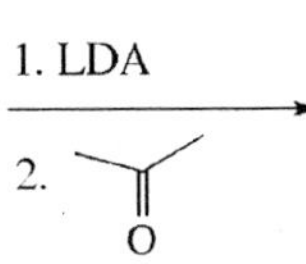

3. $POCl_3$ / pyridine

A $C_9H_{16}O_2$

^{13}C NMR:	^1H NMR:
13.9, q (2)	1.12, s, 6H
19.8, q	1.20, q, J = 7.4 Hz, 3H
24.4, q	1.62, s, 3H
47.5, s	4.07, q, J = 7.4 Hz, 2H
60.4, t	4.76, s, 2H
110.2, t	
147.7, s	
176.3, s	

WORKSHEET 3.10

From the molecular formula:

$C_6H_{12}O_2$ + C_3H_6O = $C_9H_{18}O_3$ - $C_9H_{16}O_2$ = H_2O

Fragments of the starting materials that remain:

1.20, q, J = 7.4 Hz, 3H

4.07, q, J = 7.4 Hz, 2H

Fragments that have changed in the product:

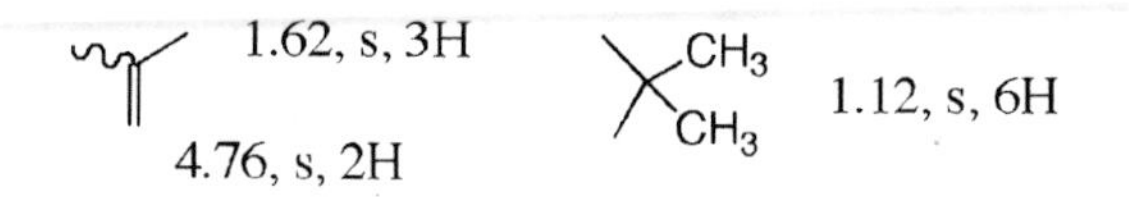

Mechanism and product:

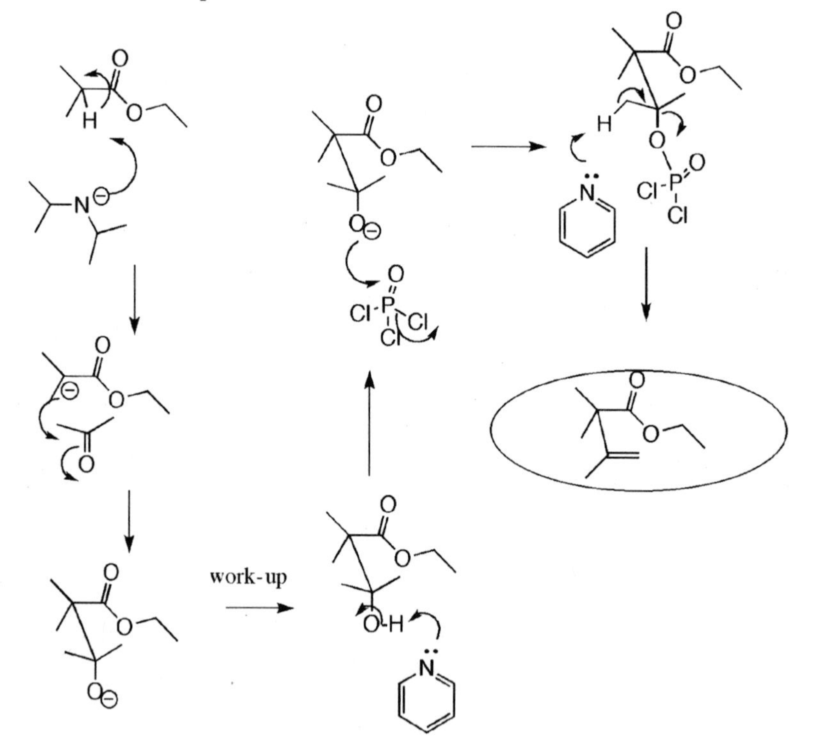

For detailed instructions for solving this problem, see http://valhalla.chem.udel.edu/SpecBook.pdf.

Problem 3.11

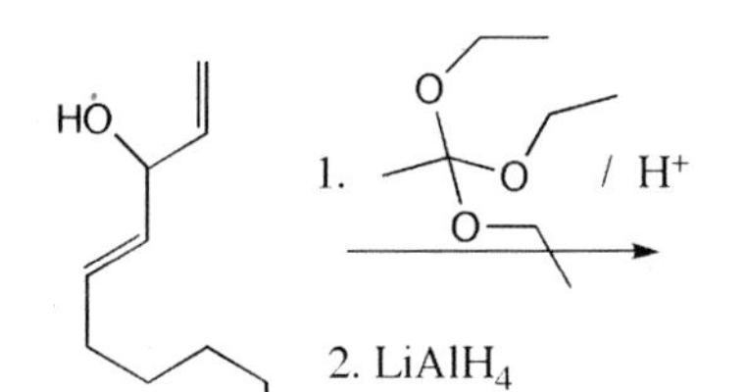

A $C_{13}H_{24}O$

^{13}C NMR:	^{1}H NMR:
14.0, q	0.88, t, J = 6.7 Hz, 3H
22.5, t	1.3–1.6, m, 10H
28.7, t	2.05, dt, J = 7.2, 6.8 Hz, 2H
28.8, t	2.15, dt, J = 7.4, 6.9 Hz, 2H
29.3, t	2.64, bs, 1H (exchanges)
31.7, t	3.66, t, J = 6.5 Hz, 2H
32.2, t	5.51, dt, J = 15.9, 7.4 Hz, 1H
32.5, t	5.62, dt, J = 16.3, 7.2 Hz, 1H
62.2, t	5.99, dd, J = 16.3, 8.2 Hz, 1H
130,0, d	6.02,dd, J = 15.9, 8.2 Hz, 1H
130.9, d	
131.0, d	
132.9, d	

WORKSHEET 3.11

From the molecular formula:

$C_{11}H_{20}O + C_8H_{18}O_3 = C_{19}H_{38}O_4 - C_{13}H_{24}O = C_6H_{14}O_3$

Fragments of the starting materials that remain:

6.02,dd, J = 15.9, 8.2 Hz, 1H
But now alkene is polarized.

0.88, t, J = 6.7 Hz, 3H

Fragments that have changed in the product:

HO

3.66, t, J = 6.5 Hz, 2H
62.2, t

Also: another alkene.

Mechanism and product:

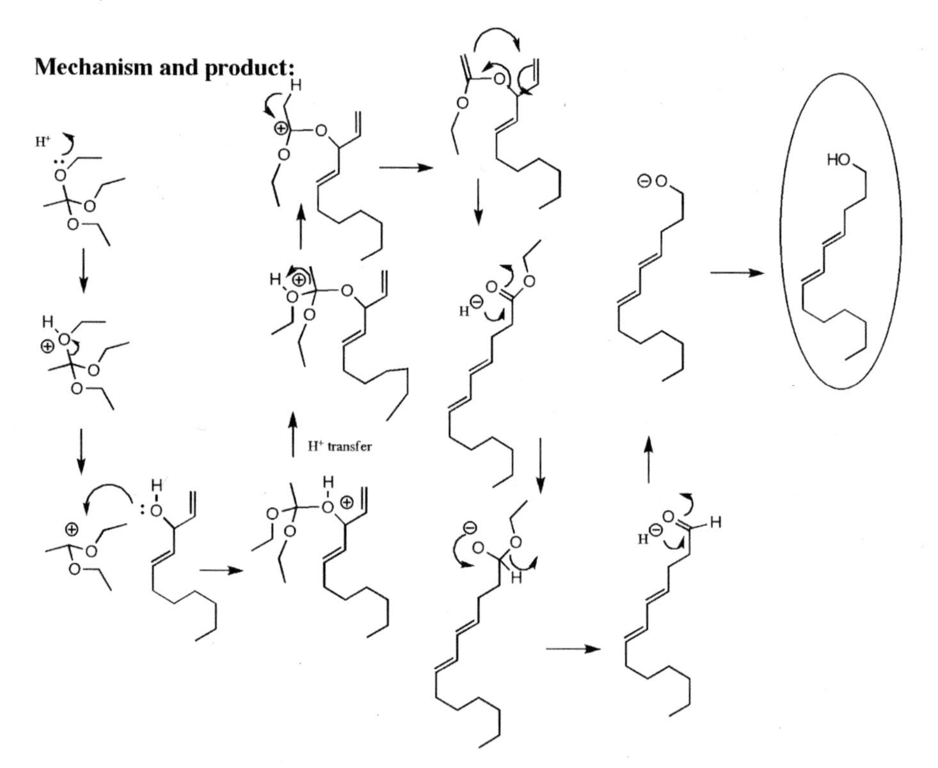

For detailed instructions for solving this problem, see
http://valhalla.chem.udel.edu/SpecBook.pdf.

Problem 3.12

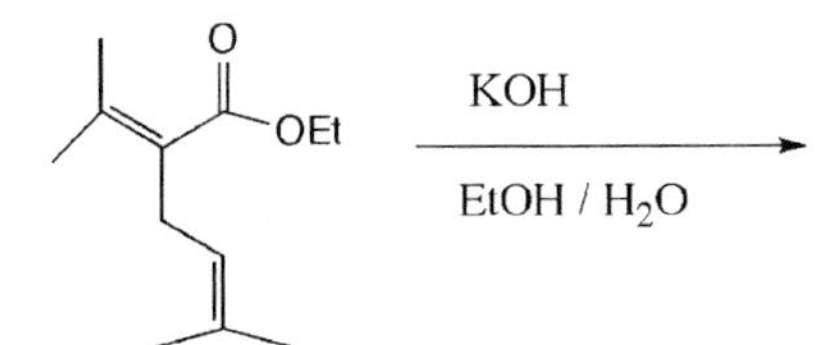

A $C_{10}H_{16}O_2$

IR: 3600–3000, 2970, 1705, 1655 cm^{-1}

^{13}C NMR:	^{1}H NMR:
17.8, q	1.62, s, 3H
20.4, q	1.68, s, 3H
25.7, q	1.77, s, 3H
28.8, t	2.28, ddd, J = 7.1, 7.6, 14.6 Hz,
53.1, d	1H
114.3, t	2.31, ddd, J = 7.1, 7.2, 14.6 Hz,
120.9, d	1H
133.8, s	3.02, dd, J = 7.2, 7.6 Hz, 1H
141.9, s	4.92, bs, 2H
179.5, s	5.05, t, J = 7.1 Hz, 1H
	11.1, bs, 1H, exchanges

WORKSHEET 3.12

From the molecular formula:

$C_{12}H_{20}O_2 + H_2O \quad = \quad C_{12}H_{22}O_3 \quad - \; C_{10}H_{16}O_2 \quad = \quad CH_3CH_2OH$

Fragments of the starting materials that remain:

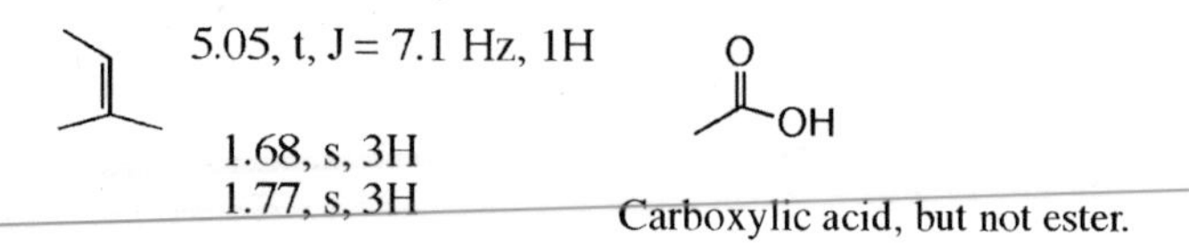

Fragments that have changed in the product:

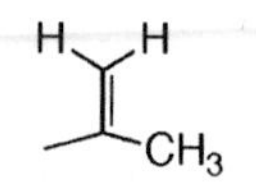

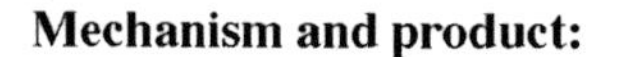

Mechanism and product:

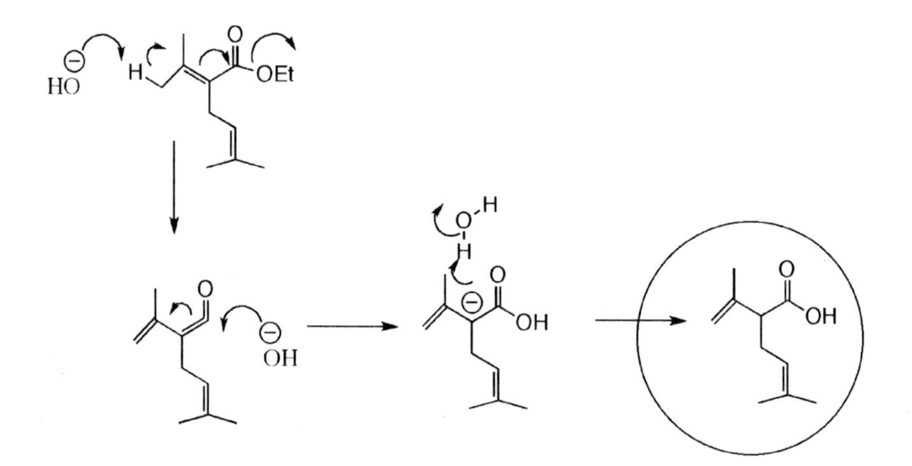

For detailed instructions for solving this problem, see http://valhalla.chem.udel.edu/SpecBook.pdf.

Problem 3.13

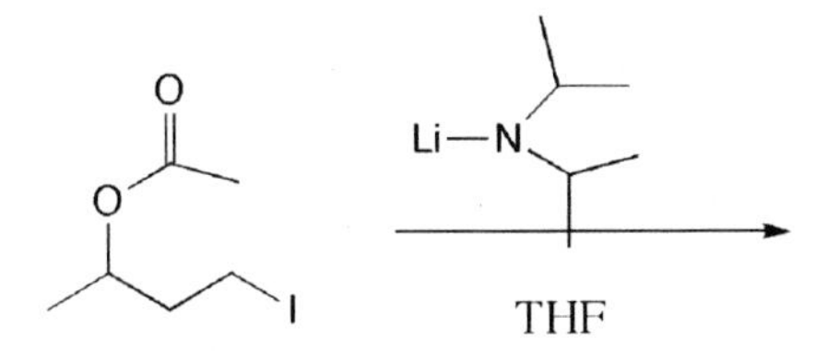

A $C_6H_{10}O_2$

IR: 1730, 1240, 1060 cm^{-1}

^{13}C NMR:	^{1}H NMR:
18.5, q	1.38, d, J = 7.2 Hz, 3H
21.7, t	1.5, m, 1H
29.2, t	1.85, m, 3H
29.6, t	2.5, m, 2H
76.9, d	4.45, m, 1H
171.9, s	

WORKSHEET 3.13

From the molecular formula:

$C_6H_{11}IO_2$ - $C_6H_{10}O_2$ = HI

Fragments of the starting materials that remain:

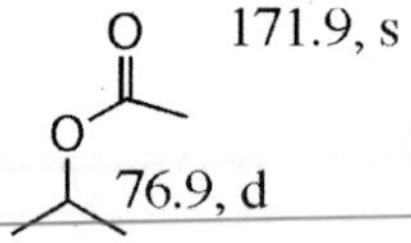

171.9, s

76.9, d

1.38, d, J = 7.2 Hz, 3H

Fragments that have changed in the product:

$-CH_2-I$ is gone.

There is a ring.

Mechanism and product:

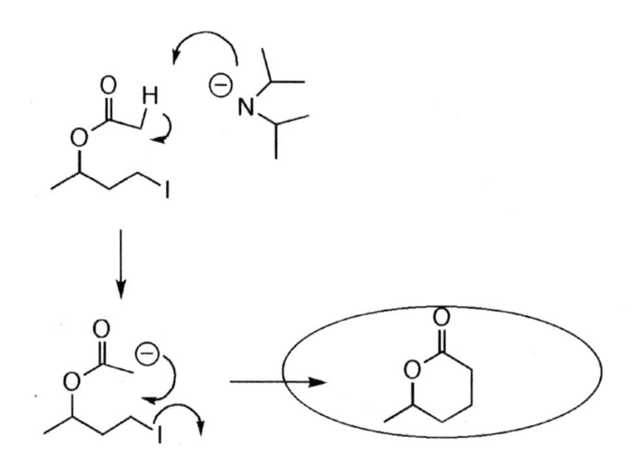

For detailed instructions for solving this problem, see http://valhalla.chem.udel.edu/SpecBook.pdf.

Problem 3.14

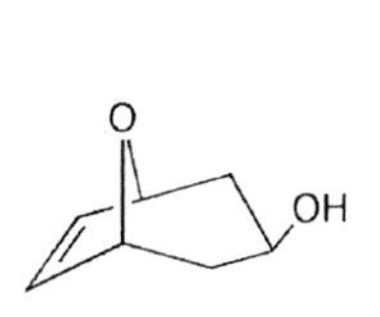

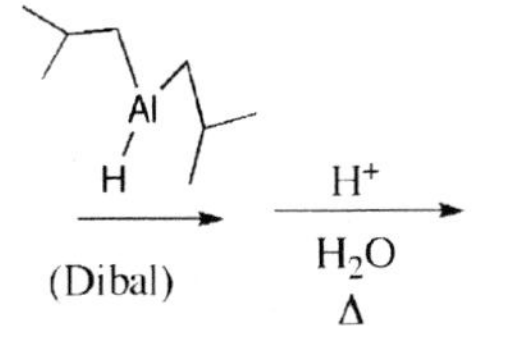

$\xrightarrow[H_2O,\ \Delta]{H^+}$ A $C_7H_{12}O_2$

^{13}C NMR:	^{1}H NMR:
37.9, t	5.70, m, 2H
51.6, t (2)	3.46, m, 2H
67.6, d (2)	2.51, bs, 2H (exchanges)
129.1, d (2)	2.21, m, 6H

WORKSHEET 3.14

From the molecular formula:

$C_7H_{10}O_2$ - $C_7H_{12}O_2$ = No change

Fragments of the starting materials that remain:

Symmetrical alkene.

Fragments that have changed in the product:

Ether is gone.

Two secondary alcohols—symmetrical.

Mechanism and product:

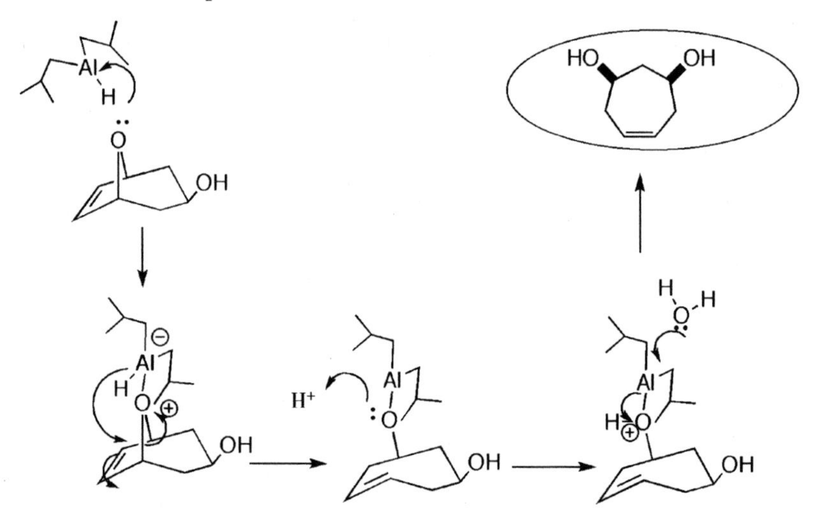

For detailed instructions for solving this problem, see http://valhalla.chem.udel.edu/SpecBook.pdf.

Problem 3.15

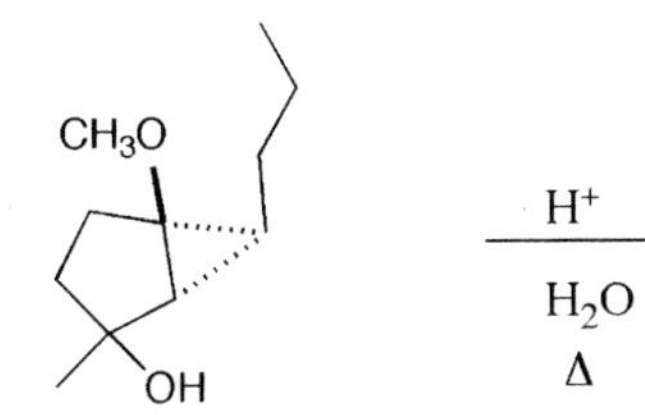

$\xrightarrow[H_2O,\ \Delta]{H^+}$ A $C_{10}H_{16}O$

^{13}C NMR:	^1H NMR:
14.0, q	0.88, t, J = 7.2 Hz, 3H
19.8, q	1.2–1.6, m, 4 H
23.0, t	1.75, s, 3H
30.8, t	2.3–2.5, m, 4H
34.7, t	2.71, brd s, 1H
37.4, t	5.37, m, 1H
48.1, d	
123.7, d	
134.2, s	
213.2, s	

WORKSHEET 3.15

From the molecular formula:

$C_{11}H_{20}O_2$ - $C_{10}H_{16}O$ = CH_3OH

Fragments of the starting materials that remain:

H_3C 0.88, t, J = 7.2 Hz, 3H

1.75, s, 3H

$-CH_3$ The methyl is still there, but now it is attached to an alkene.

Fragments that have changed in the product:

O

37.4, t

48.1, d

Only one ring.

213.2, s

Mechanism and product:

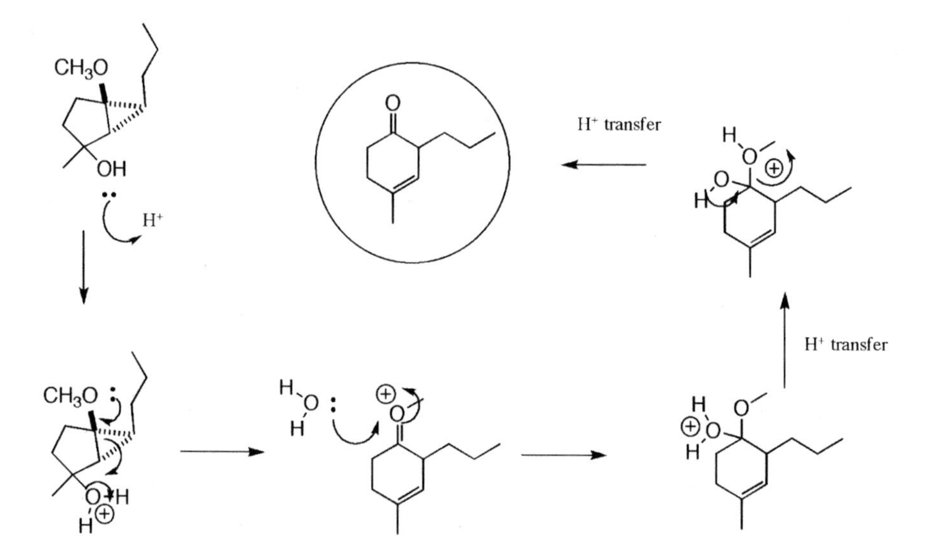

For detailed instructions for solving this problem, see http://valhalla.chem.udel.edu/SpecBook.pdf.

Problem 3.16

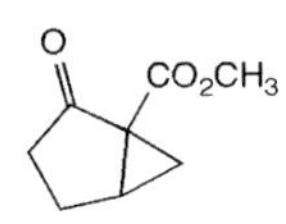

$\xrightarrow[\text{H}^+]{\text{CH}_3\text{OH}}$ **A** $C_9H_{14}O_4$

IR: 2953, 2881, 1757, 1729, 1437, 1114 cm^{-1}

^{13}C NMR:	^{1}H NMR:
24.2, t	1.65, m, 1H
38.0, t	2.09, m, 1H
41.5, d	2.27, m, 1H
52.8, q	2.35, m, 1H
58.5, q	2.81, m, 1H
59.4, d	3.05, d, J = 10.0 Hz, 1H
74.5, t	3.27, s, 3H
169.8, s	3.38, m, 2H
211.7, s	3.69, s, 3H

WORKSHEET 3.16

From the molecular formula:

$C_8H_{10}O_3$ + CH_3OH = $C_9H_{14}O_4$

Fragments of the starting materials that remain:

CO_2CH_3 3.69, s, 3H

O
H_2C —H 59.4, d—but would have to have an additional electron-withdrawing substituent.

Fragments that have changed in the product:

$H_2C{-}O{-}CH_3$ 3.27, s, 3H
74.5, t
One ring remains.

Mechanism and product:

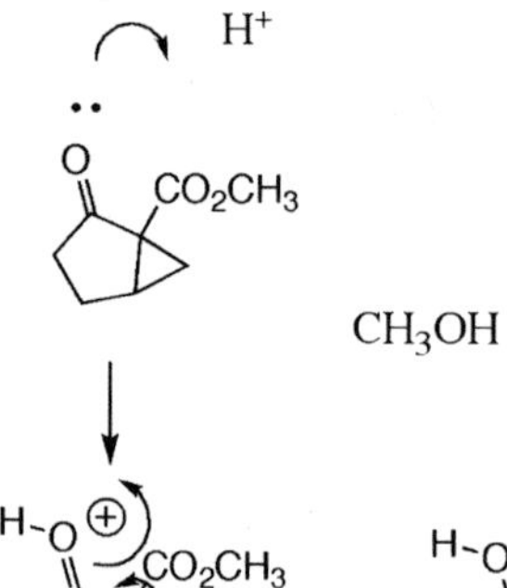

H-O ⊕
CO2CH3
CH3OH

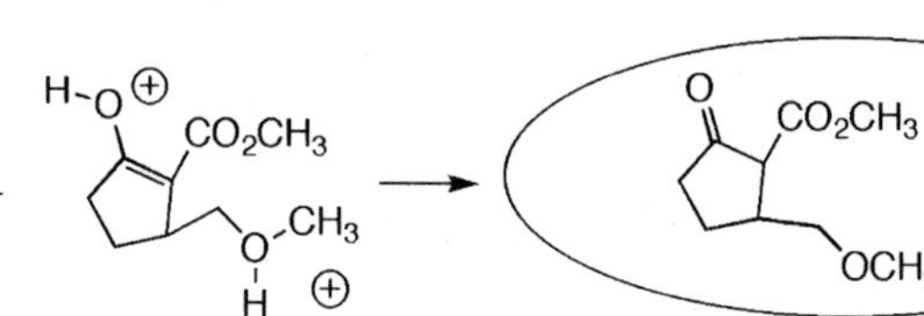

O
CO2CH3
OCH3

For detailed instructions for solving this problem, see http://valhalla.chem.udel.edu/SpecBook.pdf.

Problem 3.17

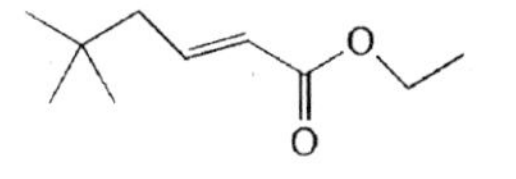

A $C_{12}H_{22}O_3$

IR: 3448, 2956, 2871, 1734, 1477 cm^{-1}

^{13}C NMR:	^{1}H NMR:
14.1, q	1.15, s, 9H
31.2, q (3)	1.25, t, J = 7.2 Hz, 3H
33.5, t	1.7, m, 2H
36.1, s	2.0, bs, 1H (exchanges)
41.6, d	3.7, t, J = 6.6 Hz, 2H
60.5, t	3.80, dt, J = 10.9, 7.1 Hz, 1H
60.6, t	4.15, q, J = 7.2 Hz, 2H
125.5, d	5.18, dd, J = 10.9, 11.5 Hz, 1H
142.7, d	5.51, d, J = 11.5 Hz, 1H
174.5, s	

WORKSHEET 3.17

From the molecular formula:

$C_{10}H_{18}O_2 \; + \; C_2H_4O \; = \; C_{12}H_{22}O_3$

Fragments of the starting materials that remain:

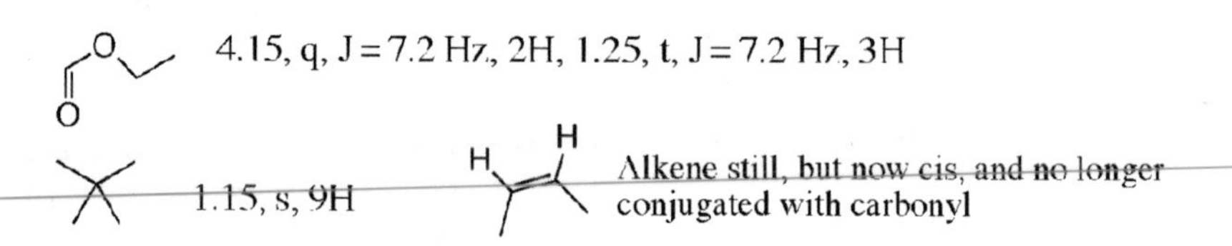

Fragments that have changed in the product:

HO

2.0, bs, 1H (exchanges)
3.7, t, J = 6.6 Hz, 2H

Mechanism and product:

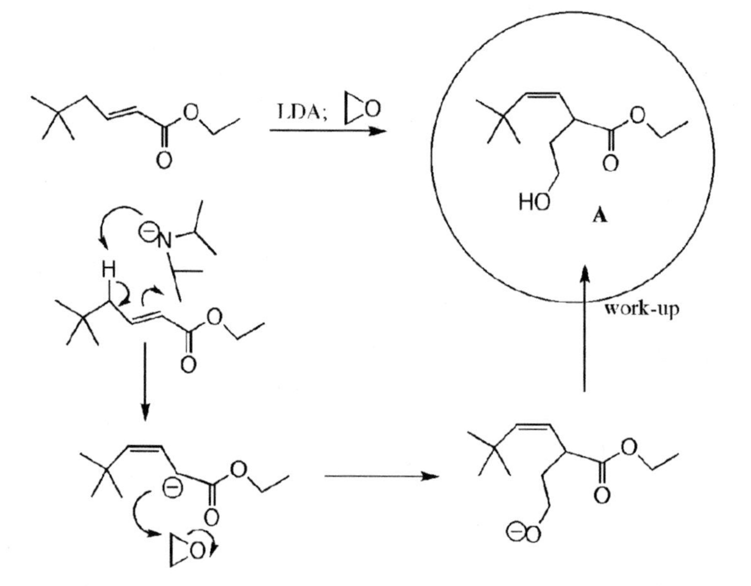

For detailed instructions for solving this problem, see http://valhalla.chem.udel.edu/SpecBook.pdf.

Problem 3.18

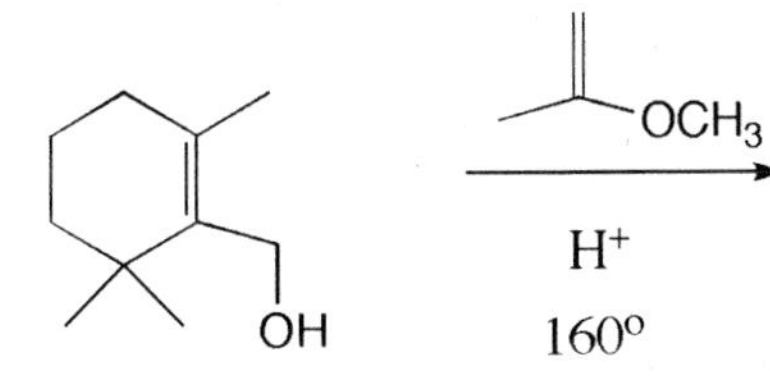

A $C_{13}H_{22}O$

IR: 1716, 1630, 905 cm^{-1}

^{13}C NMR:	**^{1}H NMR:**
18.3, t	1.12, s, 3H
29.5, q	1.14, s, 3H
30.9, q	1.2, s, 3H
31.8, q	0.8–1.7, m, 6H
32.3, q	2.08, s, 3H
36.0, s	2.62, s, 2H
38.6, t	4.84, s, 1H
38.8, s	4.97, s, 1H
40.5, t	
53.8, t	
108.3, t	
160.5, s	
207.9, s	

WORKSHEET 3.18

From the molecular formula:

$C_{10}H_{18}O$ + C_4H_8O = $C_{14}H_{26}O_2$ - $C_{13}H_{22}O$ = CH_3OH

Fragments of the starting materials that remain:

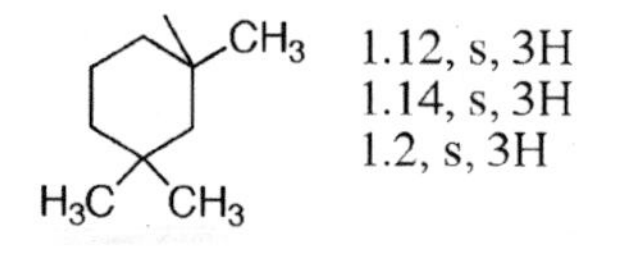

But one of the methyls used to be directly attached to an alkene—it no longer is.

Fragments that have changed in the product:

-CH_2OH gone.

The alkene is now 1,1-disubstituted 108.3, t 160.5, s.

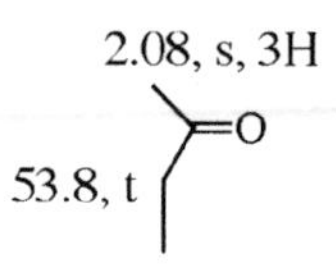

Mechanism and product:

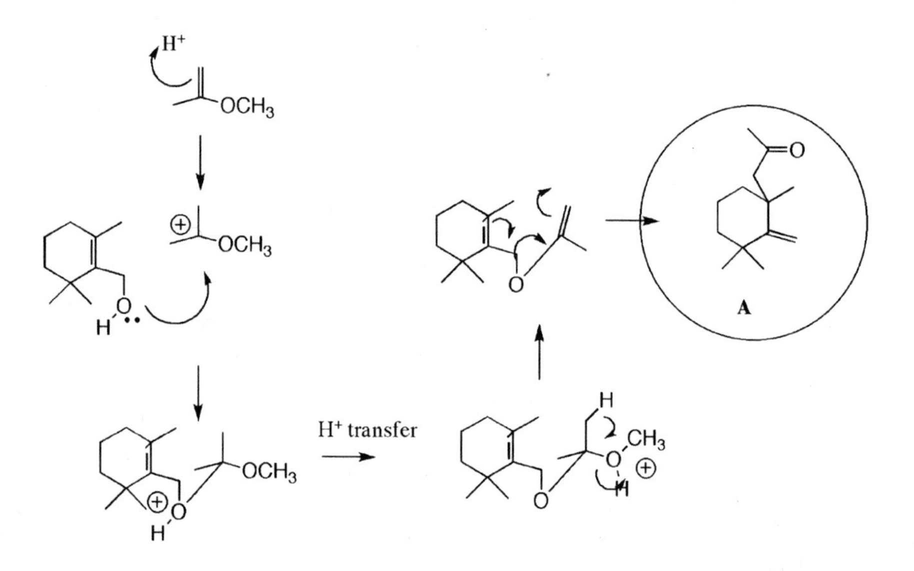

For detailed instructions for solving this problem, see http://valhalla.chem.udel.edu/SpecBook.pdf.

Problem 3.19

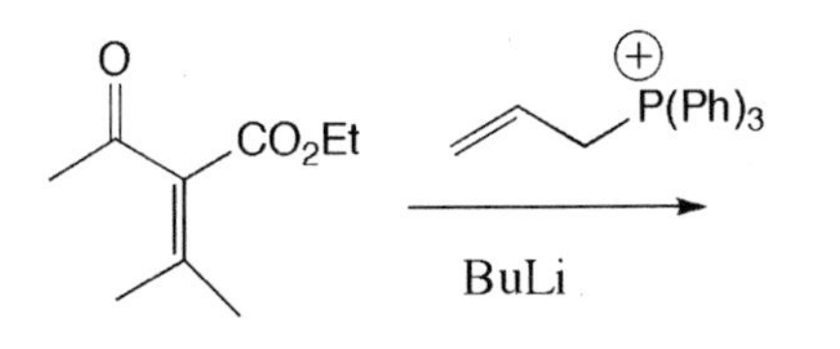

A $C_{12}H_{18}O_2$

IR: 2964, 1738, 1446 cm^{-1}

^{13}C NMR:	^{1}H NMR:
14.2, q	1.03, s, 3H
22.3, q	1.05, s, 3H
24.4, q	1.21, t, J = 7.0 Hz, 3H
28.1, q	1.75, s, 3H
34.4, s	2.74, s, 1H
56.7, d	4.10, q, J = 7.0 Hz, 2H
60.0, t	5.35, d, J = 9.6 Hz, 1H
120.3, d	5.74, m, 1H
121.3, d	5.76, m, 1H
130.8, d	
134.3, s	
171.9, s	

WORKSHEET 3.19

From the molecular formula:

$$C_9H_{14}O_3 + C_{21}H_{19}P = C_{30}H_{33}PO_3 - C_{12}H_{18}O_2 = Ph_3P{=}O$$

Fragments of the starting materials that remain:

CO_2Et 1.21, t, J = 7.0 Hz, 3H 4.10, q, J=7.0 Hz, 2H

Still three methyl groups, but now only one is directly attached to an alkene.

Fragments that have changed in the product:

We know from the appearance of $Ph_3P{=}O$, a Wittig reaction has taken place.

There are still two alkenes, one with two H's, one with only one H.

One ring. Ketone is gone.

Mechanism and product:

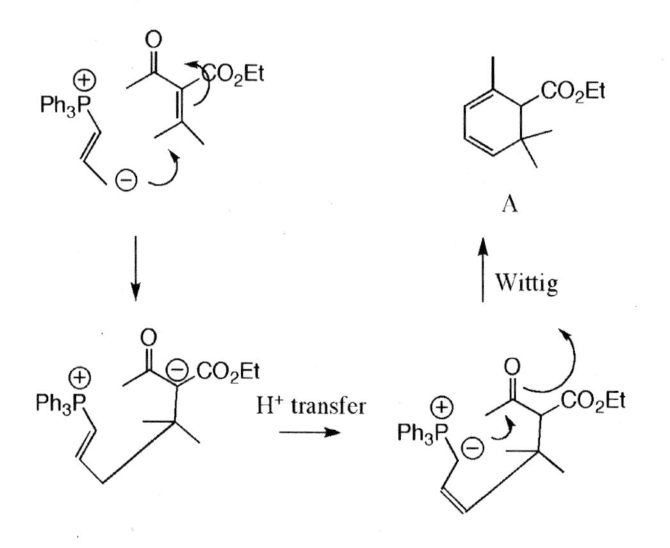

For detailed instructions for solving this problem, see http://valhalla.chem.udel.edu/SpecBook.pdf.

Problem 3.20

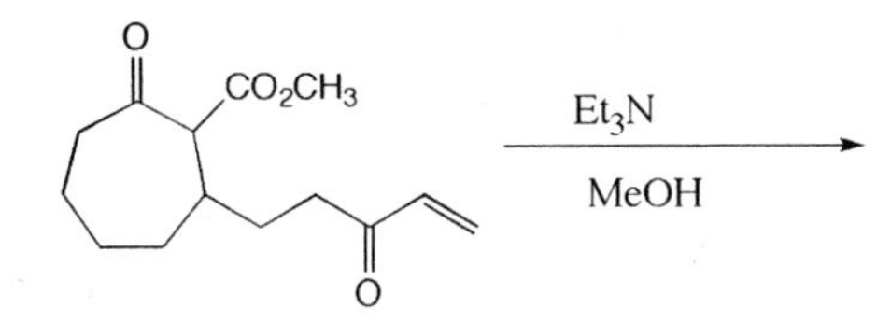

A $C_{14}H_{20}O_4$

^{13}C NMR:	**^{1}H NMR:**
24.4, t	1.36, m, 1H
28.4, t	1.56, m, 1H
29.5, t	1.71, tt, J = 11, 3 Hz, 1H
31.2, t	1.9, m, 4H
36.2, t	2.2, m, 3H
39.5, t	2.27, ddd, J = 15, 6, 3 Hz, 1H
41.5, t	2.51, ddd, J = 16, 6, 3 Hz, 1H
42.6, t	2.54, ddd, J = 15, 6, 4 Hz, 1H
46.1, d	2.57, m, 1H
52.0, q	2.61, m, 1H
66.1, s	2.65, ddd, J = 16,12,3 Hz, 1H
171.8, s	2.70, ddd, J = 13, 6, 5 Hz, 1H
209.7, s	3.83, s, 3H
212.7, s	

WORKSHEET 3.20

From the molecular formula:

$C_{14}H_{20}O_4 \quad = \quad C_{14}H_{20}O_4$

Molecule isomerized, no atoms gained or lost.

Fragments of the starting materials that remain:

CO_2CH_3 3.83, s, 3H

Two ketones still.

Fragments that have changed in the product:

Alkene gone, additional ring formed.

Mechanism and product:

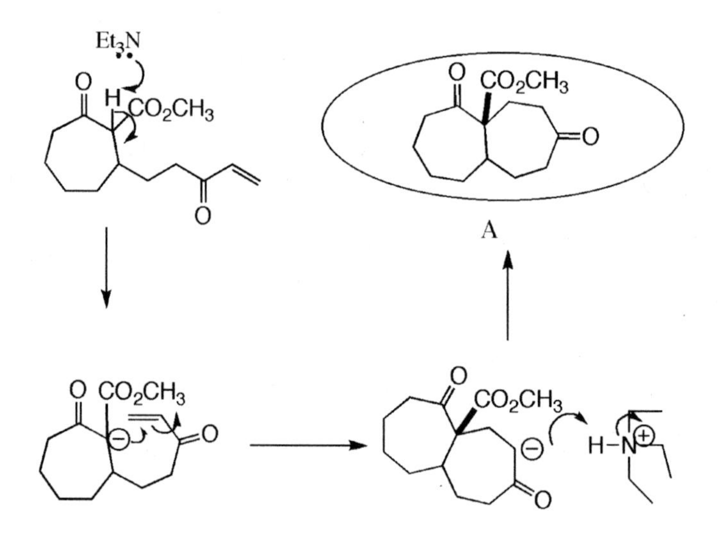

For detailed instructions for solving this problem, see http://valhalla.chem.udel.edu/SpecBook.pdf.

Problem 3.21

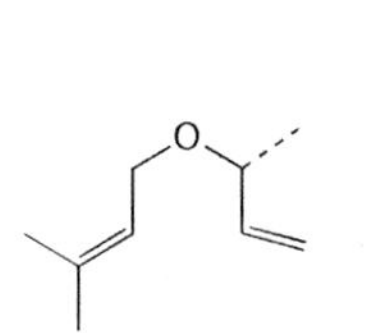

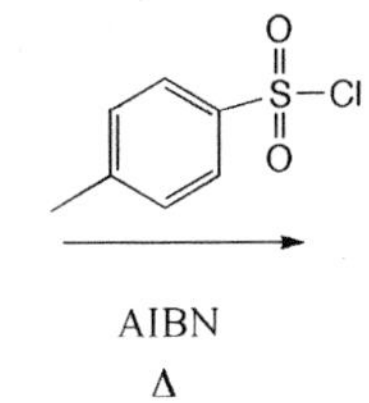

AIBN

Δ

A $C_{16}H_{23}ClSO_3$

^{13}C NMR:	^{1}H NMR:
20.6, q	1.20, d, J = 6.4 Hz, 3H
21.2, q	1.56, s, 3H
31.6, q	1.59, s, 3H
34.4, t	2.46, s, 3H
40.7, d	2.53, dddd, J = 11.0, 7.1, 1.5, 1.4 Hz, 1H
51.6, d	2.62, ddd, J = 10.2, 7.6, 7.1 Hz, 1H
55.0, t	3.15, dd, J = 14.4, 11.0, 1H
67.4, t	3.81, dd, J = 10.2, 8.3, 1H
68.3, s	3.96, dd, J = 14.4, 1.5, 1H
80.1, d	4.04, dd, J = 7.6, 8.3, 1H
127.5, d(2)	4.37, dq, J = 1.4, 6.4 Hz, 1H
129.7, d(2)	7.38, d, J = 8.2 Hz, 2H
144.5, s	7.80, d, J = 8.2 Hz, 2H
136.4, s	

WORKSHEET 3.21

From the molecular formula:

$C_9H_{16}O$ + $C_7H_7ClO_2S$ = $C_{16}H_{23}ClSO_3$

So addition product, with no loss.

Fragments of the starting materials that remain:

7.38, d, J = 8.2 Hz, 2H
7.80, d, J = 8.2 Hz, 2H

2.46, s, 3H

All three methyls are still there, but none are directly attached to alkenes.

Fragments that have changed in the product:

Both alkenes are gone.
There is an additional ring.

Mechanism and product:

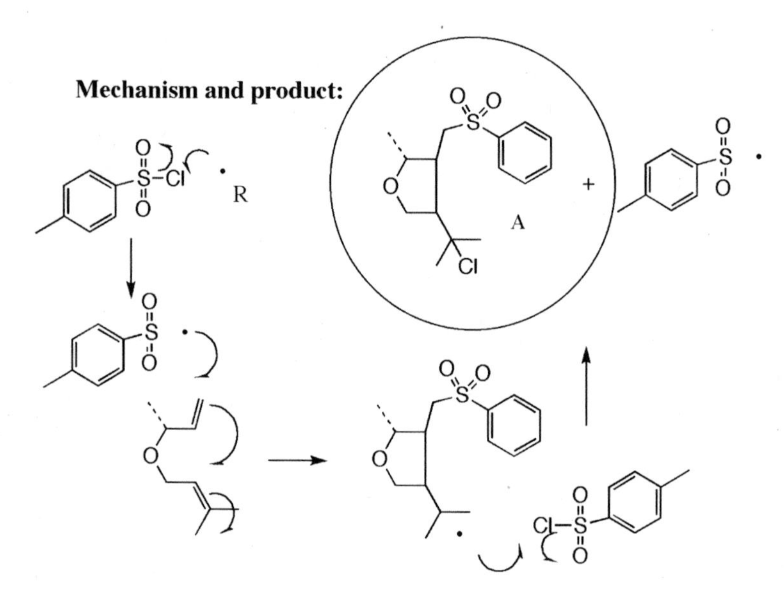

For detailed instructions for solving this problem, see http://valhalla.chem.udel.edu/SpecBook.pdf.

Problem 3.22

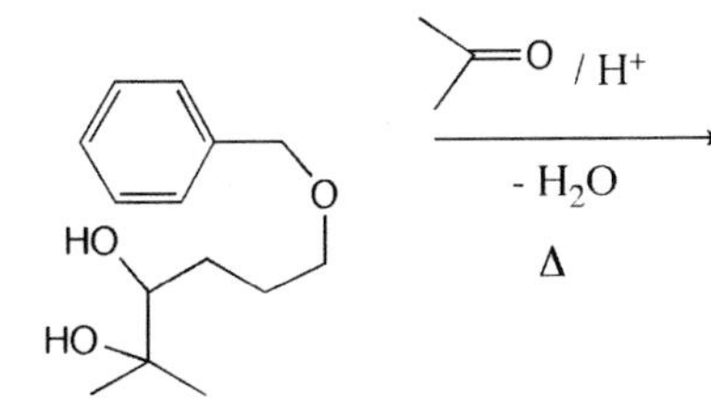

A $C_{17}H_{26}O_3$

^{13}C NMR:	^{1}H NMR:
21.7, q	1.09, s, 3H
22.8, q	1.24, s, 3H
26.0, q	1.27, s, 3H
26.8, q	1.31, s, 3H
27.1, t	1.6–1.8, m, 4H
28.5, t	3.55, t, J = 7.2 Hz, 2H
69.9, t	3.64, m, 1H
72.8, t	4.51, s, 2H
80.1, d	7.3–7.4, m, 5H
83.1, s	
106.5, s	
127.5, d (2)	
127.5, d	
128.3, d (2)	
138.5, s	

WORKSHEET 3.22

From the molecular formula:

$C_{14}H_{22}O_3$ + C_3H_6O = $C_{17}H_{28}O_4$ - $C_{17}H_{26}O_3$ = H_2O

Fragments of the starting materials that remain:

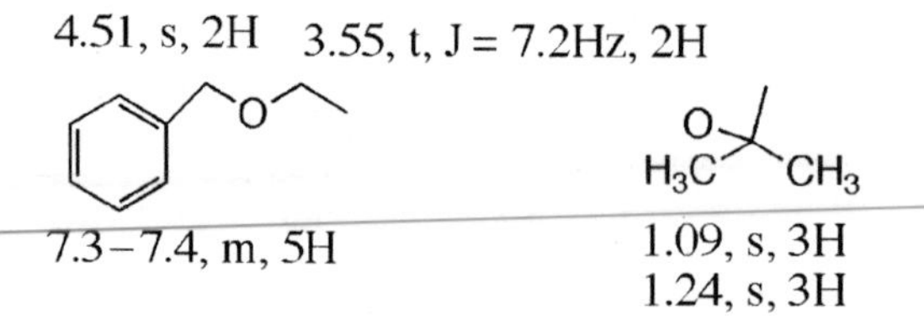

Fragments that have changed in the product:

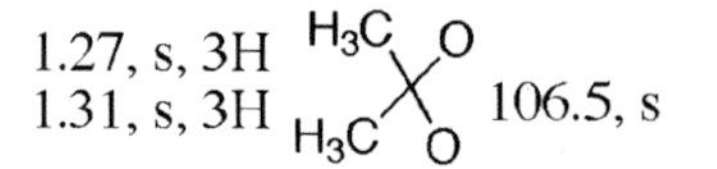

Mechanism and product:

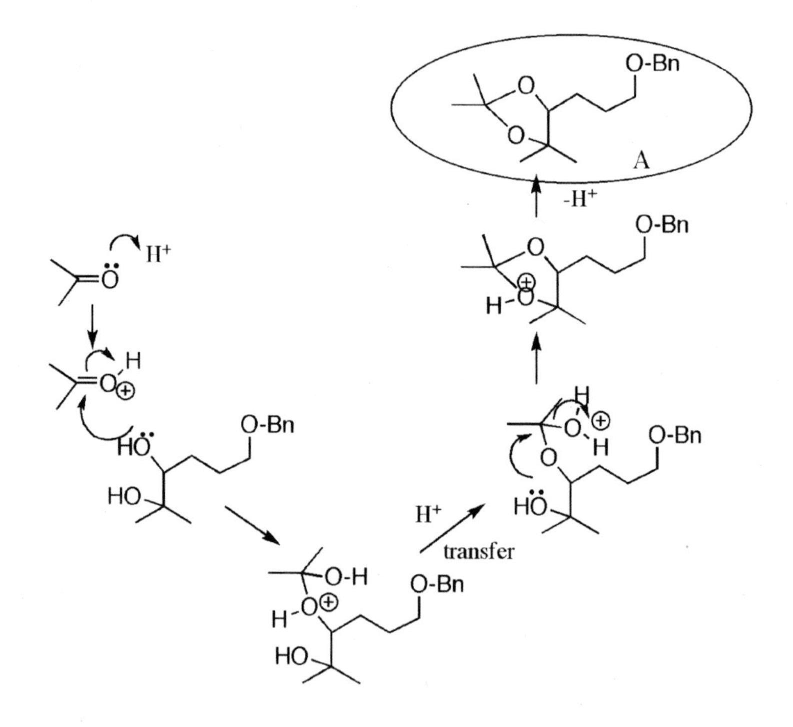

For detailed instructions for solving this problem, see http://valhalla.chem.udel.edu/SpecBook.pdf.

Problem 3.23

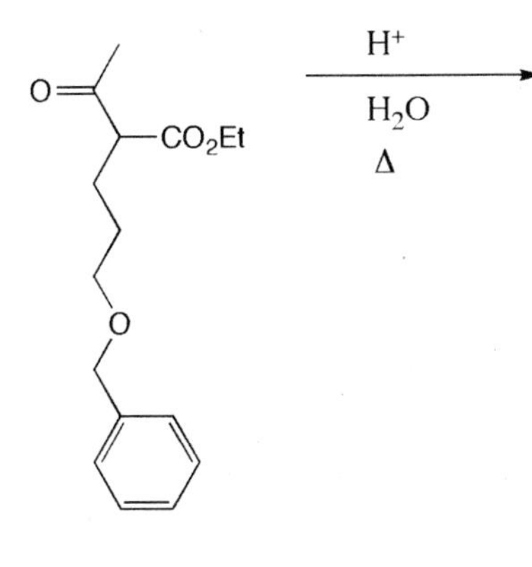

A $C_{13}H_{18}O_2$

^{13}C NMR:	^{1}H NMR:
20.5, t	1.65, m, 4H
29.1, t	2.13, s, 3H
29.8, q	2.46, t, J = 7.4 Hz, 2H
43.3, t	3.48, t, J = 6.0 Hz, 2H
69.9, t	4.50, s, 2H
72.9, t	7.3–7.4, m, 5H
127.5, d (2)	
127.6, d	
128.3, d (2)	
138.4, s	
208.9, s	

WORKSHEET 3.23

From the molecular formula:

$$C_{16}H_{22}O_4 + H_2O = C_{16}H_{24}O_5 - C_{13}H_{18}O_2 = C_3H_6O_3 = CO_2 + CH_3CH_2OH$$

Fragments of the starting materials that remain:

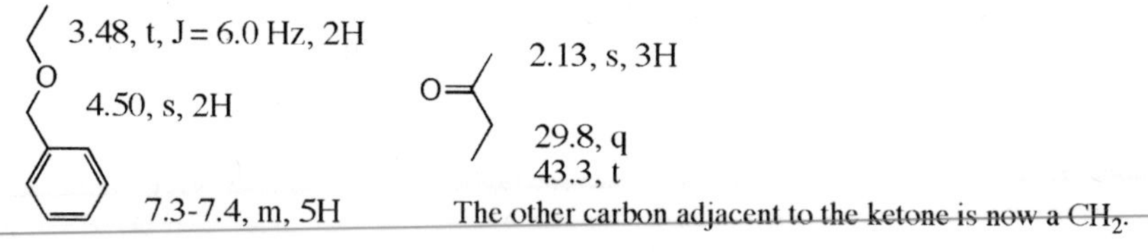

Fragments that have changed in the product:

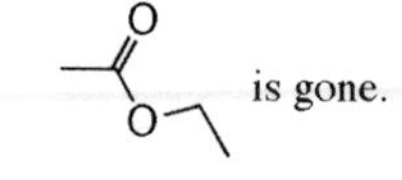

Mechanism and product:

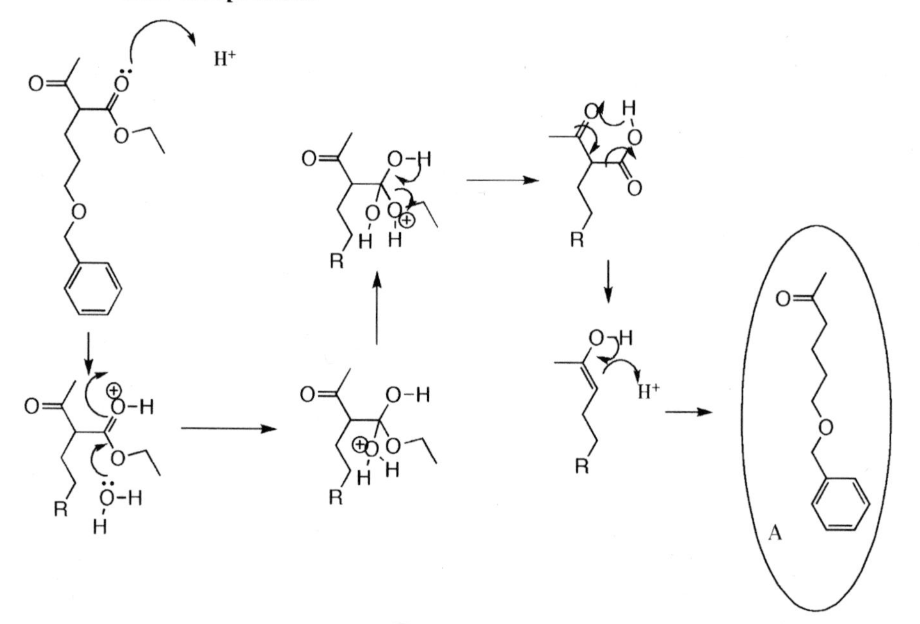

For detailed instructions for solving this problem, see http://valhalla.chem.udel.edu/SpecBook.pdf.

Problem 3.24

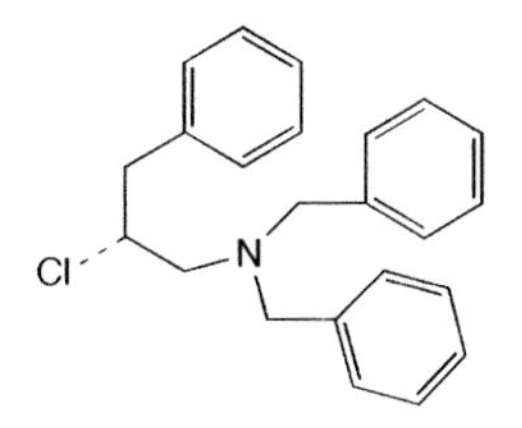

$\xrightarrow[\text{DMSO}]{\text{NaCN}}$ A $C_{24}H_{24}N_2$

^{13}C NMR:	**^{1}H NMR:**
18.9, t	2.41, dd, J = 8.3, 10.5 Hz, 1H
35,7, t	2.6, m, 2H
53.6, t (2)	3.08, dd, J = 5.5, 13.6 Hz, 1H
56.9, d	3.21, m, 1H
118.6, s	3.66, d, J = 13.7 Hz, 2H
126.6, d	3.78, d, J = 13.7 Hz, 2H
127.2, d (2)	7.3, m, 15H
128.4, d (2)	
128.6, d (4)	
128.7, d (2)	
129.0, d (4)	
138.2, s	
138.7, s (2)	

WORKSHEET 3.24

From the molecular formula:

$C_{23}H_{24}ClN$ + $NaCN$ = $C_{24}H_{24}ClN_2Na$ - $C_{24}H_{24}N_2$ = $NaCl$

Fragments of the starting materials that remain:

x 3 7.3, m, 15H   N 53.6, t (2)

Fragments that have changed in the product:

18.9, t

CH_2-CN

118.6, s

Note that direct S_N2 would lead to a methine adjacent to the nitrile–but we observe a CH_2.

Mechanism and product:

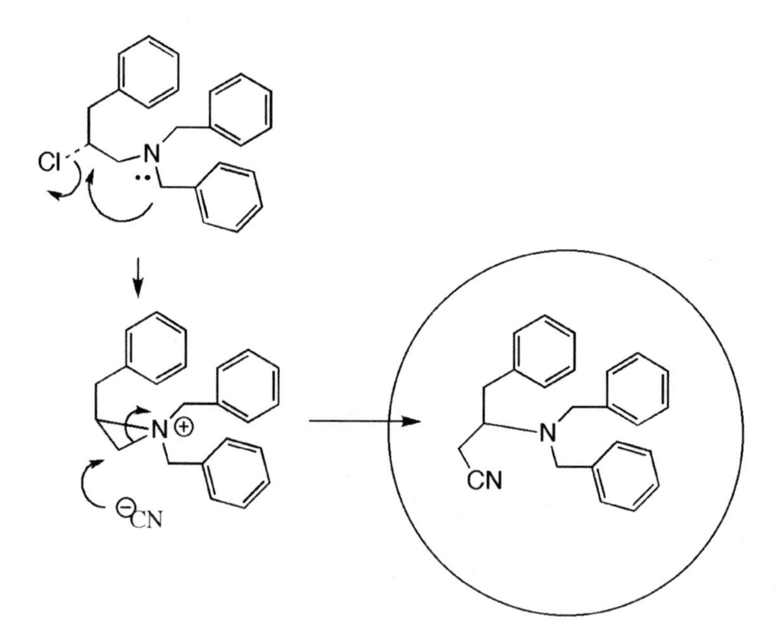

For detailed instructions for solving this problem, see http://valhalla.chem.udel.edu/SpecBook.pdf.

Problem 3.25

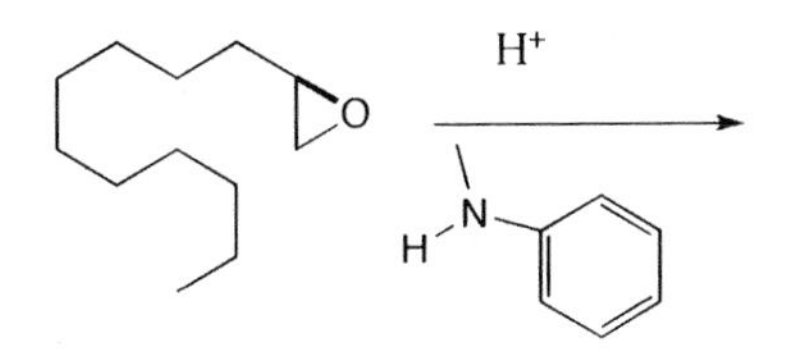

A $C_{19}H_{33}NO$

^{13}C NMR:	^{1}H NMR:
14.1, q	0.90, t, J = 6.9 Hz, 3H
22.7, t	1.28, m, 16H
25.7, t	1.5, m, 2H
29.3, t	2.2, bs, 1H (exchanges)
29.6, t	2.96, s, 3H
29.8, t	3.24, m, 2H
30.1, t	3.92, m, 1H
31.2, t	6.78, t, J = 7.5 Hz, 1H
31.9, t	6.82, d, J = 8.4 Hz, 2H
34.5, t	7.26, dd, J = 7.5, 8.4 Hz, 2H
39.3, q	
60.7, t	
69.3, d	
114.4, d (2)	
117.8, d	
129.1, d (2)	
150.5, s	

WORKSHEET 3.25

From the molecular formula:

$C_{12}H_{24}O \ + \ C_7H_9N \ = \ C_{19}H_{33}NO$

So this is a 1:1 addition product.

Fragments of the starting materials that remain:

2.96, s, 3H

6.78, t, J = 7.5 Hz, 1H
6.82, d, J = 8.4 Hz, 2H
7.26, dd, J = 7.5, 8.4 Hz, 2H

Fragments that have changed in the product:

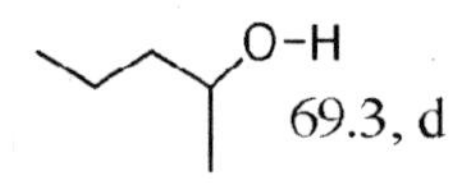

69.3, d

Mechanism and product:

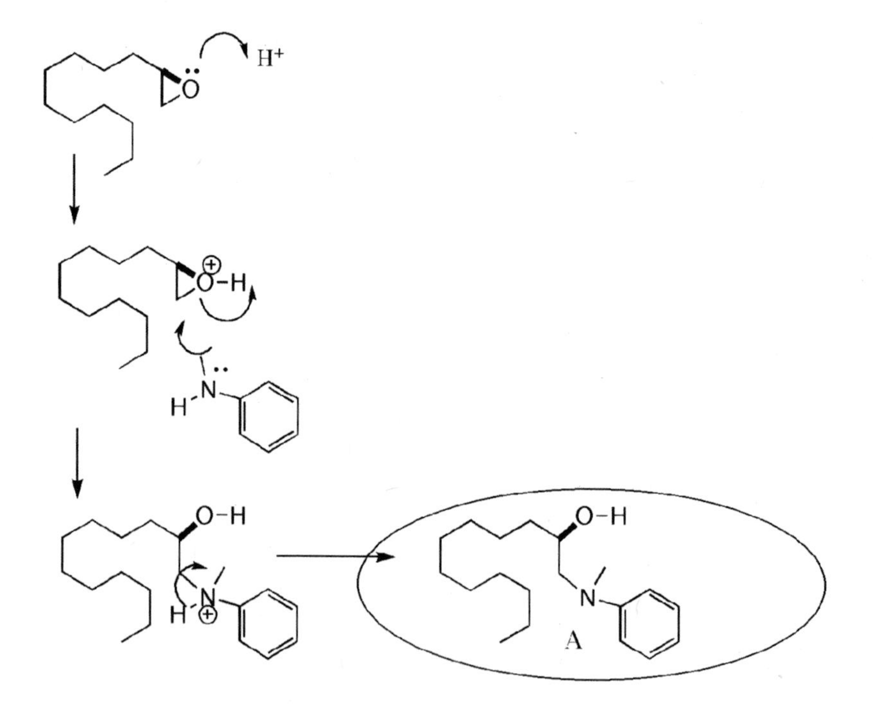

For detailed instructions for solving this problem, see http://valhalla.chem.udel.edu/SpecBook.pdf.

Problem 3.26

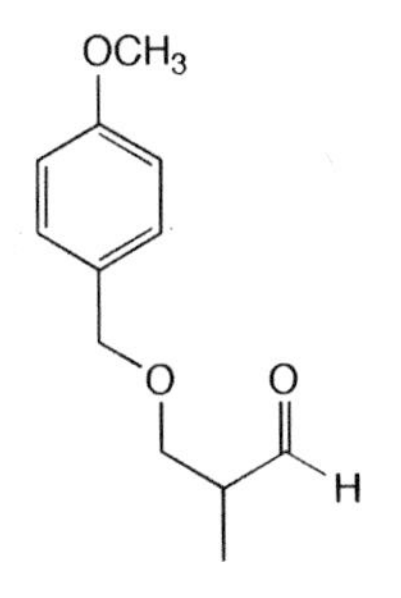

1. $Ph_3P{=}CBr_2$

2. BuLi x 2; CH_3-I

A $C_{14}H_{18}O_2$

IR: 2049, 1696, 1613 cm^{-1}

^{13}C NMR:	**^{1}H NMR:**
3.6, q	1.19, d, J = 6.9 Hz, 3H
18.1, d	1.81, s, 3H
26.7, q	2.8, m, 1H
55.2, q	3.32, dd, J = 8.9, 8.4 Hz, 1H
72.6, t	3.49, dd, J = 8.9, 6.0 Hz, 1H
74.2, t	3.82, s, 3H
77.4, s	4.51, s, 2H
81.2, s	6.90, d, J = 8.4 Hz, 2H
113.8, d (2)	7.28, d, J = 8.4 Hz, 2H
129.2, d (2)	
130.5, s	
159.2, s	

WORKSHEET 3.26

From the molecular formula:

$C_{12}H_{16}O_3$ - $C_{14}H_{18}O_2$ = C_2H_2 - O

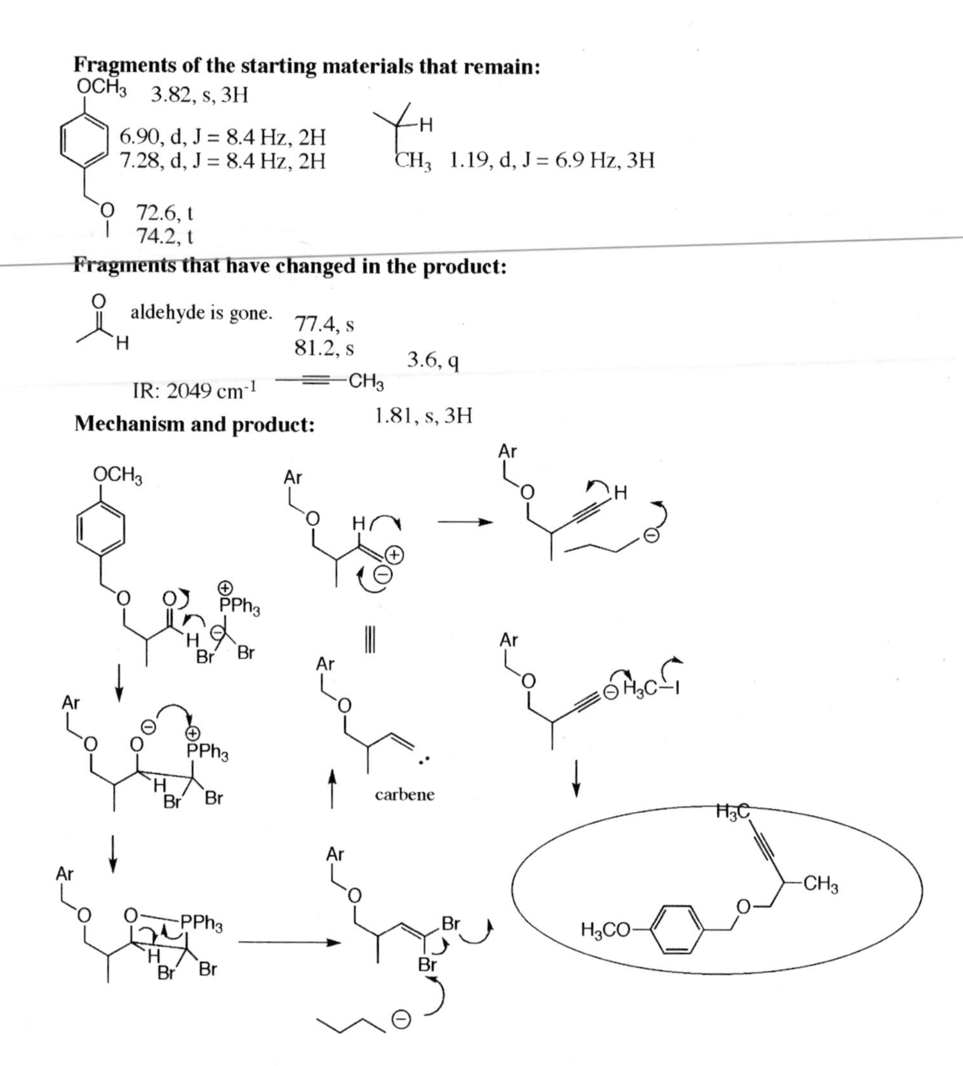

For detailed instructions for solving this problem, see http://valhalla.chem.udel.edu/SpecBook.pdf.

Problem 3.27

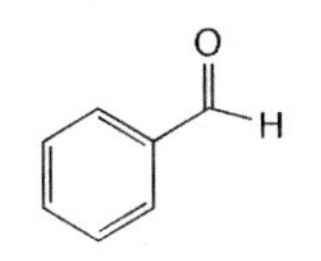

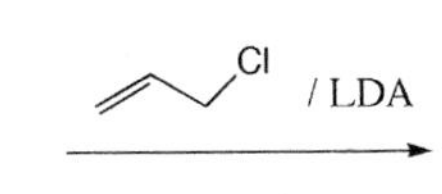

A $C_{10}H_{10}O$

IR: 1496, 1442, 1388, 1250, 1181 cm^{-1}

^{13}C NMR:	^{1}H NMR:
58.8, d	3.67, dd, J = 4.1, 4.5 Hz, 1H
59.8, d	4.25, d, J = 4.1 Hz, 1H
121.8, t	5.26, d, J = 10.2 Hz, 1H
127.7, d	5.43, d, J = 16.0 Hz, 1H
128.1, d (2)	5.66, ddd, J = 4.5, 10.2, 16.0 Hz, 1H
128.5, d (2)	7.2–7.4, m, 5H
132.2, d	
135.2, s	

WORKSHEET 3.27

From the molecular formula:

C_7H_6O + C_3H_5Cl = $C_{10}H_{11}ClO$ - $C_{10}H_{10}O$ = HCl

Fragments of the starting materials that remain:

7.2–7.4, m, 5H,
but no longer conjugated
with carbonyl.

5.26, d, J = 10.2 Hz, 1H
5.43, d, J = 16.0 Hz, 1H
5.66, ddd, J = 4.5, 10.2, 16.0 Hz, 1H

But note only one H on adjacent carbon.

Fragments that have changed in the product:

Ether, ring.

Mechanism and product:

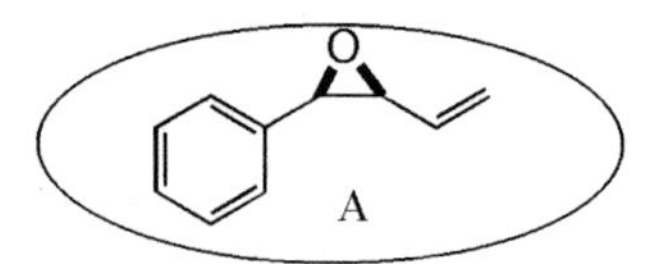

From the coupling constants,
this is probably a cis expoxide.

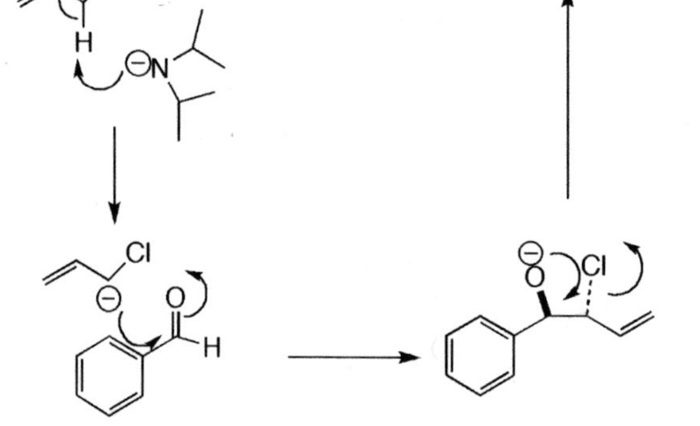

For detailed instructions for solving this problem, see
http://valhalla.chem.udel.edu/SpecBook.pdf.

Problem 3.28

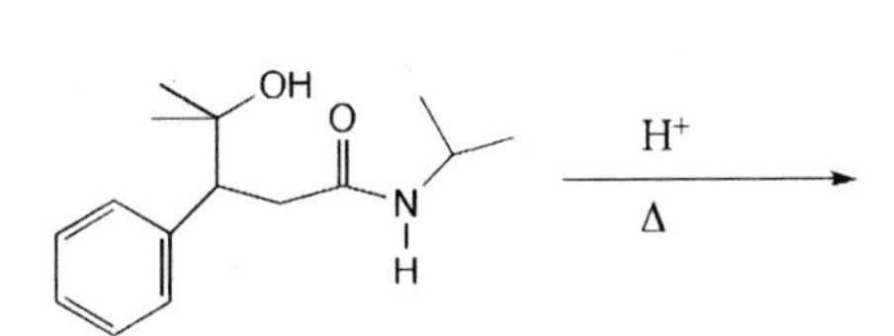

A $C_{12}H_{14}O_2$

^{13}C NMR:	^{1}H NMR:
23.1, q,	1.04, s, 3H
27.6, q	1.55, s, 3H
34.3, t	2.72, dd, J = 8.5, 17.6 Hz, 1H
51.0, d	2.94, dd, J = 9.6, 17.6 Hz, 1H
87.1, s	3.52, dd, J = 8.5, 9.6 Hz, 1H
127.6, d	7.1–7.5, m, 5H
127.7, d (2)	
128.5, d (2)	
136.5, s	
175.3, s	

WORKSHEET 3.28

From the molecular formula:

$$H^{+} + C_{14}H_{23}NO_2 \quad - C_{12}H_{15}O_2 \quad = \quad C_3H_9N$$

Fragments of the starting materials that remain:

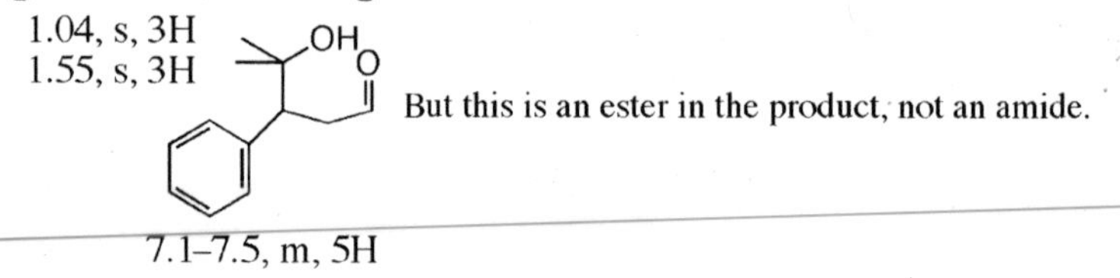

Fragments that have changed in the product:

Another ring.

Amine is gone.

Mechanism and product:

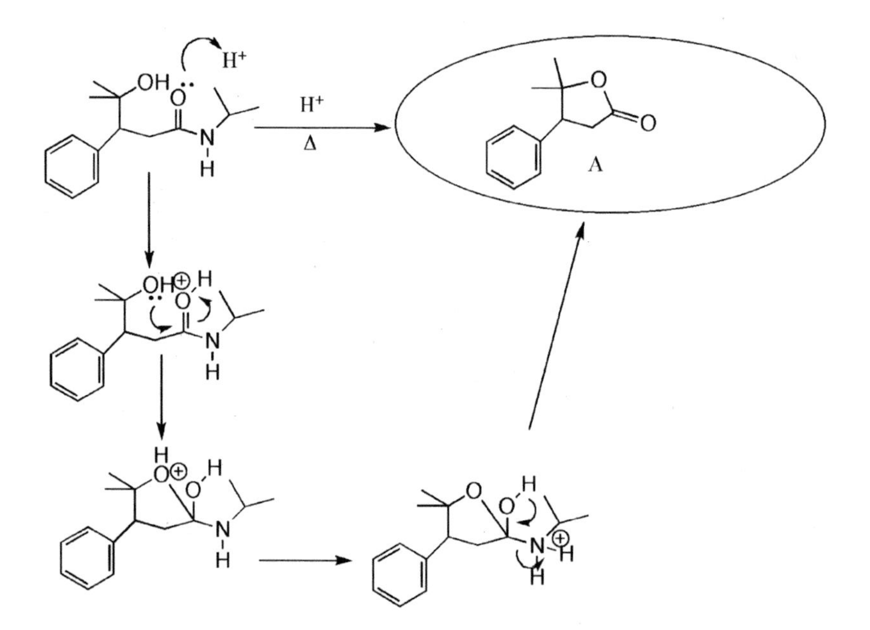

For detailed instructions for solving this problem, see
http://valhalla.chem.udel.edu/SpecBook.pdf.

Problem 3.29

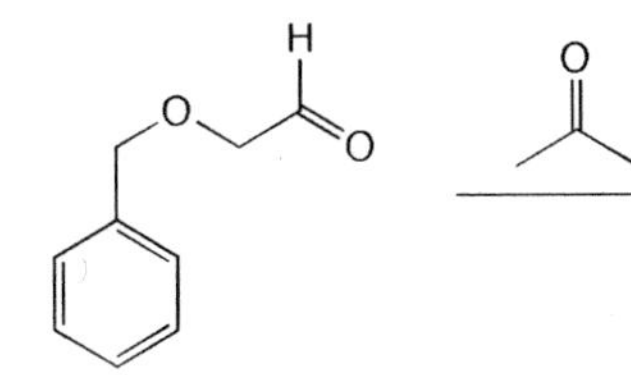

O

PPh$_3$

→ A $C_{12}H_{14}O_2$

^{13}C NMR:	**^{1}H NMR:**
27.2, q	2.27, s, 3H
68.8, t	4.20, d, J = 4.5 Hz, 2H
72.9, t	4.57, s, 2H
127.7, d (2)	6.35, d, J = 16.1 Hz, 1H
127.9, d	6.80, dt, J = 16.1, 4.5 Hz, 1H
128.5, d (2)	7.3, m, 5H
130.4, d	
137.6, s	
142.9, d	
198.0, s	

WORKSHEET 3.29

From the molecular formula:

$C_9H_{10}O_2 + C_{21}H_{19}OP = C_{30}H_{29}O_3P - C_{12}H_{14}O_2 = C_{18}H_{15}OP$

Fragments of the starting materials that remain:

68.8, t
72.9, t

Now there is only on H on the adjacent carbon

4.20, d, J = 4.5 Hz, 2H
4.57, s, 2H

O
2.27, s, 3H
CH_3

7.3, m, 5H

Fragments that have changed in the product:

Aldehyde is gone.

$Ph_3P{=}O$ is gone.

There is a trans alkene.

Mechanism and product:

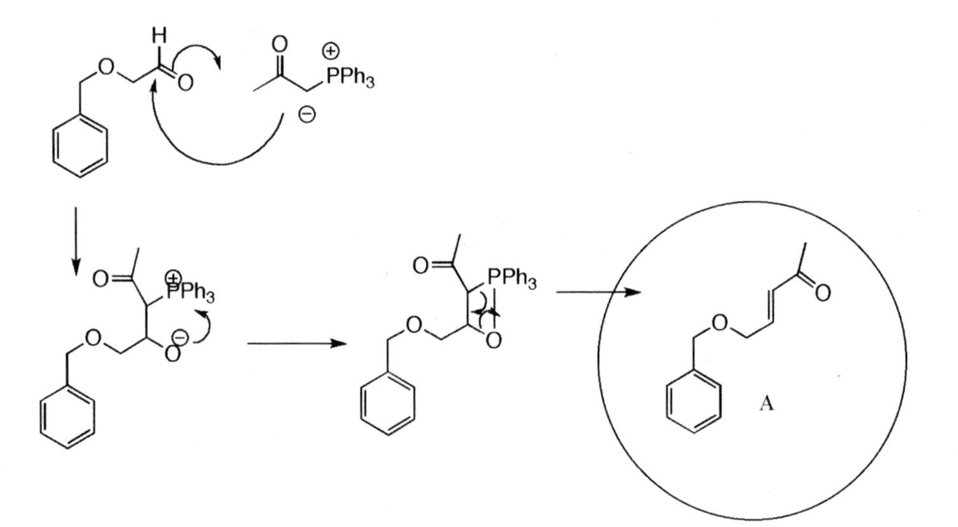

For detailed instructions for solving this problem, see http://valhalla.chem.udel.edu/SpecBook.pdf.

Problem 3.30

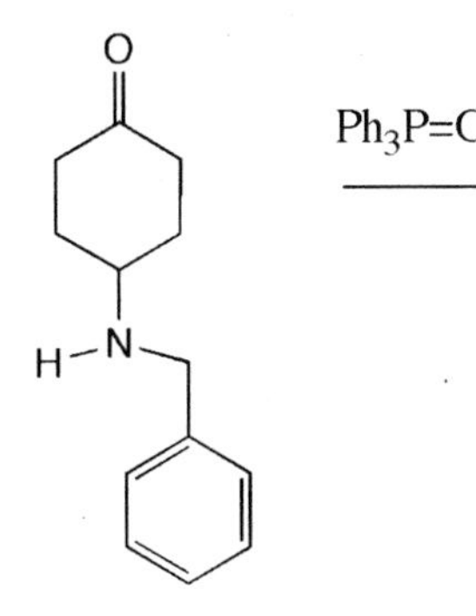

$Ph_3P{=}CH\text{-}CO_2Et$ ⟶ A $C_{17}H_{23}NO_2$

IR: 1735 cm^{-1}

^{13}C NMR:	**^{1}H NMR:**
14.2, q	1.21, t, J = 7.1 Hz, 3H
28.4, t (2)	1.3, m, 2H
33.5, t (2)	1.6, m, 2H
39.3, t	1.7, m, 4H
49.1, t	2.61, s, 2H
59.0, d	3.11, m, 1H
60.3, s	3.44, s, 2H
65.5, t	4.10, q, J = 7.1 Hz, 2H
126.6, d	7.3, m, 5H
128.1, d (2)	
128.6, d (2)	
140.4, s	
171.6, s	

WORKSHEET 3.30

From the molecular formula:

$C_{13}H_{17}NO + C_{22}H_{21}O_2P = C_{35}H_{38}NO_3P - C_{17}H_{23}NO_2 = Ph_3P{=}O$

Fragments of the starting materials that remain:

CO_2Et

4.10, q, J=7.1 Hz, 2H

1.21, t, J=7.1 Hz, 3H

H-N (benzyl) 2.61, s, 2H

7.3, m, 5H

Fragments that have changed in the product:

Ketone is gone.

This is a Wittig reaction, but instead of an alkene, we have an additional ring.

Mechanism and product:

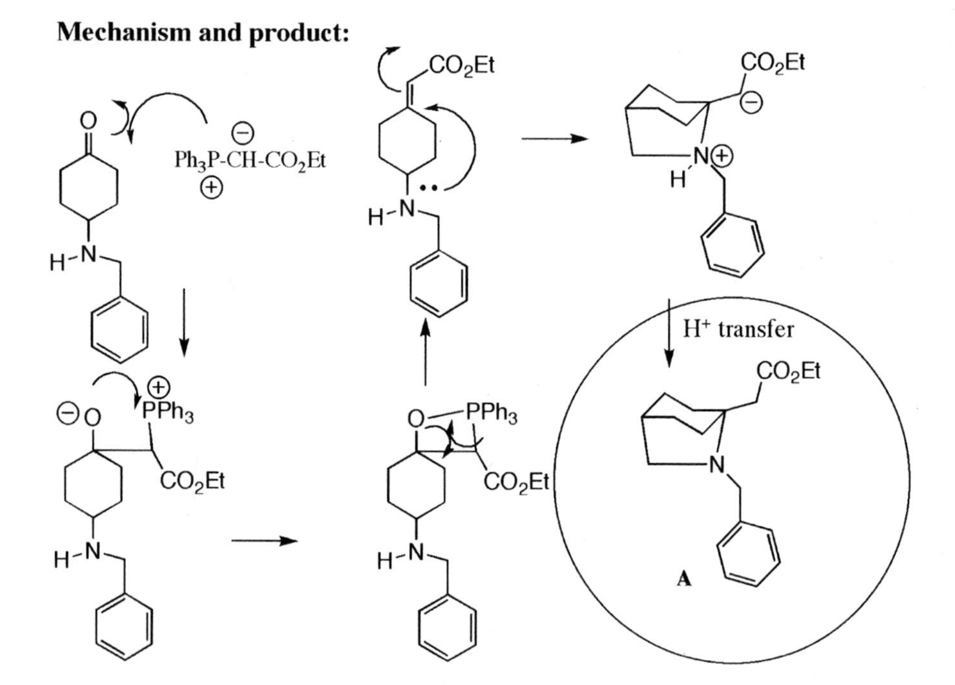

For detailed instructions for solving this problem, see http://valhalla.chem.udel.edu/SpecBook.pdf.

Problem 3.31

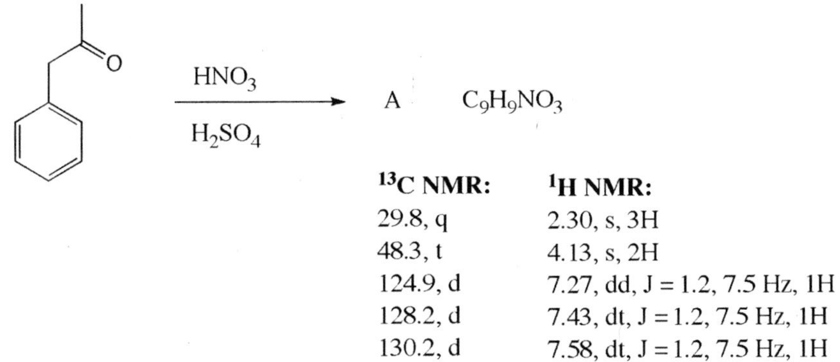

^{13}C NMR:	^{1}H NMR:
29.8, q	2.30, s, 3H
48.3, t	4.13, s, 2H
124.9, d	7.27, dd, J = 1.2, 7.5 Hz, 1H
128.2, d	7.43, dt, J = 1.2, 7.5 Hz, 1H
130.2, d	7.58, dt, J = 1.2, 7.5 Hz, 1H
133.4, d	8.08, dd, J = 1.2, 7.5 Hz, 1H
133.5, s	
203.5, s	
148.4, s	

WORKSHEET 3.31

From the molecular formula:

$C_9H_{10}O$ + HNO_3 = $C_9H_{11}NO_4$ - $C_9H_9NO_3$ = H_2O

Fragments of the starting materials that remain:

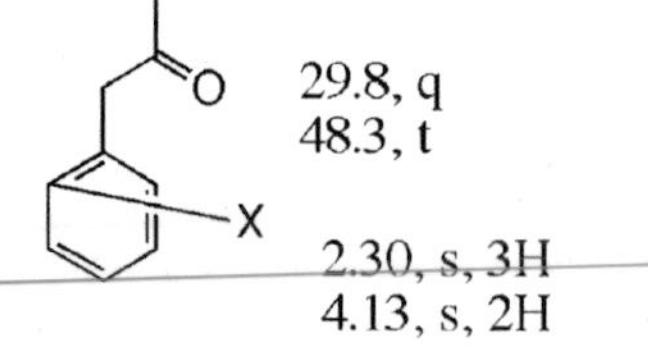

Fragments that have changed in the product:

The aromatic ring is now disubstituted—an NO_2 group has been attached.
It is not symmetrical.

Mechanism and product:

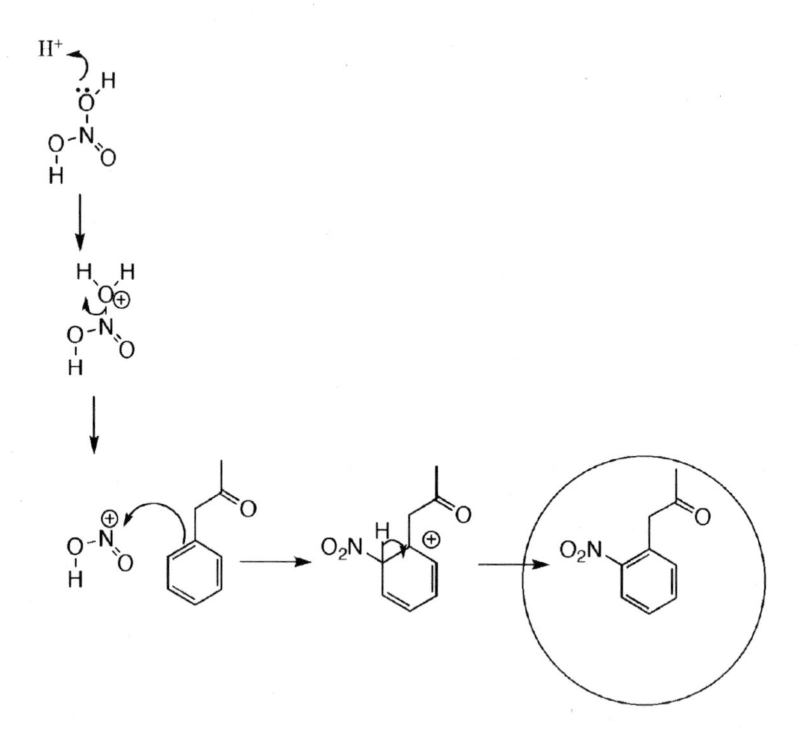

For detailed instructions for solving this problem, see
http://valhalla.chem.udel.edu/SpecBook.pdf.

Problem 3.32

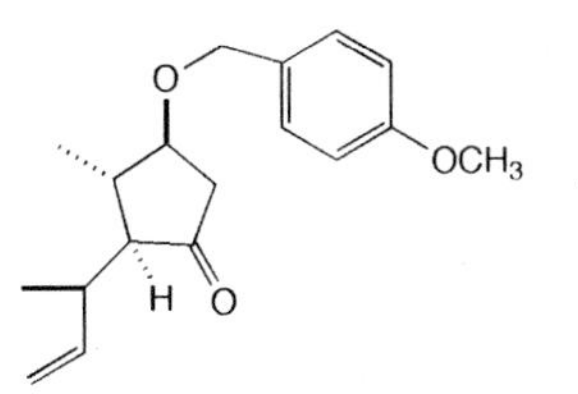

MCPBA → A $C_{18}H_{24}O_4$

IR: 2972, 2936, 1744, 1613, 1514, 1248 cm^{-1}

^{13}C NMR:	^{1}H NMR:
12.6, q	1.03, d, J = 6.7 Hz, 3H
14.7, q	1.06, d, J = 6.9 Hz, 3H
35.6, t	1.9, m, 1H
37.0, d	2.5, m, 1H
39.4, d	2.59, dd, 6.5, 16.9 Hz, 1H
55.4, q	2.83, dd, 5.5, 16.9 Hz, 1H
70.7, t	3.48, dd, J = 1.0, 6.5 Hz, 1H
76.1, d	3.80, s, 3H
85.2, d	3.86, dd, J = 2.8, 10.0 Hz, 1H
114.0, d, (2)	4.39, d, J = 11.4 Hz, 1H
115.3, t	4.55, d, J = 11.4 Hz, 1H
129.5, d (2)	4.97, d, J = 17.3 Hz, 1H
129.7, s	5.05, d, J = 10.0 Hz, 1H
140.7, d	5.93, ddd, J = 7.6, 10.0, 17.3 Hz, 1H
159.5, s	6.88, d, J = 8.6 Hz, 2H
170.8, s	7.24, d, J = 8.6 Hz, 2H

WORKSHEET 3.32

From the molecular formula:

$C_{18}H_{24}O_3 + O = C_{18}H_{24}O_4$

Fragments of the starting materials that remain:

4.97, d, J = 17.3 Hz, 1H
5.05, d, J = 10.0 Hz, 1H
5.93, ddd, J = 7.6, 10.0, 17.3 Hz, 1H

6.88, d, J = 8.6 Hz, 2H
7.24, d, J = 8.6 Hz, 2H

Fragments that have changed in the product:

There is a new oxygenated methine.
76.1, d
85.2, d

The carbonyl is now an ester
170.8, s.

Mechanism and product:

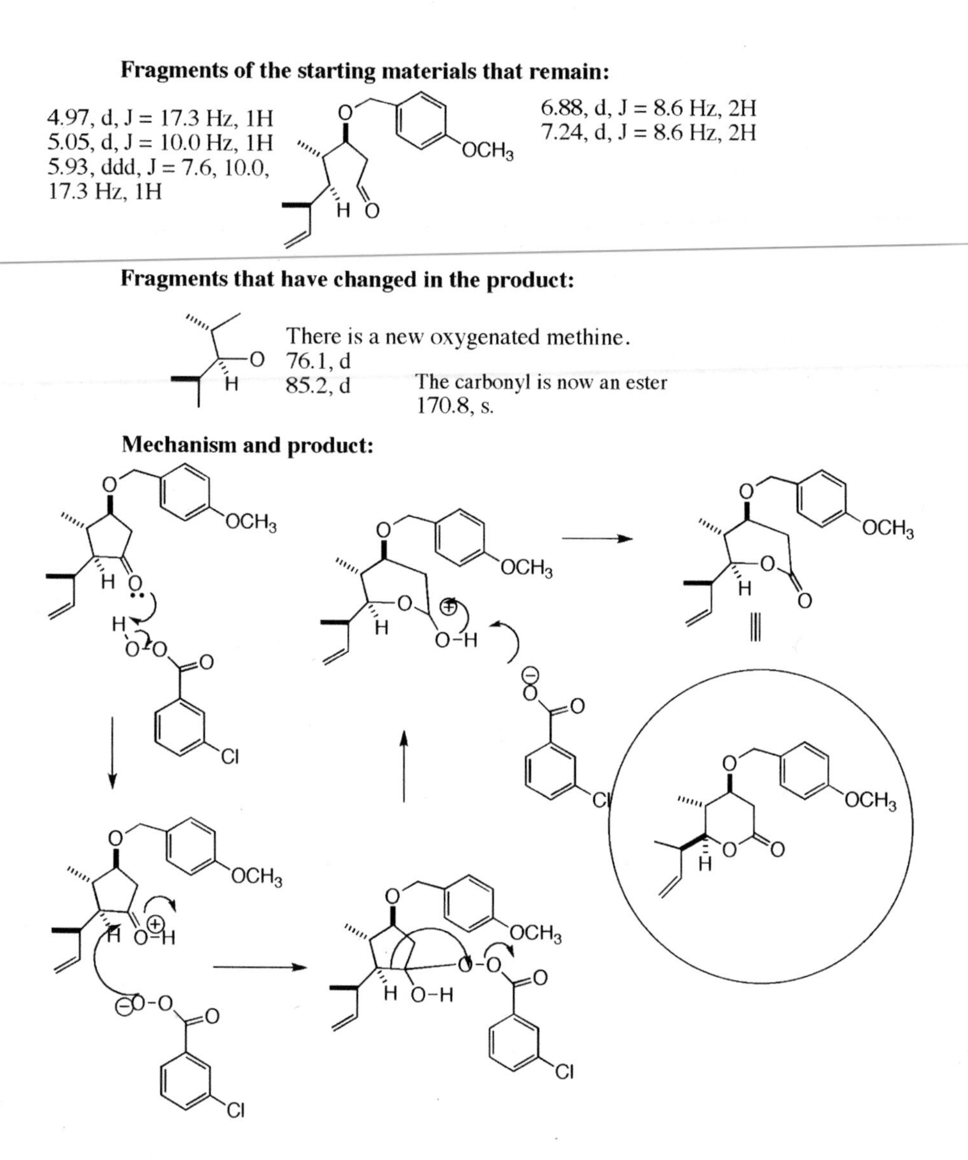

For detailed instructions for solving this problem, see http://valhalla.chem.udel.edu/SpecBook.pdf.

Problem 3.33

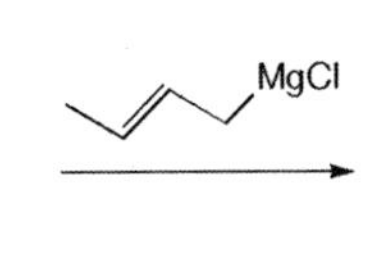

A $C_{11}H_{14}O$

^{13}C NMR:	^{1}H NMR:
16.9, q	0.85, d, J = 6.8 Hz, 3H
46.7, d	2.29, bs, 1H (exchanges)
78.3, d	2.45, ddq. J = 7.2, 7.8, 6.8 Hz, 1H
116.7, t	4.33, d, J = 7.8 Hz, 1H
126.8, d (2)	5.15, d, J = 10.2 Hz, 1H
127.5, d	5.21, d, J = 15.9 Hz, 1H
128.1, d (2)	5.80, ddd, J = 15.9, 10.2, 7.2 Hz, 1H
140.6, d	7.3, m, 5H
142.4, s	

WORKSHEET 3.33

From the molecular formula:

C_7H_6O + C_4H_7MgCl = $C_{11}H_{13}MgClO$ + H_2O - $C_{11}H_{14}O$ = MgClOH

Fragments of the starting materials that remain:

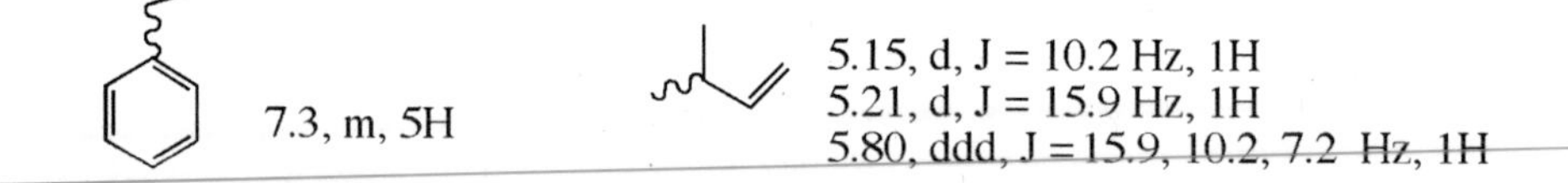

Fragments that have changed in the product:

Aldehyde is gone.

Secondary alcohol has appeared. 78.3, d

Mechanism and product:

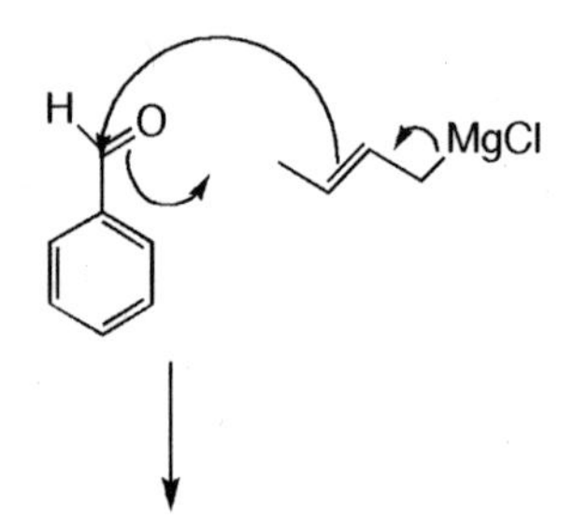

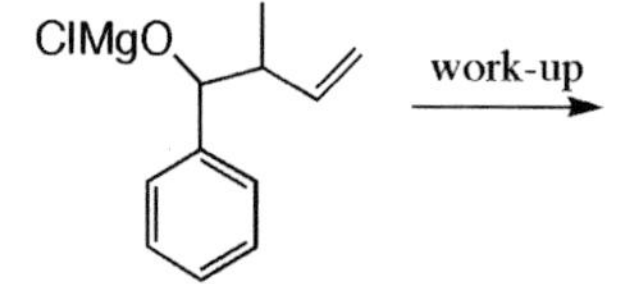

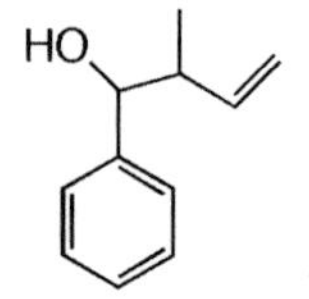

For detailed instructions for solving this problem, see http://valhalla.chem.udel.edu/SpecBook.pdf.

Problem 3.34

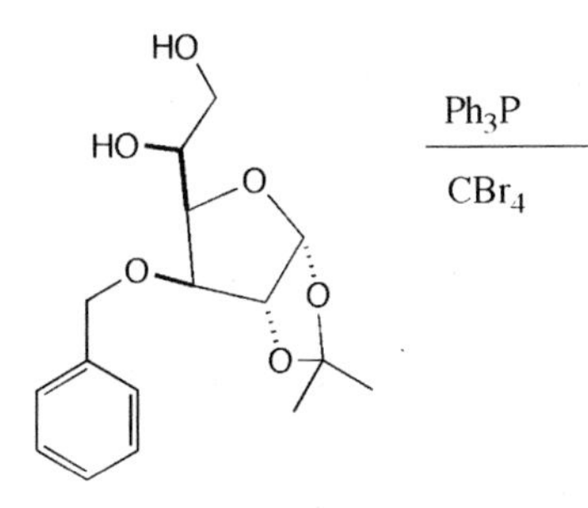

A $C_{16}H_{21}BrO_5$

^{13}C NMR:	**^{1}H NMR:**
26.4, q	1.31, s, 3H
27.0, q	1.46, s, 3H
38.8, t	2.38, bs, 1H (exchanges)
68.5, d	3.58, dd, J = 5.7, 10.5 Hz, 1H
72.6, t	3.71, dd, J = 2.7, 10.5 Hz, 1H
81.1, d	4.02, m, 3H
82.0, d	4.58, d, J = 11.7 Hz, 1H
82.6, d	4.64, d, J = 3.7 Hz, 1H
105.6, d	4.72, d, J = 10.7 Hz, 1H
112.3, s	5.89, d, J = 3.7 Hz, 1H
128.2, d	7.4, m, 5H
128.5, d (2)	
128.9, d (2)	
137.9, s	

WORKSHEET 3.34

From the molecular formula:

$C_{16}H_{22}O_6 + Br_2 + Ph_3P \;=\; C_{16}H_{22}Br_2O_6 + Ph_3P \;-\; C_{16}H_{21}BrO_5 \;=\; HBr + Ph_3P{=}O$

Fragments of the starting materials that remain:

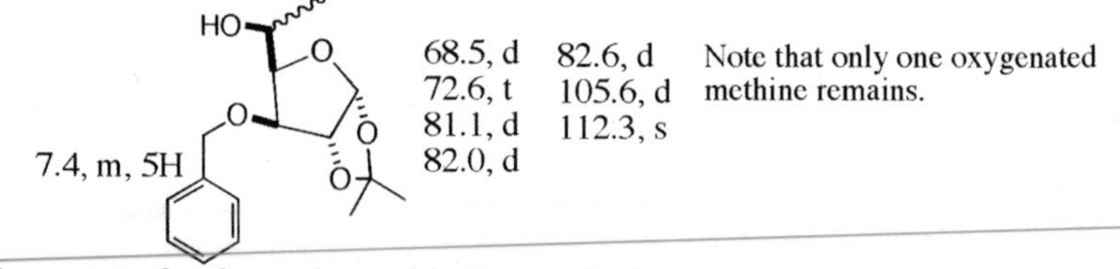

Fragments that have changed in the product:

Primary alcohol is replaced by CH_2-Br.

Mechanism and product:

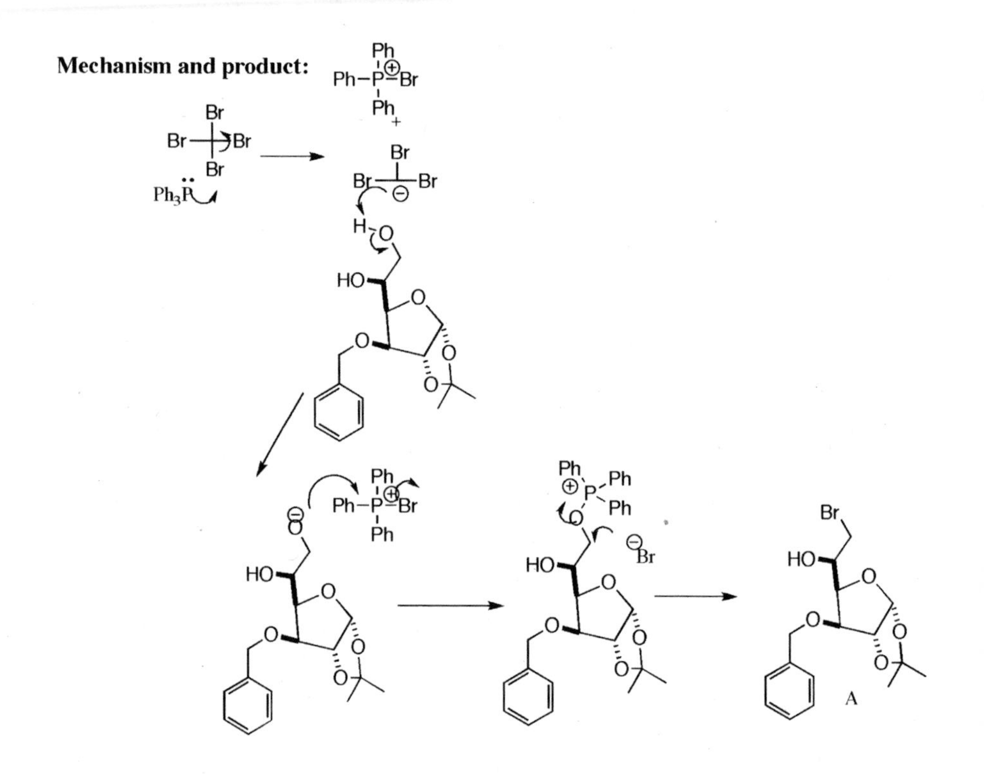

For detailed instructions for solving this problem, see http://valhalla.chem.udel.edu/SpecBook.pdf.

Problem 3.35

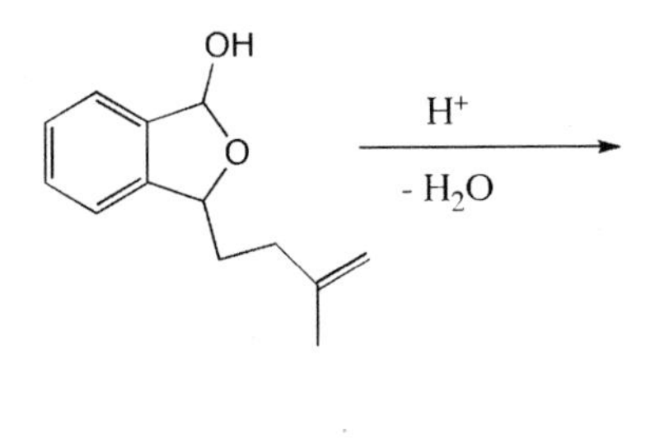

A $C_{13}H_{14}O$

^{13}C NMR:	^{1}H NMR:
31.5, t	1.7, m, 2H
36.1, t	2.1, m, 2H
45.7, t	2.45, dd, J = 0.9, 14.4 Hz, 1H
80.3, d	2.98, dd, J = 6.6, 14.4 Hz, 1H
82.3, d	4.55, dd, J = 1.2, 5.4 Hz, 1H
114.6, t	4.68,dd, J = 0.9, 6.6 Hz, 1H
120.3, d	5.44, d, J = 5.7 Hz, 1H
120.4, d	5.48, d, J = 5.7 Hz, 1H
127.4, d	7.3, m, 4H
127.5, d	
142.7, s	
143.5, s	
146.6, s	

WORKSHEET 3.35

From the molecular formula:

$C_{13}H_{16}O_2$ - H_2O = $C_{13}H_{14}O$

Fragments of the starting materials that remain:

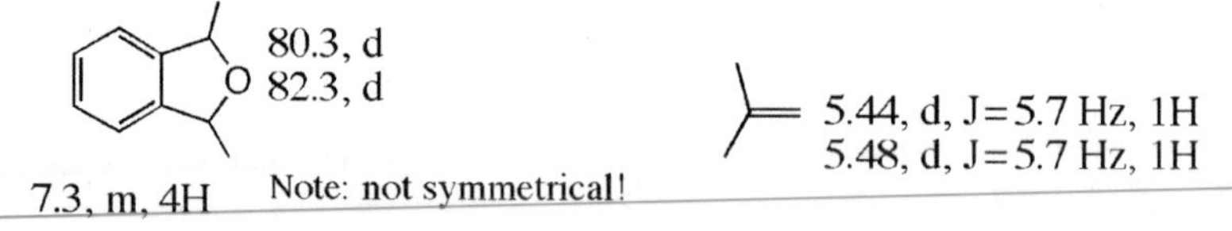

Fragments that have changed in the product:

The OH is gone.

Now there are two ethers, and an additional ring has been created.

Mechanism and product:

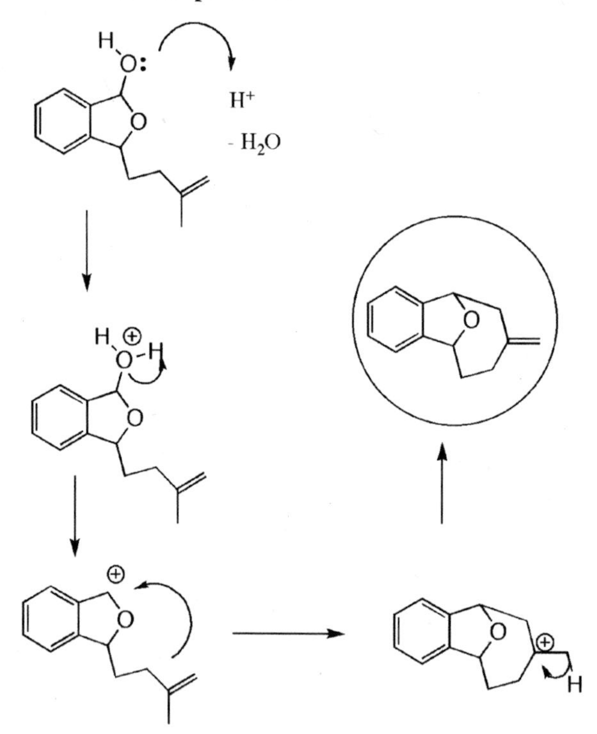

For detailed instructions for solving this problem, see
http://valhalla.chem.udel.edu/SpecBook.pdf.

Problem 3.36

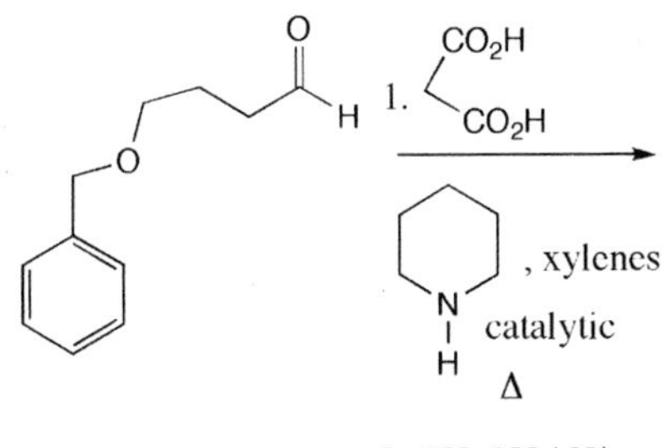

2. CH_3OH / H^+

A $C_{14}H_{18}O_3$

IR: 3030, 2950, 1765, 1454, 1164, 1023 cm^{-1}

^{13}C NMR:	^{1}H NMR:
32.9, t	2.3, m, 2H
37.9, t	3.05, d, J = 5.9 Hz, 2H
51.7, q	3.51, t. J = 6.7 Hz, 2H
69.6, t	3.68, s, 3H
72.9, t	4.51, s, 2H
123.6, d	5.6, m, 2H
127.5, d	7.33, m, 5H
127.6, d (2)	
128.2, d (2)	
131.0, d	
138.4, s	
172.3, s	

WORKSHEET 3.36

From the molecular formula:

$C_{11}H_{14}O_2$ + $C_3H_4O_4$ + CH_3OH = $C_{15}H_{22}O_7$ - $C_{14}H_{18}O_3$ = $CO_2 + H_2O \times 2$

Fragments of the starting materials that remain:

7.33, m, 5H

69.6, t
72.9, t

Fragments that have changed in the product: Aldehyde gone.

Mechanism and product: Methyl ester, alkene added. 3.68, s, 3H; 5.6, m, 2H

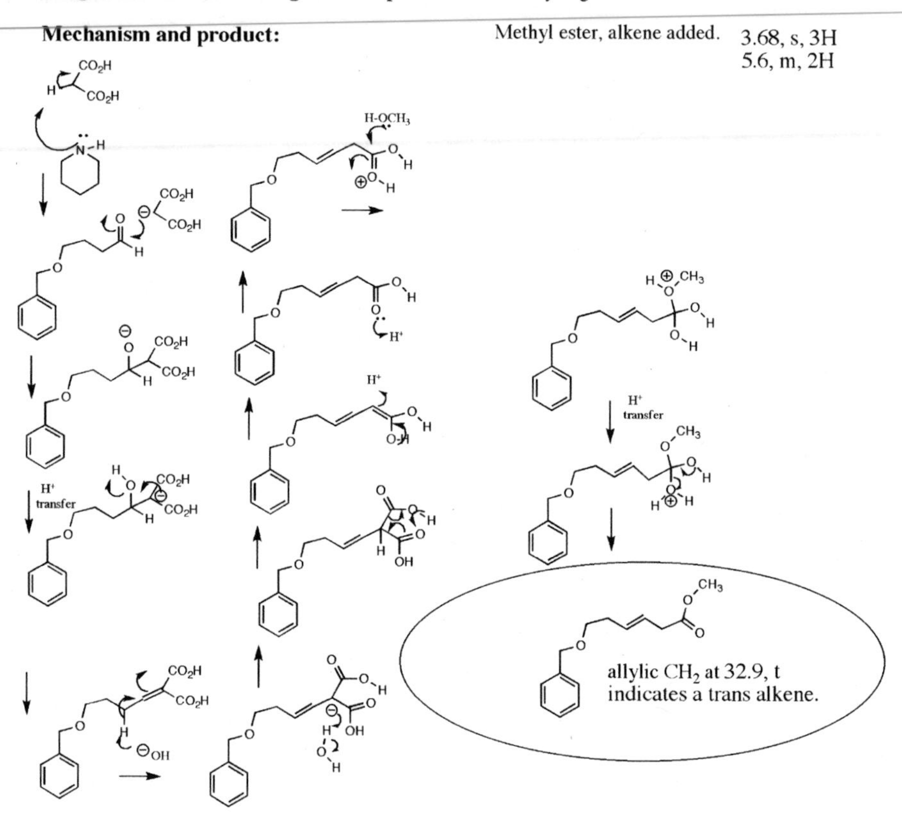

For detailed instructions for solving this problem, see http://valhalla.chem.udel.edu/SpecBook.pdf.

Problem 3.37

Br

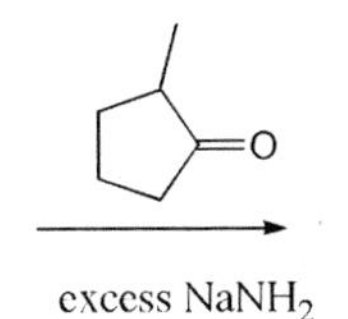

A $C_{12}H_{14}O$

IR: 3060, 2920, 1675, 1600, 1450, 1405, 1375, 1355, 1315, 1285, 1250, 1220, 1180, 1110, 1020, 930, 820, 760 cm^{-1}

^{13}C NMR:	^{1}H NMR:
19.3, q	1.39, d, J = 6.9 Hz, 3H
20.4, t	1.4–1.6, m, 2H
34.2, t	1.8–2.0, m, 2H
34.4, d	2.5–2.7, m, 2H
41.2, t	3.07, m, 1H
125.1, d	7.22–7.55, m, 2H
126.3, d	7.43, ddd, J = 7.5, 7.5, 1.5 Hz, 1H
127.7, d	7.50, d, J = 7.5 Hz, 1H
131.8, d	
139.3, s	
143.0, s	
208.1, s	

WORKSHEET 3.37

From the molecular formula:

$C_6H_5Br + C_6H_{10}O = C_{12}H_{15}BrO - C_{12}H_{14}O = HBr$

Fragments of the starting materials that remain:

Fragments that have changed in the product:

Benzene is still there, but now disubstituted, with the ketone directly attached 7.50, d, J=7.5 Hz, 1H.

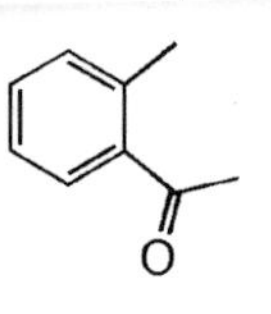

Mechanism and product:

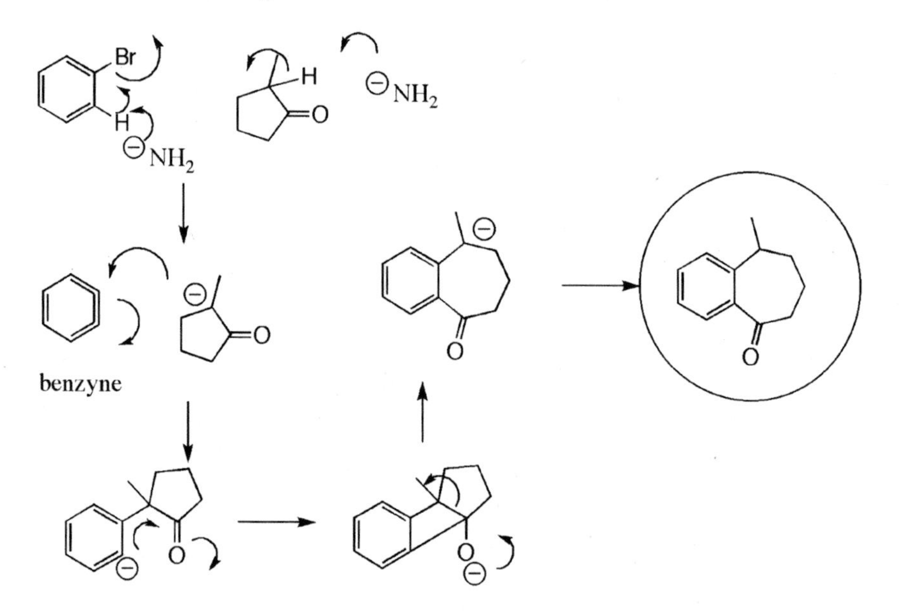

For detailed instructions for solving this problem, see http://valhalla.chem.udel.edu/SpecBook.pdf.

Problem 3.38

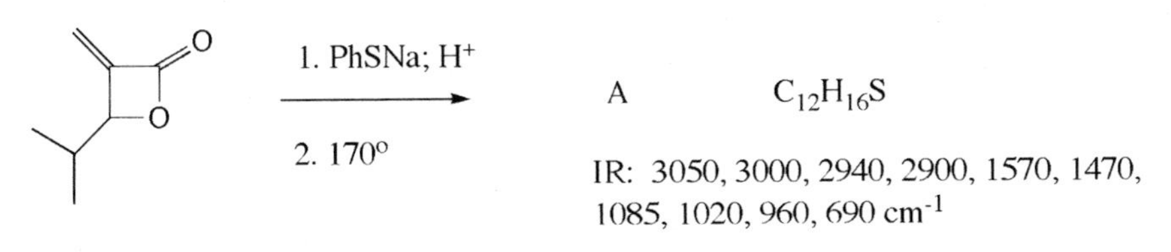

IR: 3050, 3000, 2940, 2900, 1570, 1470, 1085, 1020, 960, 690 cm^{-1}

^{13}C NMR:	^{1}H NMR:
22.2, q (2)	0.85, d, J = 6.8 Hz, 6H
29.7, d	2.48, m, 1H
36.7, t	3.40, d, J = 5.2 Hz, 2H
127.5, d (2)	5.3, m, 2H
128.6, d (2)	7.3, m, 5H
129.0, d	
130.2, d	
137.0, s	
141.5, d	

WORKSHEET 3.38

From the molecular formula:

$C_7H_{10}O_2$ + C_6H_6S = $C_{13}H_{16}O_2S$ - $C_{12}H_{16}S$ = CO_2

Fragments of the starting materials that remain:

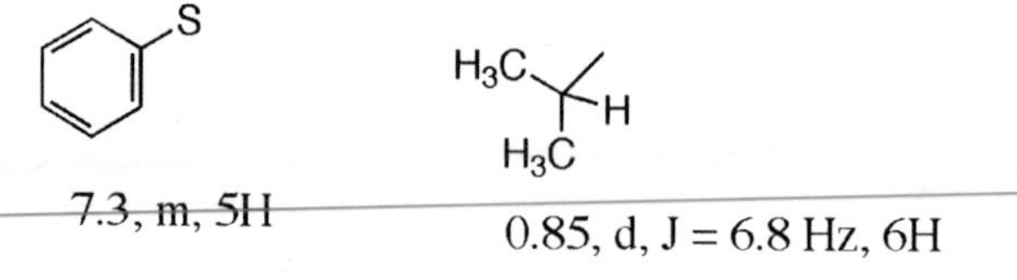

7.3, m, 5H

0.85, d, J = 6.8 Hz, 6H

Fragments that have changed in the product:

Disubstituted alkene
5.3, m, 2H.

Mechanism and product:

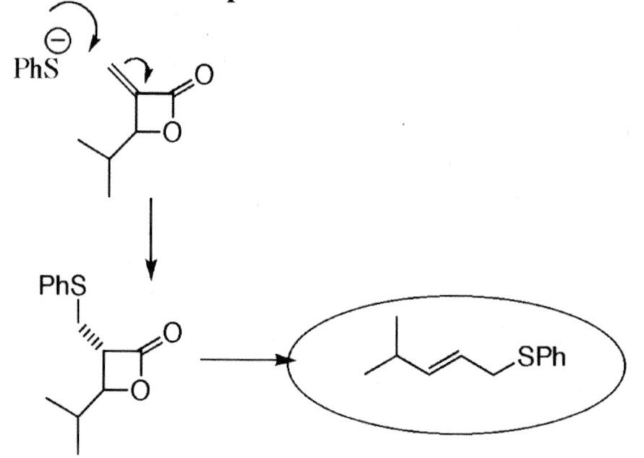

For detailed instructions for solving this problem, see
http://valhalla.chem.udel.edu/SpecBook.pdf.

Problem 3.39

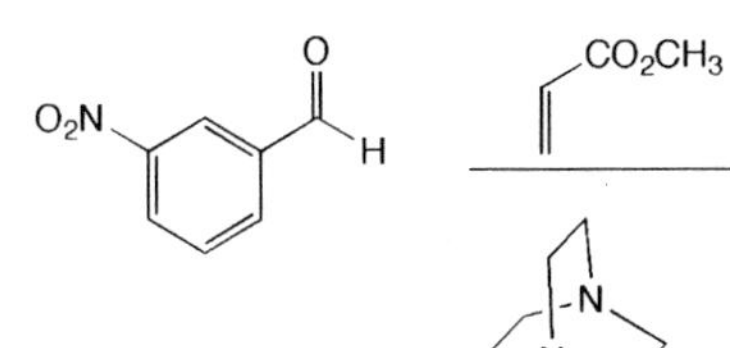

(catalytic)

A $C_{11}H_{11}NO_5$

IR: 3488, 1716, 1531, 1351 cm^{-1}

^{13}C NMR:	^1H NMR:
52.4, q	3.58, bs, 1H (exchanges)
72.4, d	3.70, s, 3H
121.8, t	5.62, s, 1H
122.9, d	5.94, s, 1H
127.2, d	6.38, s, 1H
129.5, d	7.50, dd, J = 7.7, 8.0 Hz, 1H
133.0, d	7.72, m, 1H
141.2, s	8.21, dd, J = 1.1, 1.3 Hz, 1H
143.9, s	8.10, ddd, J = 1.1, 1.2, 8.0 Hz, 1H
148.4, s	
166.5, s	

WORKSHEET 3.39

From the molecular formula:

$C_7H_5NO_3 + C_4H_6O_2 \;=\; C_{11}H_{11}NO_5$

Fragments of the starting materials that remain:

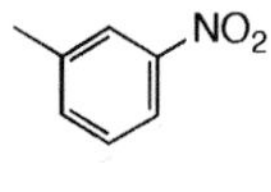

7.50, dd, J = 7.7, 8.0 Hz, 1H
7.72, m, 1H
8.21, dd, J = 1.1, 1.3 Hz, 1H
8.10, ddd, J = 1.1, 1.2, 8.0 Hz, 1H

CO_2CH_3 3.70, s, 3H
CH_2 5.94, s, 1H
6.38, s, 1H

Fragments that have changed in the product:

Aldehyde is gone.
Secondary alcohol has appeared 72.4, d.

Mechanism and product:

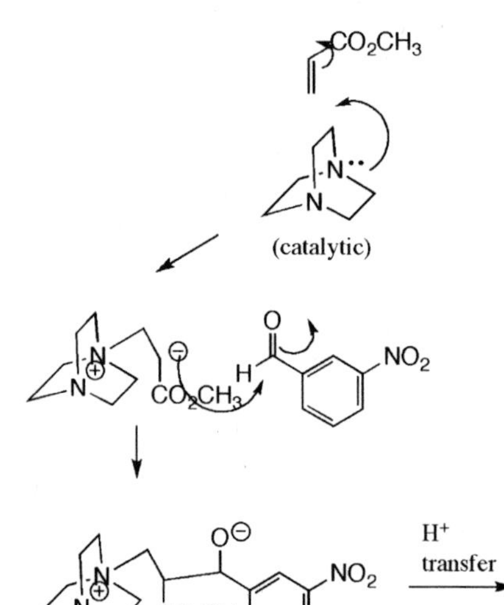

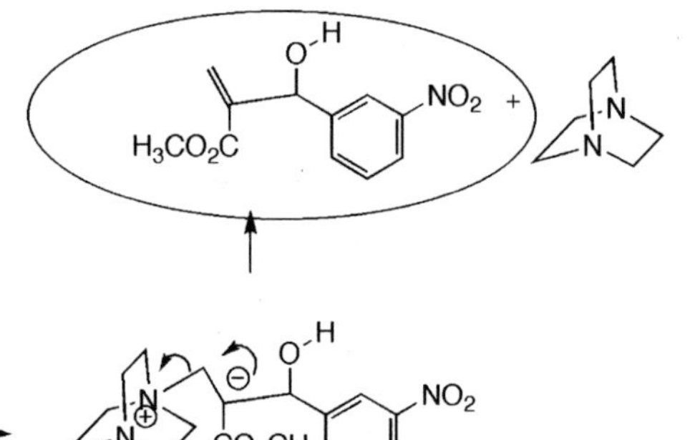

O⊖ NO_2 CO_2CH_3 H^+ transfer CO_2CH_3 NO_2 O-H

For detailed instructions for solving this problem, see http://valhalla.chem.udel.edu/SpecBook.pdf.

Problem 3.40

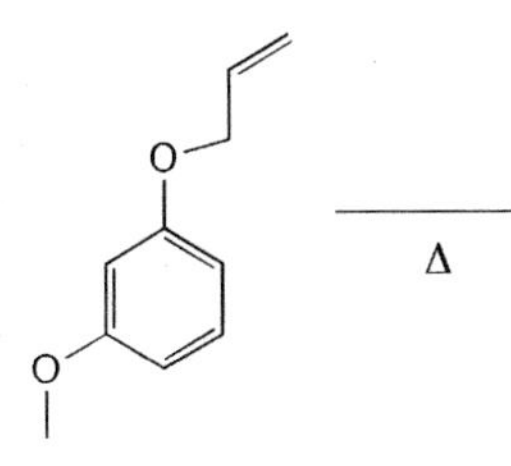

$\xrightarrow{\Delta}$ A $C_{10}H_{12}O_2$

IR: 3443, 3078, 1637, 1596, cm^{-1}

^{13}C NMR:	**^{1}H NMR:**
27.3, t	3.46, d, J = 5.8 Hz, 2H
55.8, q	3.80, s, 3H
103.3, s	5.07, d, J = 10.9 Hz, 1H
108.8, d	5.10, d, J = 17.4 Hz, 1H
113.6, d	5.98, ddt, J = 10.9, 17.4, 5.8 Hz, 1H
115.3, t	6.48, d, J = 7.5 Hz, 1H
127.5, d	6.49, d, J = 7.5 Hz, 1H
136.3, d	7.07, t, J = 7.5 Hz, 1H
155.1, s	9.92 Hz, 1H, s
158.2, s	

WORKSHEET 3.40

From the molecular formula:

$$C_{10}H_{12}O_2 \quad = \quad C_{10}H_{12}O_2$$

Fragments of the starting materials that remain:

Benzene ring is still there, but now trisubstituted.

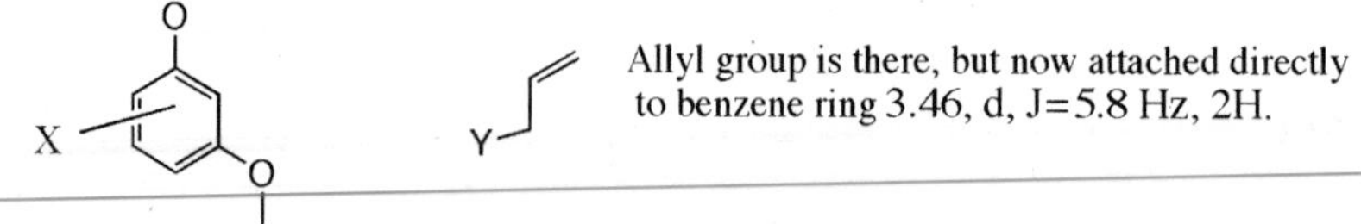

Allyl group is there, but now attached directly to benzene ring 3.46, d, J=5.8 Hz, 2H.

Fragments that have changed in the product:

Allyl ether is gone.

OH has appeared. This is directly attached to benzene ring (a phenol) 9.92 Hz, 1H, s.

Mechanism and product:

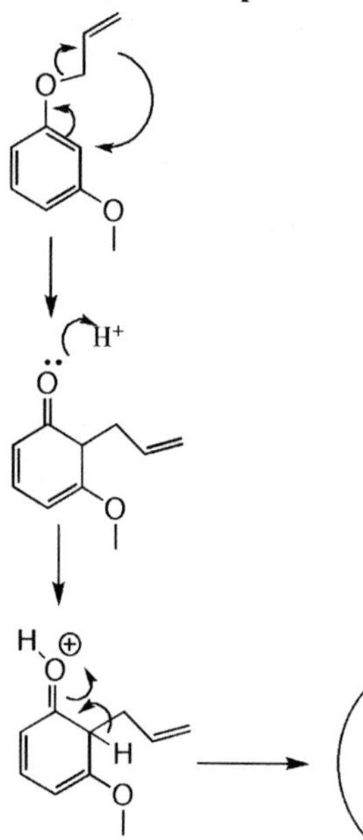

Note that the upfield ring carbon C-2 to both oxygens is at 103.3, singlet, so the carbon chain must be attached there.

For detailed instructions for solving this problem, see http://valhalla.chem.udel.edu/SpecBook.pdf.

Problem 3.41

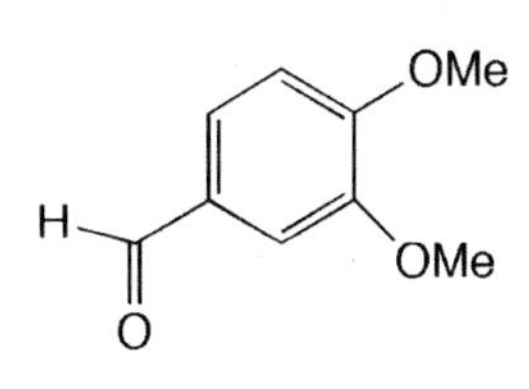

$\xrightarrow[\text{AcOH}]{\text{Br}_2}$ A $C_9H_9BrO_3$

^{13}C NMR:	^{1}H NMR:
56.1, q	3.90, s, 3H
56.4, q	3.94, s, 3H
110.4, d	7.03, s, 1H
115.4, d	7.40, s, 1H
120.3, s	10.17, s, 1H
126.5, s	
148.7, s	
154.5, s	
190.7, d	

WORKSHEET 3.41

From the molecular formula:

$C_9H_{10}O_3$ + Br_2 = $C_9H_{10}Br_2O_3$ - $C_9H_9BrO_3$ = HBr

Fragments of the starting materials that remain:

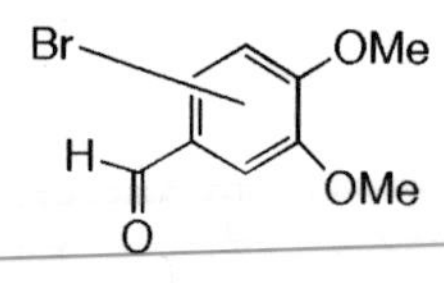

Clearly, the starting material is monobrominated. The question is, which of the three positions has been brominated?

Fragments that have changed in the product:

The H on the ring ortho to the aldehyde is shifted downfield. There is only one, so the Br must be adjacent to the aldehyde.

The benzene carbons C-2 to the ethers are shifted upfield. They are both doublets, 110.4, d; 115.4, d so the Br is not there.

Mechanism and product:

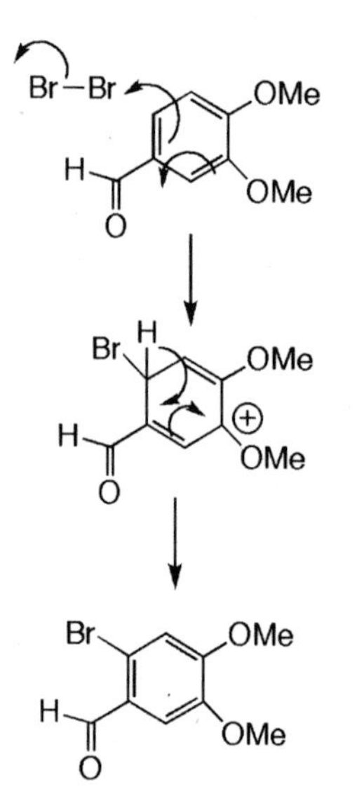

For detailed instructions for solving this problem, see http://valhalla.chem.udel.edu/SpecBook.pdf.

Problem 3.42

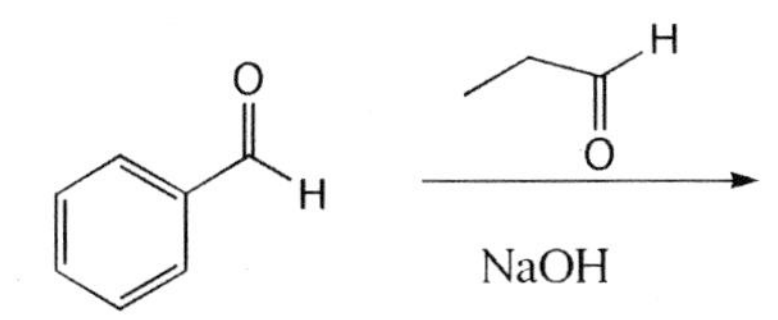

A $C_{10}H_{12}O_2$

^{13}C NMR:	^{1}H NMR:
7.4, q	1.22, d, J = 7.3 Hz, 3H
53.0, d	2.7, m, 1H
72.5, d	2.8, bs, 1H (exchanges)
125.1, d (2)	5.25, d, J = 3.8 Hz, 1H
125.8, d (2)	7.3, m, 5H
128.3, d	9.79, d, J = 1.1 Hz, 1H
141.5, s	
204.4, d	

WORKSHEET 3.42

From the molecular formula:

C_7H_6O + C_3H_6O = $C_{10}H_{12}O_2$

Fragments of the starting materials that remain:

9.79, d, J=1.1 Hz, 1H

7.3, m, 5H no longer
polarized directly attached aldehdye.

Fragments that have changed in the product:

Aromatic aldehyde is gone.

Mechanism and product:

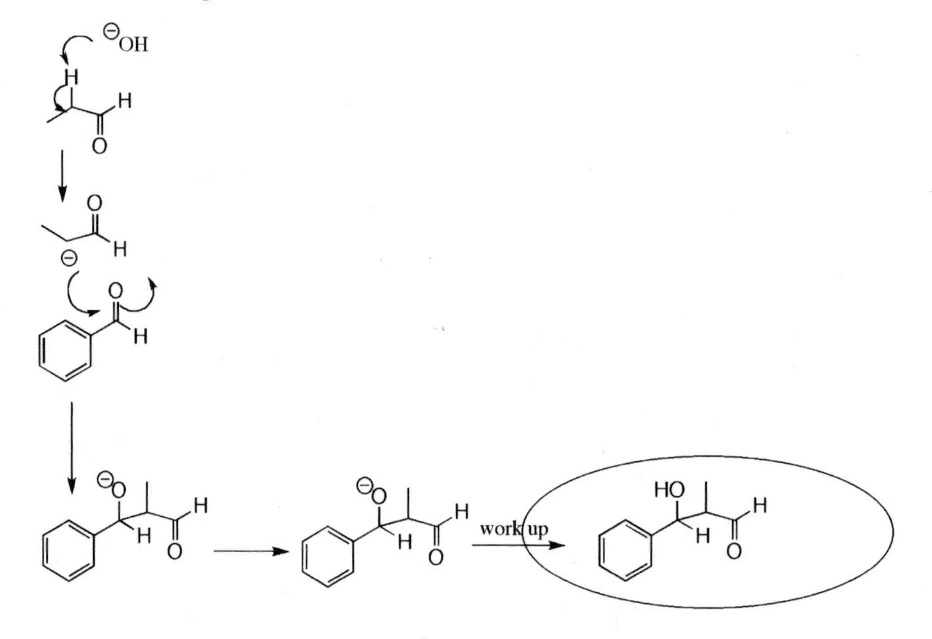

For detailed instructions for solving this problem, see
http://valhalla.chem.udel.edu/SpecBook.pdf.

Problem 3.43

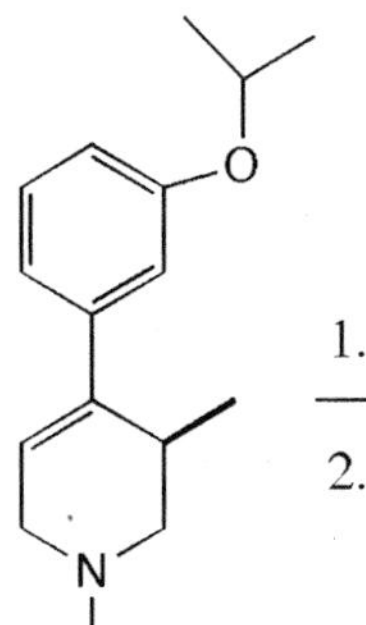

1. n-BuLi
2. CH_3-I

A $C_{17}H_{25}NO$

IR: 3009, 2977, 1643, 1604, 1577 cm^{-1}

^{13}C NMR:	^{1}H NMR:
15.0, q	0.59, d, J = 7.0 Hz, 3H
22.1, q	1.33, d, J = 6.1 Hz, 6H
22.2, q	1.43, s, 3H
28.0, q	1.9, m, 1H
38.8, q	2.47, m, 1H
40.1, d	2.67, s, 3H
42.5, t	2.7, m, 1H
52.8, s	4.34, d, J = 7.9 Hz, 1H
69.7, d	4.52, septet, J = 6.1 Hz, 1H
106.2, d	5.97, d, J = 7.9 Hz, 1H
112.7, d	6.71, d, J = 8.2 Hz, 1H
117.3, d	6.9, m, 2H
121.2, d	7.17, t, J = 8.2 Hz, 1H
127.8, d	
135.2, d	
148.0, s	
157.0, s	

WORKSHEET 3.43

From the molecular formula:

$C_{16}H_{23}NO$ + CH_3I = $C_{17}H_{26}INO$ - $C_{17}H_{25}NO$ = HI

Fragments of the starting materials that remain:

O

6.71, d, J = 8.2 Hz, 1H
6.9, m, 2H
7.17, t, J = 8.2 Hz, 1H

N CH_2 CH_3 42.5, t—but there is only one, not two.

2.67, s, 3H

CH_3 H

0.59, d, J = 7.0 Hz, 3H

Fragments that have changed in the product:

CH_3

1.43, s, 3H

This is a polarized alkene.
4.34, d, J = 7.9 Hz, 1H
5.97, d, J = 7.9 Hz, 1H

H H X

Mechanism and product:

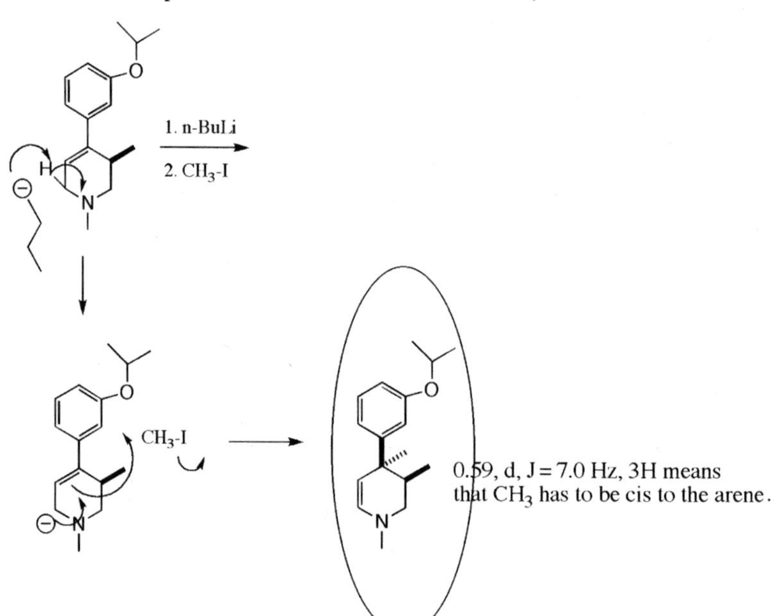

For detailed instructions for solving this problem, see http://valhalla.chem.udel.edu/SpecBook.pdf.

Problem 3.44

1. NBS, DME / H_2O; 2. reflux 18 h → A $C_8H_{10}O_2$

IR: 1765 cm^{-1}

^{13}C NMR:	^1H NMR:
16.5, q	1.63, s, 3H
36.1, t	2.10, dd, J = 16.7, 4.4 Hz, 1H
36.5, t	2.15, dd, J = 18.5, 5.2 Hz, 1H
43.9, d	2.59, dd, J = 16.7, 7.9 Hz, 1H
90.7, d	2.76, dd, J = 18.5, 10.7 Hz, 1H
123.7, d	3.10, m, 1H
148.6, s	5.38, m, 1H
178.0, s	5.42, m, 1H

WORKSHEET 3.44

From the molecular formula:

$C_8H_{10}O$ + HOBr = $C_8H_{11}BrO_2$ - $C_8H_{10}O_2$ = HBr

Fragments of the starting materials that remain:

CH_3 1.63, s, 3H

Fragments that have changed in the product:

O H 90.7, d

O 178.0, s

Note 1765 cm^{-1} five-membered ring lactone.

Mechanism and product:

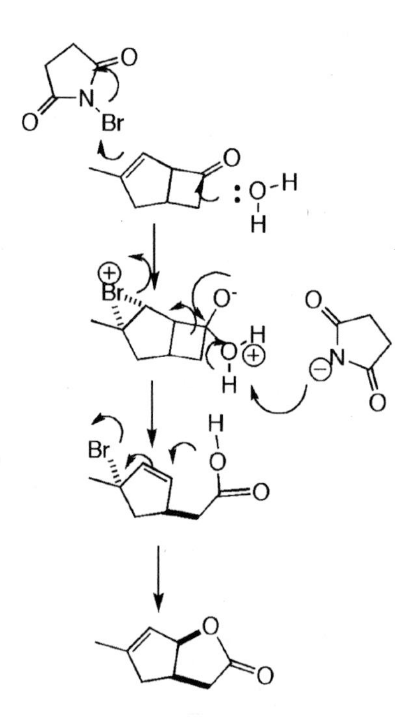

For detailed instructions for solving this problem, see http://valhalla.chem.udel.edu/SpecBook.pdf.

Problem 3.45

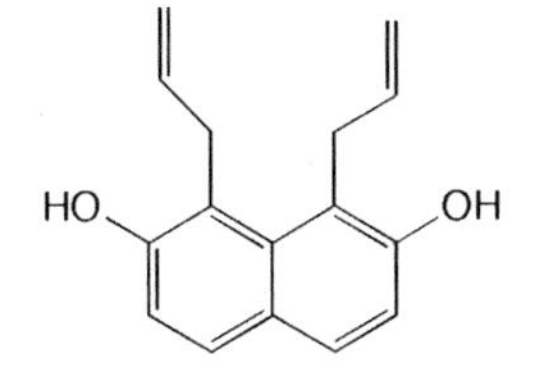

$\xrightarrow[\text{NaH}]{\text{CH}_3\text{-I}}$ A $\quad C_{18}H_{20}O_2$

^{13}C NMR:	**^{1}H NMR:**
25.8, q	1.58, s, 3H
31.9, t	2.87, dd, J = 8.0, 13.9 Hz, 1H
44.8, t	2.99, dd, J = 6.3, 13.9 Hz, 1H
52.3, s	3.72, d, J = 4.5 Hz, 2H
55.4, q	3.85, s, 3H
108.8, d	4.72, d, J = 10.4 Hz, 1H
114.9, t	4.82, d, J = 17.2 Hz, 1H
117.1, t	4.92, d, J = 17.4 Hz, 1H
121.7, s	5.02, d, J = 10.2 Hz, 1H
123.6, d	5.24, m, 1H
128.1, d	5.93, m, 1H
130.3, d	6.04, d, J = 9.8 Hz, 1H
133.4, d	6.86, d, J = 8.4 Hz, 1H
136.4, s	7.20, d, J = 8.4 Hz, 1H
144.4, s	7.35, d, J = 9.8 Hz, 1H
147.1, d	
160.3, s	
204.2, s	

WORKSHEET 3.45

From the molecular formula:

$C_{16}H_{16}O_2$ + CH_3I x 2 = $C_{18}H_{22}I_2O_2$ - $C_{18}H_{20}O_2$ = HI x 2

Fragments of the starting materials that remain:

One benzene ring.

Two allyl groups.

Fragments that have changed in the product:

A ketone with an alkene conjugated to it has appeared.
There is now a methyl ether,3.85, s, 3H, and a methyl group attached to a carbon with no H's, 1.58, s, 3H.

Mechanism and product:

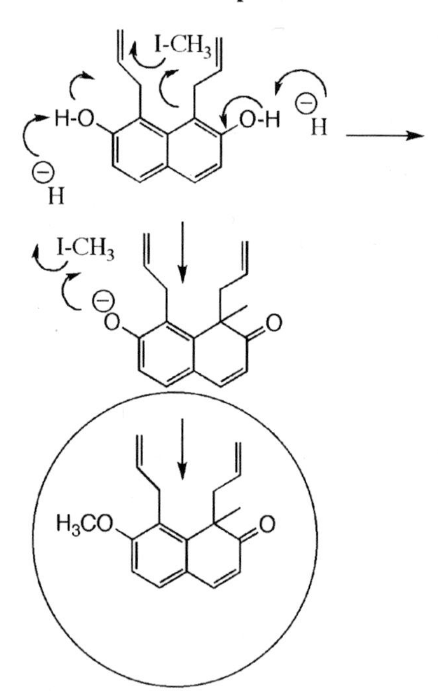

For detailed instructions for solving this problem, see http://valhalla.chem.udel.edu/SpecBook.pdf.

Problem 3.46

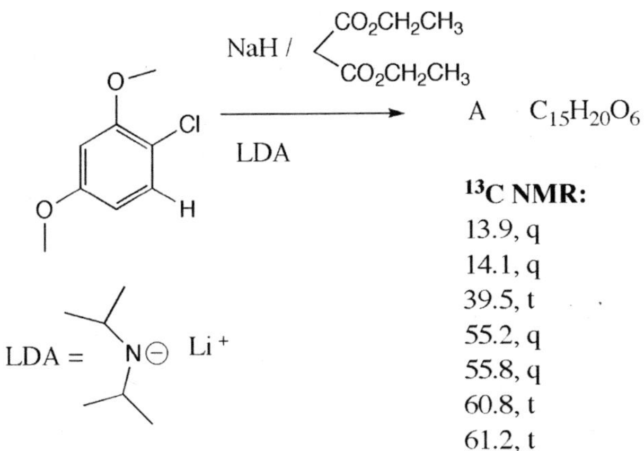

^{13}C NMR:	^{1}H NMR:
13.9, q	1.24, t, J = 7.1 Hz, 3H
14.1, q	3.66, s, 2H
39.5, t	3.81, s, 3H
55.2, q	3.82, s, 3H
55.8, q	4.14, q, J = 7.1 Hz, 2H
60.8, t	4.34, q, J = 7.4 Hz, 2H
61.2, t	6.40, bs, 2H
97.7, d	
107.3, d	
116.2, s	
134.8, s	
158.8, s	
161.4, s	
167.1, s	
170.6, s	

WORKSHEET 3.46

From the molecular formula:

$C_8H_9ClO_2$ + $C_7H_{12}O_4$ = $C_{15}H_{21}ClO_6$ - $C_{15}H_{20}O_6$ = HCl

Fragments of the starting materials that remain:

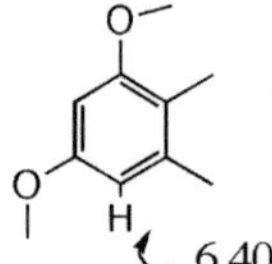

6.40, bs, 2H—so not shifted by carbonyl.

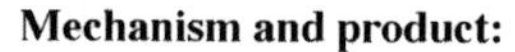

Fragments that have changed in the product:

$CO_2CH_2CH_3$
$CO_2CH_2CH_3$
Doubly shifted CH_2 gone.

Mechanism and product:

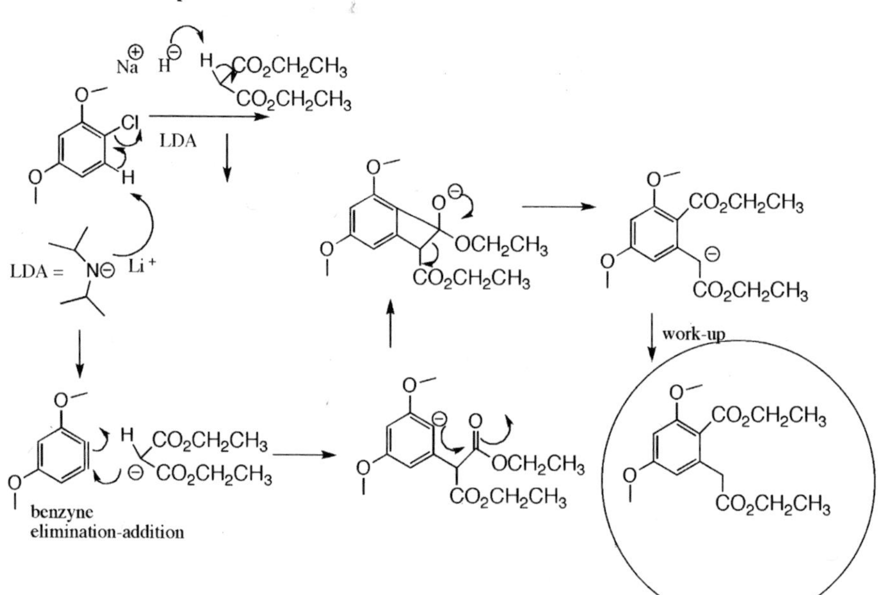

For detailed instructions for solving this problem, see
http://valhalla.chem.udel.edu/SpecBook.pdf.

Problem 3.47

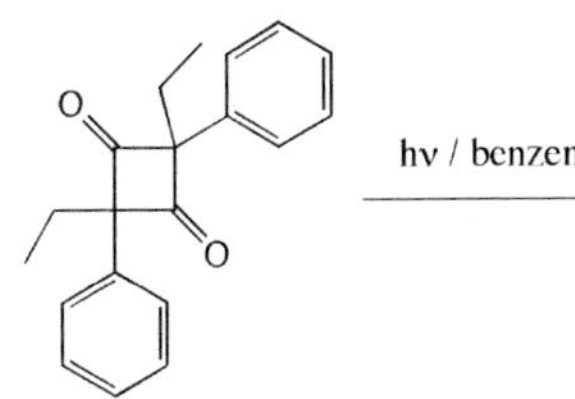

A $C_{19}H_{20}O$

IR: 3061, 3021, 2967, 2934, 2876, 1745, 1445, 1149 cm^{-1}

^{13}C NMR:	^{1}H NMR:
	0.68, t, J = 7.2 Hz, 3H
9.7, q	0.75, t, J = 7.2 Hz, 3H
11.3, q	1.7, m, 2H
25.8, t	2.1, m, 1H
32.1, t	2.5, m, 1H
54.1, d	3.37, t, J = 6.6 Hz, 1H
63.1, s	7.30–7.17, m, 6H
124.6, d	7.42–7.37, m, 3H
126.8, d	
127.3, d	
127.7, d	
128.2, d	
128.3, d (2)	
128.4, d (2)	
140.0, s	
141.7, s	
143.0, s	
219.8, s	

WORKSHEET 3.47

From the molecular formula:

$C_{20}H_{20}O_2$ - $C_{19}H_{20}O$ = CO

Fragments of the starting materials that remain:

There are still two benzene rings, but one is now disubstituted.
There are still two ethyl groups.
The one remaining ketone 1745 cm^{-1} must be a five-membered ring.

Fragments that have changed in the product:

Mechanism and product:

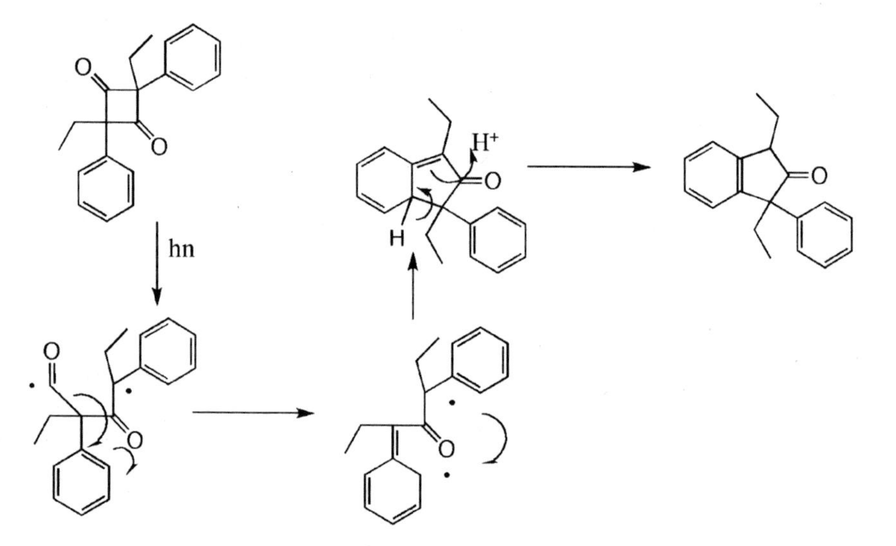

For detailed instructions for solving this problem, see http://valhalla.chem.udel.edu/SpecBook.pdf.

Problem 3.48

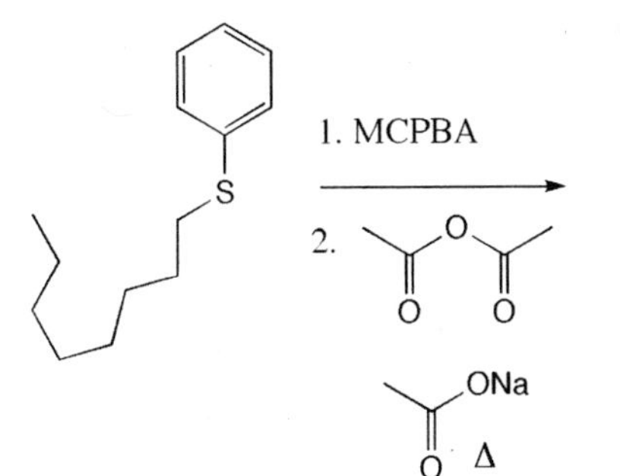

A $C_{16}H_{24}O_2S$

IR: 1748 cm^{-1}

^{13}C NMR:	^{1}H NMR:
13.9, q	0.90, t, J = 6.9 Hz, 3H
21.0, q	1.26, m, 8H
22.5, t	1.40, m, 2H
25.6, t	1.64, dt, J = 6.7, 7.5 Hz, 2H
28.8, t	2.04, s, 3H
28.9, t	6.11, t, J = 6.7 Hz, 1H
31.6, t	7.3, m, 5H
34.5, t	
80.4, d	
128,3, d	
128.9, d (2)	
131.8, d (2)	
133.7, s	
170.1, s	

WORKSHEET 3.48

From the molecular formula:

$C_{16}H_{24}O_2S$ - $C_{14}H_{22}S$ = $C_2H_2O_2$ acetate?

Fragments of the starting materials that remain:

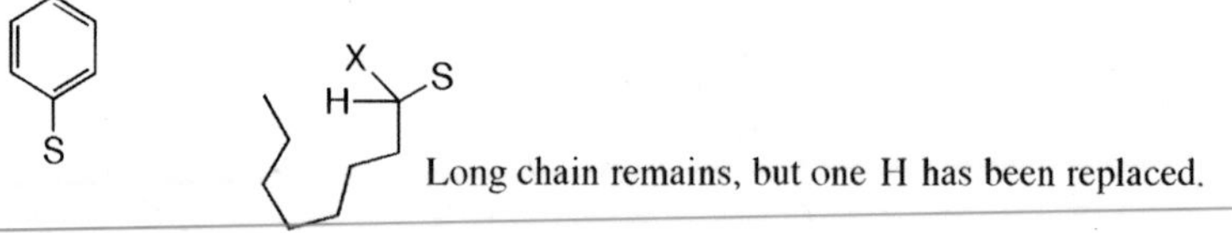

Fragments that have changed in the product:

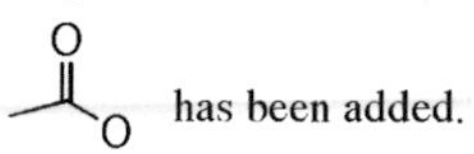

Mechanism and product:

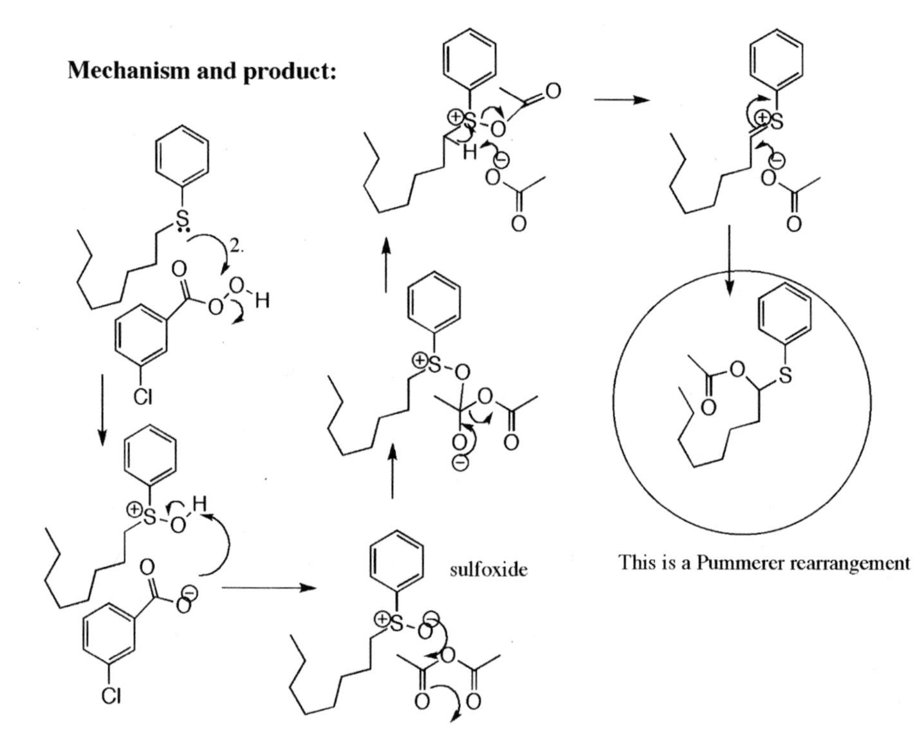

For detailed instructions for solving this problem, see http://valhalla.chem.udel.edu/SpecBook.pdf.

Problem 3.49

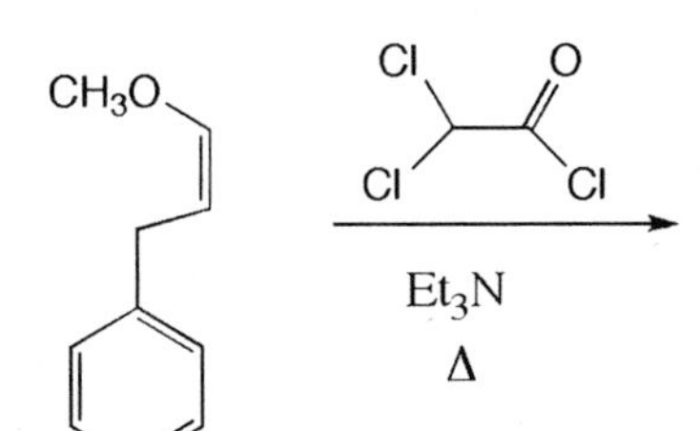

A $C_{12}H_{12}Cl_2O_2$

IR: 1808 cm^{-1}

^{13}C NMR:	**^{1}H NMR:**
33.6, t	3.2–3.4, m, 3H
54.3, q	3.30, s, 3H
59.9, d	4.37, d, J = 9.3 Hz, 1H
77.0, d	7.2, m, 5H
88.0, s	
126.2, d (2)	
128.1, d	
128.5, d (2)	
137.9, s	
195.2, s	

WORKSHEET 3.49

From the molecular formula:

$C_{10}H_{12}O$ + C_2HCl_3O = $C_{12}H_{13}Cl_3O_2$ - $C_{12}H_{12}Cl_2O_2$ = HCl

Fragments of the starting materials that remain:

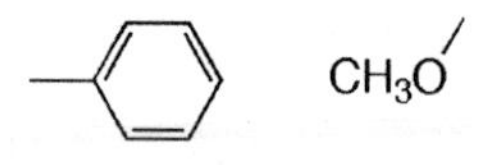

Fragments that have changed in the product:

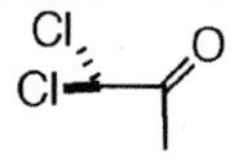

Carbonyl is still there, but now it is a ketone 195.2, s.

Alkene is gone.
There is an additional ring.

Mechanism and product:

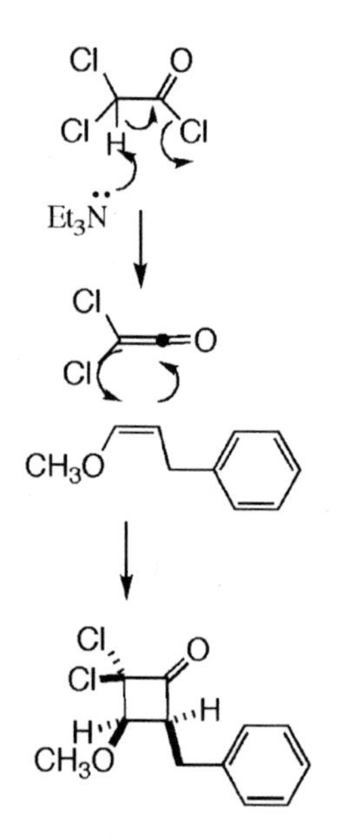

For detailed instructions for solving this problem, see http://valhalla.chem.udel.edu/SpecBook.pdf.

Problem 3.50

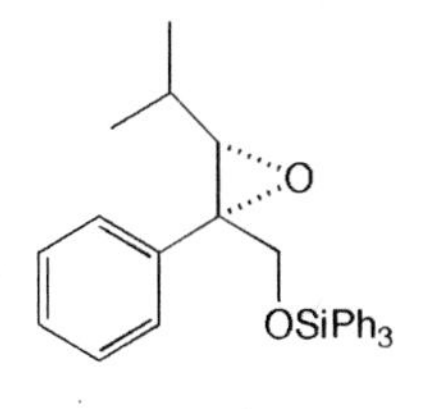

$(ArO)_2Al\text{-}CH_3$ → A $C_{30}H_{30}O_2Si$

$[\alpha]_D = -28.5$

^{13}C NMR:

18.0, q
18.6, q
29.2, d
62.5, s
64.5, t
127.4, d (6)
128.1, d (2)
128.6, d (3)
128.9, d
130.4, d (6)
133.7, d (2)
135.6, s (3)
136.8, s
203.9, d

^{1}H NMR:

0.80, d, J = 6.8 Hz, 3H
0.88, d, J = 6.8 Hz, 3H
2.75, septet, J = 6.8 Hz, 1H
4.24, d, J = 10.4 Hz, 1H
4.27, d, J = 10.4 Hz, 1H
7.4, m, 20H
9.79, s, 1H

WORKSHEET 3.50

From the molecular formula:

$C_{30}H_{30}O_2Si$ = $C_{30}H_{30}O_2Si$ isomer

Fragments of the starting materials that remain:

Three symmetrical phenyls.

One symmetrical phenyl.

0.80, d, J = 6.8 Hz, 3H
0.88, d, J = 6.8 Hz, 3H

4.24, d, J = 10.4 Hz, 1H
4.27, d, J = 10.4 Hz, 1H

OSiPh3

Fragments that have changed in the product:

Epoxide gone.

Aldehyde has appeared 203.9, d 9.79, s, 1H.

Mechanism and product:

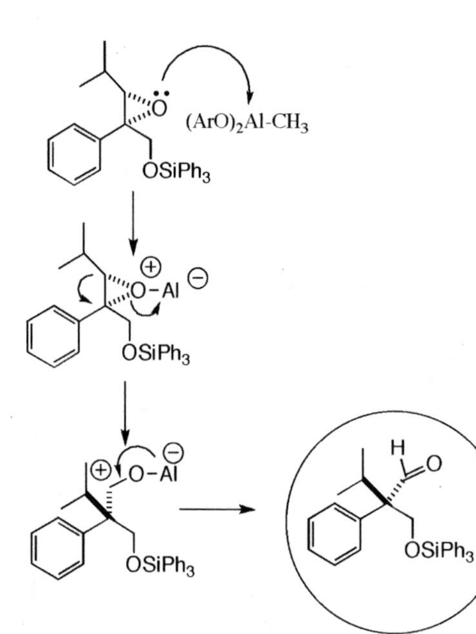

For detailed instructions for solving this problem, see
http://valhalla.chem.udel.edu/SpecBook.pdf.

Section IV
Tables of Spectroscopic Data

^{13}C NMR Tables

All tables, unless otherwise noted, are from R.M. Silverstein, G.C. Bassler, and T.C. Morrill, Spectrometric Identification of Organic Compounds, *5th ed. (New York: John Wiley and Sons, 1991). Reprinted with permission of John Wiley and Sons, Inc.*

Table C.1 Chemical Shifts of Cycloalkanes (ppm from TMS)

C_3H_6	−2.9	C_7H_{14}	28.4
C_4H_8	22.4	C_8H_{16}	26.9
C_5H_{10}	25.6	C_9H_{18}	26.1
C_6H_{12}	26.9	$C_{10}H_{20}$	25.3

Table C.2 Chemical Shifts for Saturated Heterocyclics (ppm from TMS, neat)

Unsubstituted

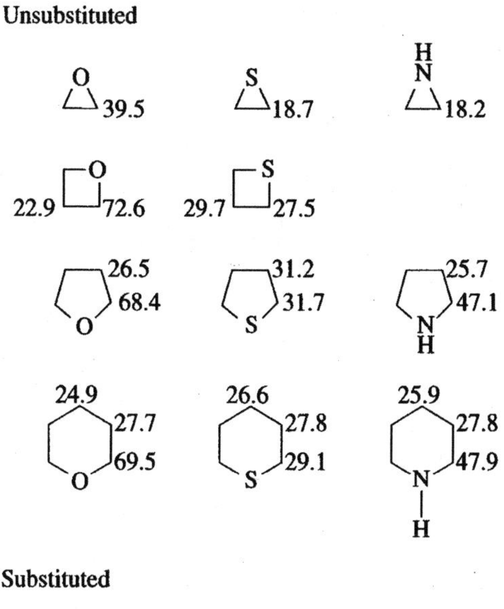

Substituted

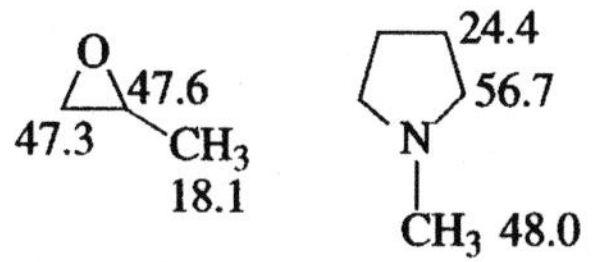

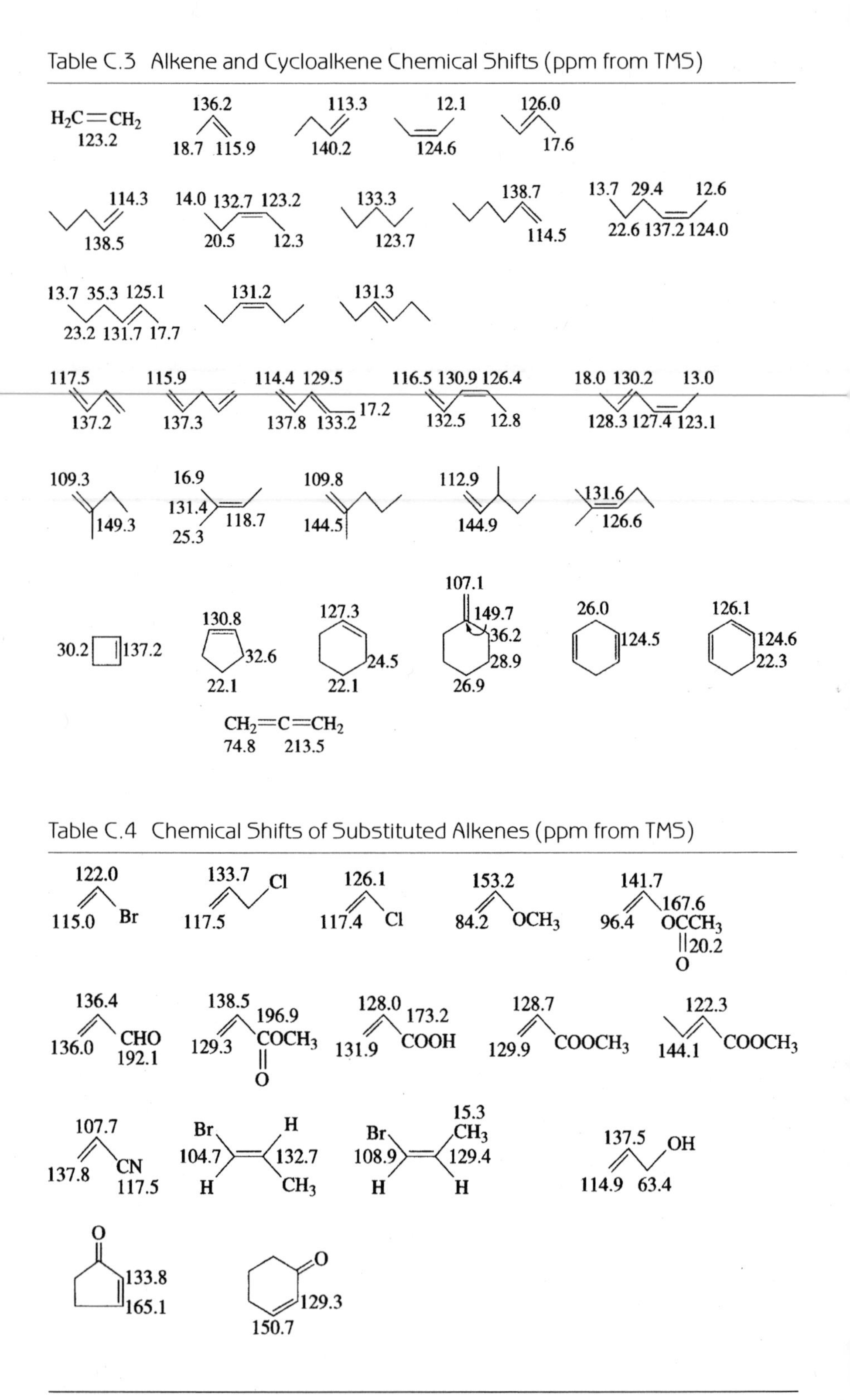

Table C.3 Alkene and Cycloalkene Chemical Shifts (ppm from TMS)

Table C.4 Chemical Shifts of Substituted Alkenes (ppm from TMS)

Table C.5 Alkyne Chemical Shifts (ppm)

COMPOUND	C-1	C-2	C-3	C-4	C-5	C-6
1-Butyne	67.0	84.7				
2-Butyne		73.6				
1-Hexyne	67.4	82.8	17.4	29.9	21.2	12.9
2-Hexyne	1.7	73.7	76.9	19.6	21.6	12.1
3-Hexyne	14.4	12.0	79.9			

Table C.6 Chemical Shifts of Alcohols (neat, ppm from TMS)

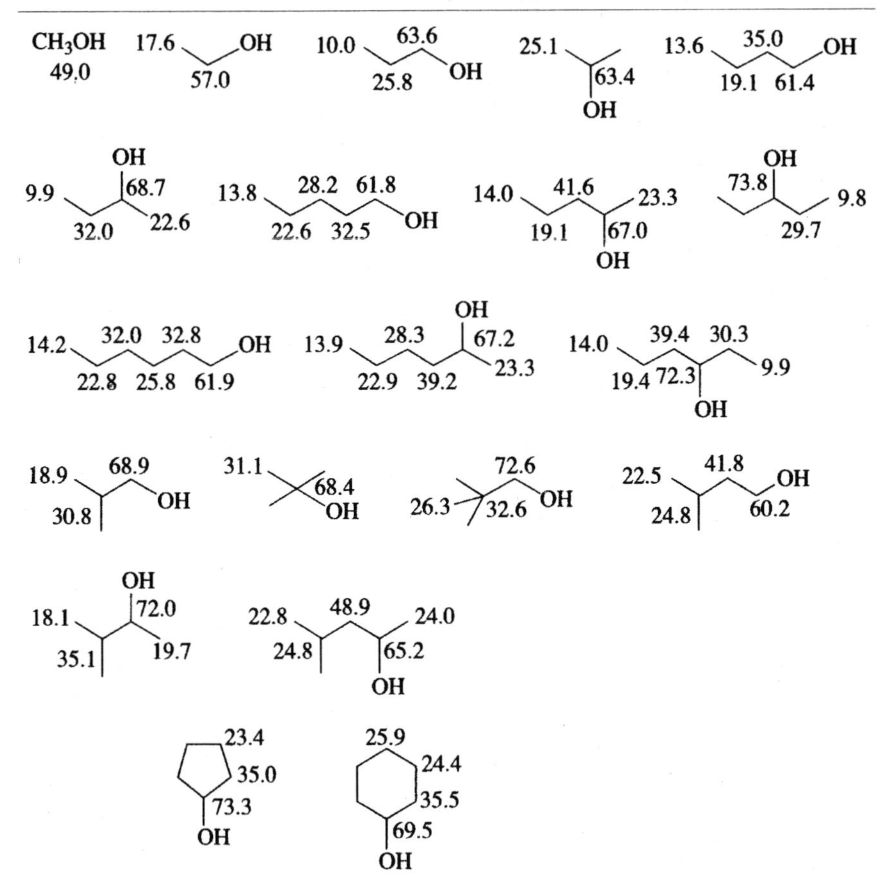

Table C.7 Chemical Shifts of Ethers, Acetals, and Epoxides (ppm from TMS)

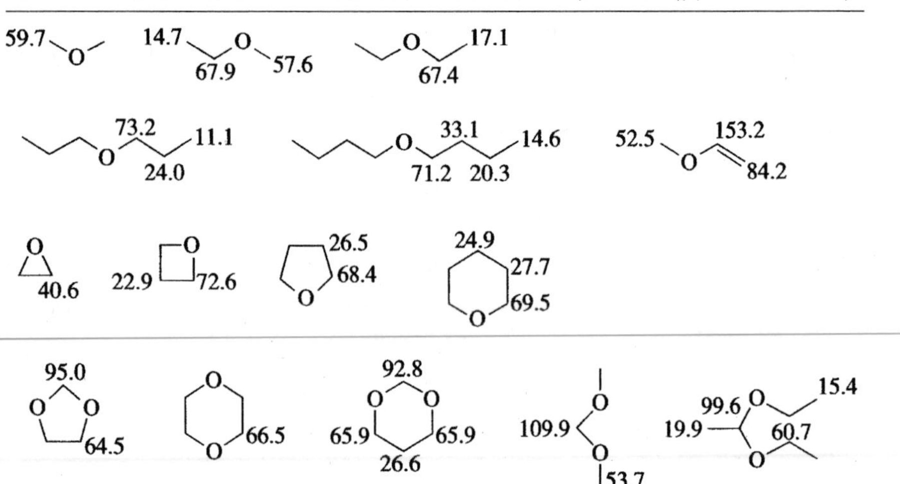

Table C.8 Shift Positions of the C=O Group and Other Carbon Atoms of Ketones and Aldehydes (ppm from TMS)

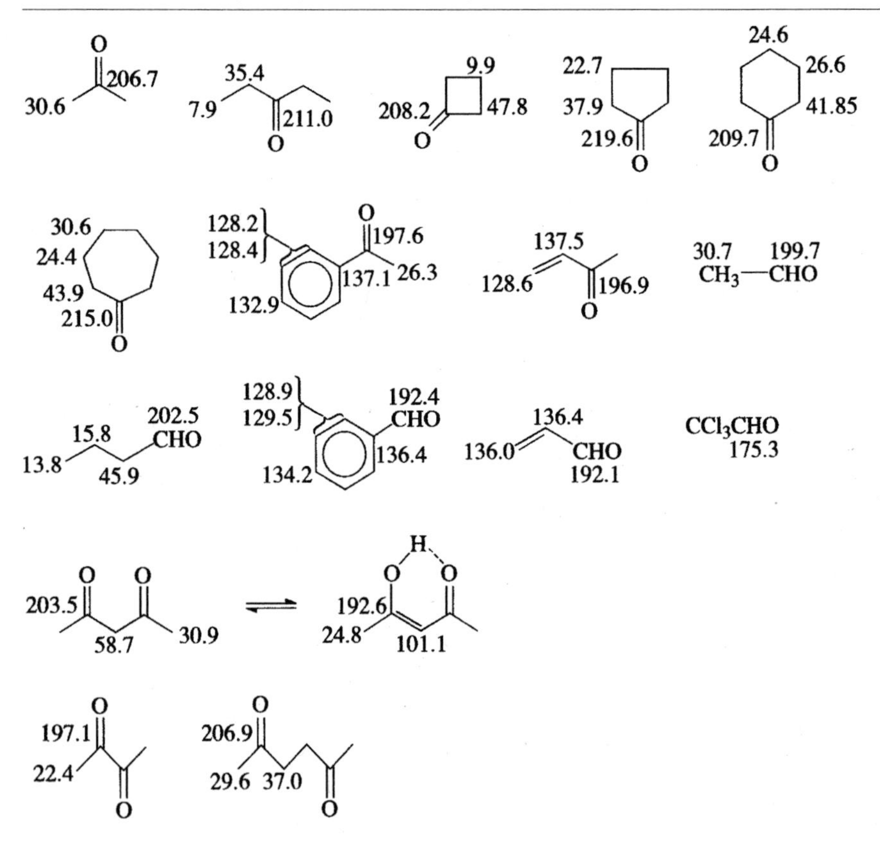

Table C.9 Shift Positions for the C=O Group and Other Carbon Atoms of Carboxylic Acids, Esters, Lactones, Chlorides, Anhydrides, Amides, Carbamates, and Nitriles (ppm from TMS)

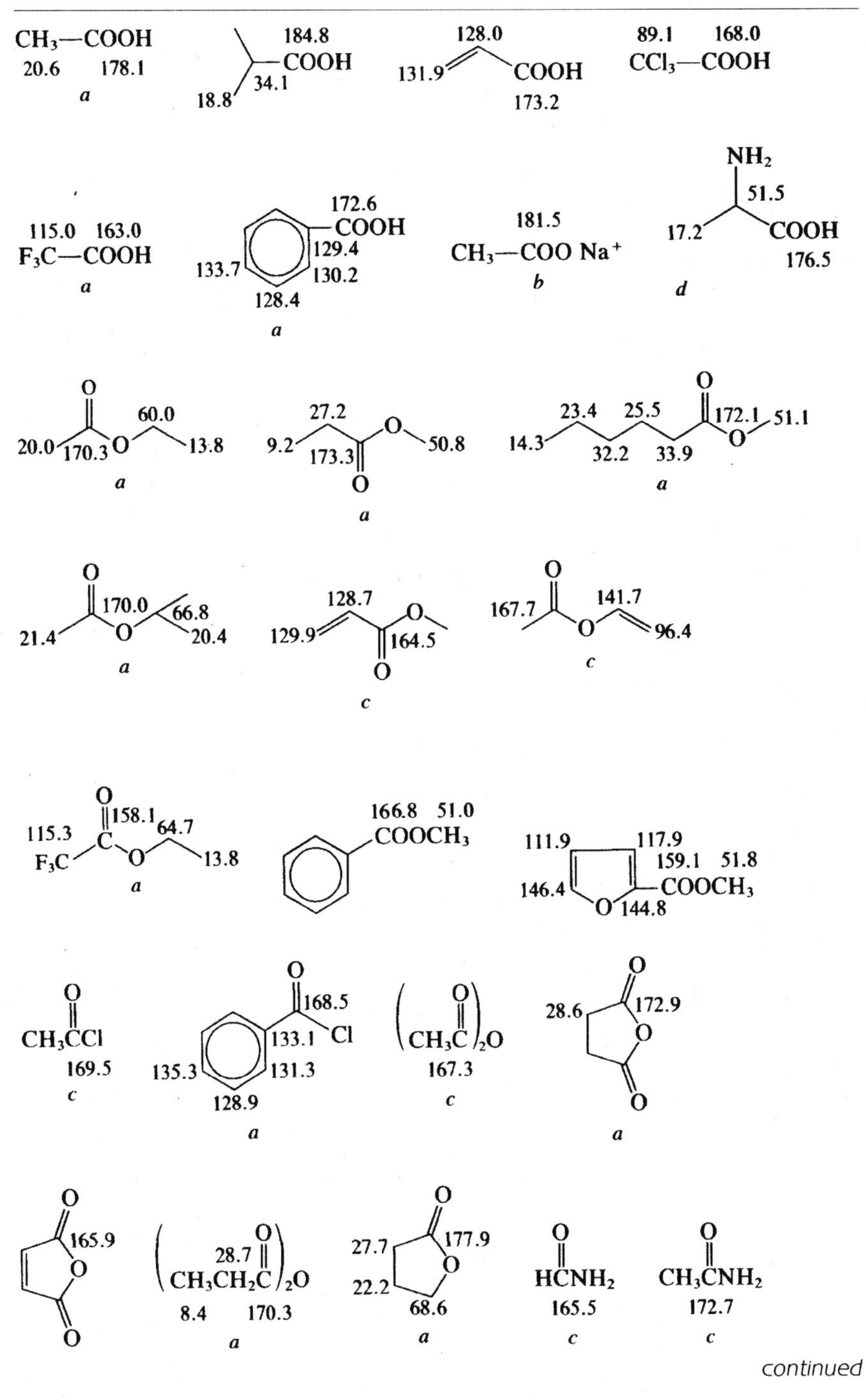

continued

Table C.9 *continued*

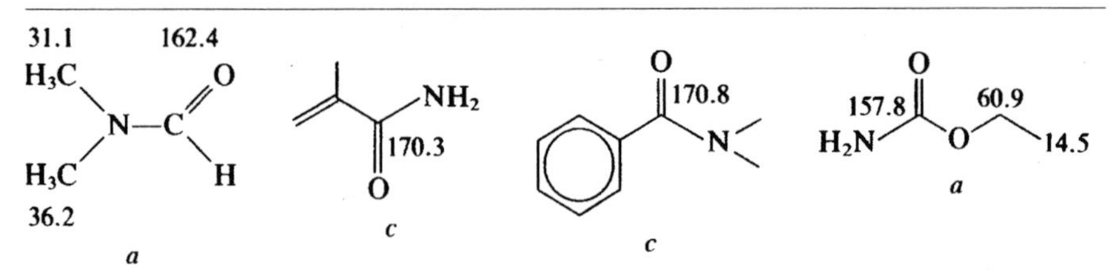

Table C.10 The ^{13}C Shifts for Some Linear and Branched-Chain Alkanes (ppm from TMS)

COMPOUND	C-1	C-2	C-3	C-4	C-5
Methane	−2.3				
Ethane	5.7				
Propane	15.8	16.3	15.8		
Butane	13.4	25.2	25.2		
Pentane	13.9	22.8	34.7	22.8	13.9
Hexane	14.1	23.1	32.2	32.2	23.1
Heptane	14.1	23.2	32.6	29.7	32.6
Octane	14.2	23.2	32.6	29.9	29.9
Nonane	14.2	23.3	32.6	30.0	30.3
Decane	14.2	23.2	32.6	31.1	30.5
Isobutane	24.5	25.4			
Isopentane	22.2	31.1	32.0	11.7	
Isohexane	22.7	28.0	42.0	20.9	14.3
Neopentane	31.7	28.1			
2,2-Dimethylbutane	29.1	30.6	36.9	8.9	
3-Methylpentane	11.5	29.5	36.9	(18.8, 3-CH_3)	
2,3-Dimethylbutane	19.5	34.3			
2,2,3-Trimethylbutane	27.4	33.1	38.3	16.1	
2,3-Dimethylpentane	7.0	25.3	36.3	(14.6, 3-CH_3)	

Table C.11 Incremental Substituent Effects (ppm) on Replacement of H by Y in Alkanes. Y is Terminal or Internal[a] (+ Downfield, − Upfield)

	α		β		γ
Y	TERMINAL	INTERNAL	TERMINAL	INTERNAL	
CH_3	+ 9	+ 9	+ 6	+ 6	−2
$CH{=}CH_2$	+20		+ 6		−0.5
C CH	+ 4.5		+ 5.5		−3.5
COOH	+21	+16	+3	+ 2	−2
COO^-	+25	+20	+5	+ 3	−2
COOR	+20	+17	+3	+ 2	−2
COCl	+33	28		+ 2	
$CONH_2$	+22		+ 2.5		−0.5
COR	+30	+24	+ 1	+ 1	−2
CHO	+31		0		−2
Phenyl	+23	+17	+ 9	+ 7	−2
OH	+48	+41	+10	+ 8	−5
OR	+58	+51	+ 8	+ 5	−4
OCOR	+51	+45	+ 6	+ 5	−3
NH_2	+29	+24	+11	+10	−5
NH_3	+26	+24	+ 8	+ 6	−5
NHR	+37	+31	+ 8	+ 6	−4
NR_2	+42		+ 6		−3
NR_3	+31		+ 5		−7
NO_2	+63	+57	+ 4	+ 4	
CN	+ 4	+ 1	+ 3	+ 3	−3
SH	+11	+11	+12	+11	−4
SR	+20		+ 7		−3
F	+68	+63	+ 9	+ 6	−4
Cl	+31	+32	+11	+10	−4
Br	+20	+25	+11	+10	−3
I	− 6	+ 4	+11	+12	−1

[a] Add these increments to the shift values of the appropriate carbon atom in Table C.10.

Source: F. W. Wehrli, A. P. Marchand, and S. Wehrli, *Interpretation of Carbon-13 NMR Spectra*. 2nd ed., (London: Heyden, 1983).

Table C.12 Incremental Shifts of the Aromatic Carbon Atoms of Monosubstituted Benzenes (ppm from Benzene at 128.5 ppm. + Downfield, − Upfield). Carbon Atoms of Substituents ppm from TMS[a]

SUBSTITUENT	C-1 (ATTACHMENT)	C-2	C-3	C-4	C OF SUBSTITUENT (ppm from TMS)
H	0.0	0.0	0.0	0.0	
CH_3	9.3	+0.7	−0.1	−2.9	21.3
CH_2CH_3	+15.6	−0.5	0.0	−2.6	29.2 (CH_2), 15.8 (CH_3)
$CH(CH_3)_2$	+20.1	−2.0	0.0	−2.5	34.4 (CH), 24.1 (CH_3)
$C(CH_3)_3$	+22.2	−3.4	−0.4	−3.1	34.5 (C), 31.4 (CH_3)
$CH{=}CH_2$	+9.1	−2.4	+0.2	−0.5	137.1 (CH), 113.3 (CH_2)
$C{\equiv}CH$	−5.8	+6.9	+0.1	+0.4	84.0 (C), 77.8 (CH)
C_6H_5	+12.1	−1.8	−0.1	−1.6	
CH_2OH	+13.3	−0.8	−0.6	−0.4	64.5
$CH_2OC(=O)CH_3$	+7.7	~0.0	~0.0	~0.0	20.7 (CH_3), 66.1 (CH_2), 170.5 (C=O)
OH	+26.6	−12.7	+1.6	−7.3	
OCH_3	+31.4	−14.4	+1.0	−7.7	54.1
OC_6H_5	+29.0	−9.4	+1.6	−5.3	
$OC(=O)CH_3$	+22.4	−7.1	−0.4	−3.2	23.9 (CH_3), 169.7 (C=O)
$C(=O)H$	+8.2	+1.2	+0.6	+5.8	192.0
$C(=O)CH_3$	+7.8	−0.4	−0.4	+2.8	24.6 (CH_3), 195.7 (C=O)
$C(=O)C_6H_5$	+9.1	+1.5	−0.2	+3.8	196.4 (C=O)
$C(=O)CF_3$	−5.6	+1.8	+0.7	+6.7	
$C(=O)OH$	+2.9	+1.3	+0.4	+4.3	168.0
$C(=O)OCH_3$	+2.0	+1.2	−0.1	+4.8	51.0 (CH_3), 166.8 (C=O)
$C(=O)Cl$	+4.6	+2.9	+0.6	+7.0	168.5
$C{\equiv}N$	−16.0	+3.6	+0.6	+4.3	119.5
NH_2	+19.2	−12.4	+1.3	−9.5	
$N(CH_3)_2$	+22.4	−15.7	+0.8	−11.8	40.3
$NHC(=O)CH_3$	+11.1	−9.9	+0.2	−5.6	

continued

Table C.12 *continued*

SUBSTITUENT	C-1 (ATTACHMENT)	C-2	C-3	C-4	C OF SUBSTITUENT (ppm from TMS)
NO_2	+19.6	−5.3	+0.9	+6.0	
N=C=O	+5.7	−3.6	+1.2	−2.8	129.5
F	+35.1	−14.3	+0.9	−4.5	
Cl	+6.4	+0.2	+1.0	−2.0	
Br	−5.4	+3.4	+2.2	−1.0	
I	−32.2	+9.9	+2.6	−7.3	
CF_3	+2.6	−3.1	+0.4	+3.4	
SH	+2.3	+0.6	+0.2	−3.3	
SCH_3	+10.2	−1.8	+0.4	−3.6	15.9
SO_2NH_2	+15.3	−2.9	+0.4	+3.3	
$Si(CH_3)_3$	+13.4	+4.4	−1.1	−1.1	

[a] See D. E. Ewing, Org. Magn. Reson., 12, 499 (1979) for chemical shifts of 709 monosubstituted benzenes.

^{1}H NMR Tables

Table H.1 Chemical Shifts of Protons on a Carbon Atom Adjacent (α Position) to a Functional Group in Aliphatic Compounds (M—Y)

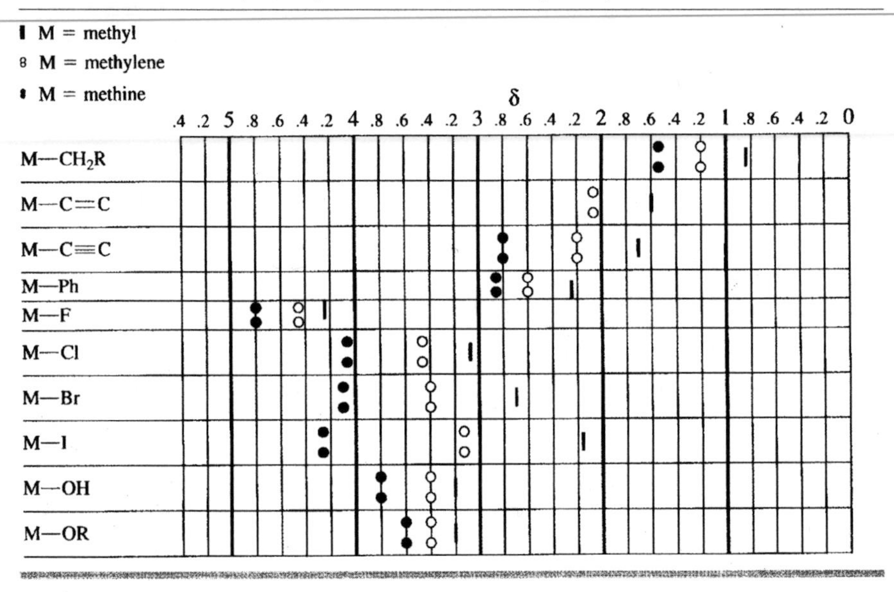

continued

Table H.1 *continued*

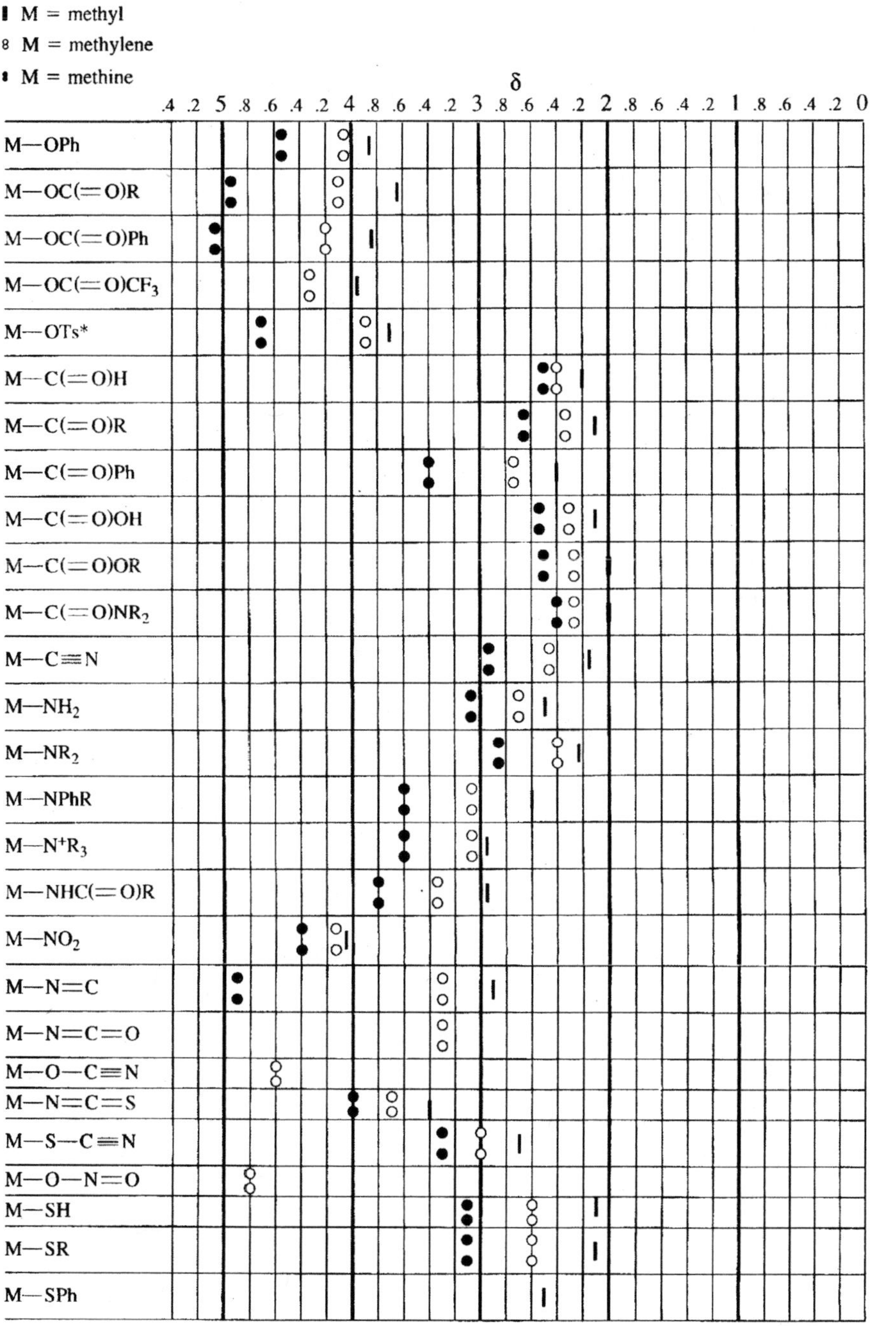

▮ M = methyl
8 M = methylene
▪ M = methine

δ (scale: 5.4 to 0)

Group	Methine (M)	Methylene (M)	Methyl (M)
M—OPh	~4.55	~4.05	~3.85
M—OC(=O)R	~4.95	~4.10	~3.65
M—OC(=O)Ph	~5.05	~4.20	~3.85
M—OC(=O)CF_3		~4.30	~3.95
M—OTs*	~4.70	~3.90	~3.70
M—C(=O)H	~2.50	~2.40	~2.20
M—C(=O)R	~2.65	~2.35	~2.10
M—C(=O)Ph	~3.40	~2.75	~2.40
M—C(=O)OH	~2.50	~2.30	~2.10
M—C(=O)OR	~2.50	~2.25	~2.00
M—C(=O)NR_2	~2.40	~2.25	~2.00
M—C≡N	~2.90	~2.45	~2.15
M—NH_2	~3.05	~2.65	~2.45
M—NR_2	~2.80	~2.35	~2.20
M—NPhR	~3.55	~3.05	~2.55
M—N^+R_3	~3.55	~3.05	~2.90
M—NHC(=O)R	~3.75	~3.30	~2.90
M—NO_2	~4.35	~4.10	~4.00
M—N=C	~4.85	~3.25	~2.85
M—N=C=O		~3.25	
M—O—C≡N		~4.55	
M—N=C=S	~3.95	~3.65	~3.35
M—S—C≡N	~3.25	~2.95	~2.65
M—O—N=O		~4.75	
M—SH	~3.05	~2.55	~2.05
M—SR	~3.05	~2.55	~2.05
M—SPh			~2.45

continued

Table H.1 *continued*

▮ M = methyl
8 M = methylene
⁝ M = methine

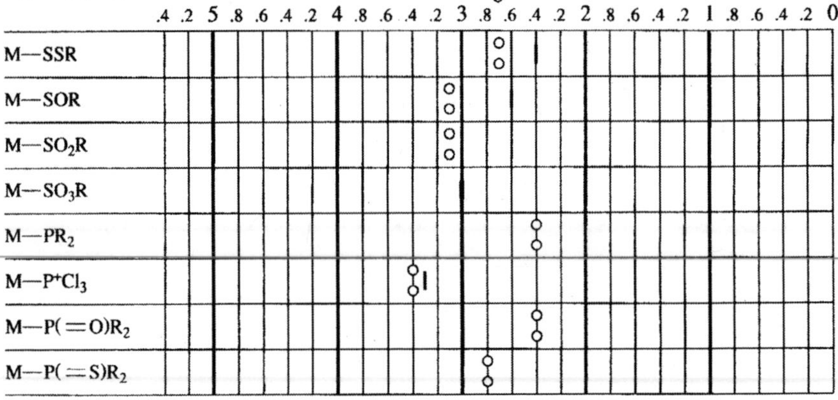

δ

.4 .2 5 .8 .6 .4 .2 4 .8 .6 .4 .2 3 .8 .6 .4 .2 2 .8 .6 .4 .2 1 .8 .6 .4 .2 0

M—SSR
M—SOR
M—SO_2R
M—SO_3R
M—PR_2
M—P^+Cl_3
M—P(=O)R_2
M—P(=S)R_2

*OTs is —O—S(=O)(=O)—C_6H_4—CH_3

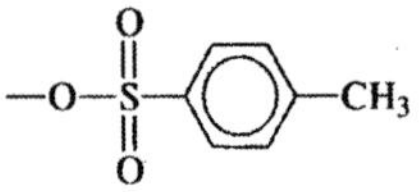

Table H.2 Proton Spin-Coupling Constants

TYPE	J_{ab} (Hz)	J_{ab} TYPICAL	TYPE	J_{ab} (Hz)	J_{ab} TYPICAL
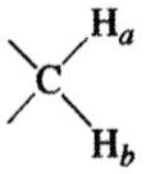 $>C(H_a)(H_b)$	0–30	12–15	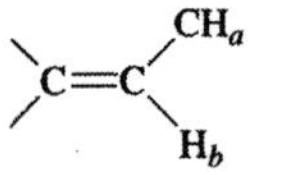 $>C=C(CH_a)$—H_b	4–10	7
CH_a—CH_b (free rotaion)	6–8	7	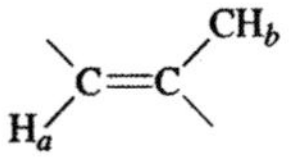 H_a—C=C—CH_b	0–3	1.5
CH_a—C—CH_b	0–1	0			
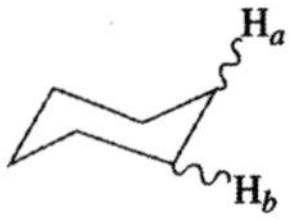 (cyclohexane H_a, H_b)			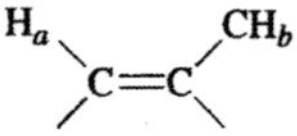 H_a—C=C—CH_b	0–3	2
			C=CH_a—CH_b=C	9–13	10
ax–ax	6–14	8–10	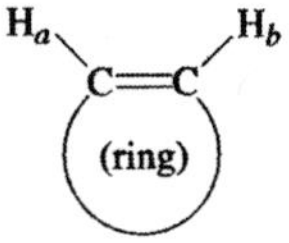 H_a—C=C—H_b (ring)	3 member	0.5–2.0
ax–eq	0–5	2–3		4 member	2.5–4.0
eq–eq	0–5	2–3		5 member	5.1–7.0
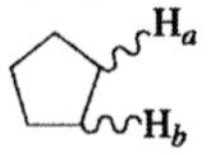 (cyclopentane H_a, H_b) (cis or trans)	*cis* 5–10			6 member	8.8–11.0
	trans 5–10			7 member	9–13
				8 member	10–13

continued

Table H.2 *continued*

TYPE	J_{ab} (Hz)	J_{ab} TYPICAL
Cyclobutane with H_a, H_b (cis or trans)	*cis* 4–12 *trans* 2–10	
Cyclopropane with H_a, H_b (cis or trans)	*cis* 7–13 *trans* 4–9	
CH_a—OH_b (no exchange)	4–10	5
>CH_a—C(=O)H_b	1–3	2–3
C=CH_a—C(=O)H_b	5–8	6
H_a(C=C)H_b (trans)	12–18	17
>C=C(H_a)(H_b)	0–3	0–2
H_a(C=C)H_b (cis)	6–12	10
CH_a(C=C)CH_b (cis)	0–3	1–2

TYPE		J_{ab} (Hz)	J_{ab} TYPICAL
CH_a—C≡CH_b		2–3	
—CH_a—C≡C—CH_b—		2–3	
Epoxide, H_a and H_b on same carbon			6
Epoxide, H_a and H_b cis			4
Epoxide, H_a and H_b trans			2.5
Benzene ring, H_a, H_b	*J* (ortho)	6–10	9
	J (meta)	1–3	3
	J (para)	0–1	~0
Pyridine (positions 2–6, N)	*J* (2–3)	(5–6)	5
	J (3–4)	(7–9)	8
	J (2–4)	(1–2)	1.5
	J (3–5)	(1–2)	1.5
	J (2–5)	(0–1)	1
	J (2–6)	(0–1)	~0
Furan (positions 2–5, O)	*J* (2–3)	1.3–2.0	1..8
	J (3–4)	3.1–3.8	3.6
	J (2–4)	0–0	~0
	J (2–5)	1–2	1.5
Thiophene (positions 2–5, S)	*J* (2–3)	4.9–6.2	5.4
	J (3–4)	3.4–5.0	4.0
	J (2–4)	1.2–1.7	1.5
	J (2–5)	3.2–3.7	3.4

Table H.3 Chemical Shifts in Alicylic Rings

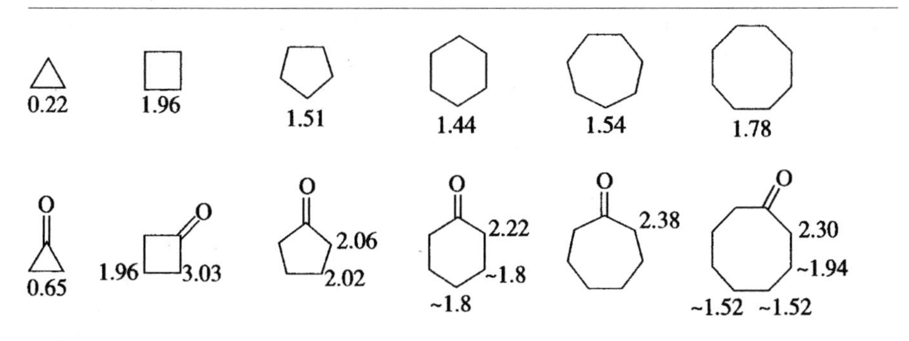

Table H.4 Chemical Shifts of Protons on Monosubstituted Benzene Rings

	9	.8	.6	.4	.2	8	.8	.6	.4	.2	7	.8	.6	.4	.2	6			δ	
Benzene									⁝											
CH_3 (omp)										⁝										
CH_3CH_2 (omp)										⁝										
$(CH_3)_2CH$ (omp)										⁝										
$(CH_3)_3C$ o, m, p									⁝	⁝⁝										
$C{=}CH_2$ (omp)									⁝											
$C{\equiv}CH$ o, (mp)								⁝	⁝											
Phenyl o, m, p								⁝	⁝	⁝										
CF_3 (omp)								⁝												
CH_2Cl (omp)									⁝											
$CHCl_2$ (omp)									⁝											
CCl_3 o, (mp)					⁝			⁝												
CH_2OH (omp)										⁝										
CH_2OR (omp)									⁝											
$CH_2OC({=}O)CH_3$ (omp)									⁝											
CH_2NH_2 (omp)									⁝											
F m, p, o									⁝	⁝	⁝									
Cl (omp)									⁝											
Br o, (pm)								⁝	⁝											
I o, p, m							⁝		⁝	⁝										
OH m, p, o										⁝	⁝	⁝								
OR m, (op)									⁝		⁝									
$OC({=}O)CH_3$ (mp), o									⁝	⁝										
OTs[a] (mp), o									⁝	⁝										
$CH({=}O)$ o, p, m						⁝	⁝	⁝												
$C({=}O)CH_3$ o, (mp)						⁝		⁝												
$C({=}O)OH$ o, p, m						⁝		⁝	⁝											
$C({=}O)OR$ o, p, m					⁝			⁝⁝												
$C({=}O)Cl$ o, p, m					⁝			⁝⁝												
$C{\equiv}N$								⁝												
NH_2 m, p, o										⁝	⁝		⁝							
$N(CH_3)_2$ m(op)										⁝		⁝								
$NHC({=}O)R$ o								⁝												
NH_3^+ o							⁝													
NO_2 o, p, m				⁝			⁝	⁝												
SR (omp)									⁝											
$N{=}C{=}O$ (omp)										⁝										
	9	.8	.6	.4	.2	8	.8	.6	.4	.2	7	.8	.6	.4	.2	6			δ	

[a] OTs = *p*-Toluenesulfonyloxy group.

Table H.5 Chemical Shifts of Alkyne Protons

HC≡CR	1.73–1.88
HC≡C—COH	2.23
HC≡C—C≡CR	1.95
HC≡CH	1.80
HC≡CAR	2.71–3.37
HC≡C—C≡CR	2.60–3.10

Table H.6 Calculation of ^{1}H NMR Chemical Shifts for Alkenes

See Figure 3.10 for more information

	Z				Z		
SUBSTITUENT R	GEM	CIS	TRANS	SUBSTITUENT R	GEM	CIS	TRANS
—H	0	0	0				
—Alkyl	0.44	−0.26	−0.29	—C(H)=O	1.03	0.97	1.21
—Alkyl-ring[a]	0.71	−0.33	−0.30				
—CH_20, —CH_2I	0.67	−0.02	−0.07				
—CH_2S	0.53	−0.15	−0.15				
—CH_2Cl, —CH_2Br	0.72	0.12	0.07	—C(N)=O	1.37	0.93	0.35
—CH_2N	0.66	−0.05	−0.23				
—C≡C	0.50	0.35	0.10	—C(Cl)=O	1.10	1.41	0.99
—C≡N	0.23	0.78	0.58	—OR, R:aliph	1.18	−1.06	−1.28
—C=C	0.98	−0.04	−0.21	—OR, R:conj[b]	1.14	−0.65	−1.05
—C=C conj[b]	1.26	0.08	−0.01	—OCOR	2.09	−0.40	−0.67
—C=O	1.10	1.13	0.81	—Aromatic	1.35	0.37	−0.10
—C=O conj[b]	1.06	1.01	0.95	—Cl	1.00	0.19	0.03
—COOH	1.00	1.35	0.74	—Br	1.04	0.40	0.55
—COOH conj[b]	0.69	0.97	0.39	—N(R)R R: aliph	0.69	−1.19	−1.31
—COOR	0.84	1.15	0.56	—N(R)R R:conj[b]	2.30	−0.73	−0.81
—COOR conj[b]	0.68	1.02	0.33				
				—SR	1.00	−0.24	−0.04
				—SO_2	1.58	1.15	0.95

[a]Alkyl ring indicates that the double bond is part of the ring R(C=C)

[b]The *Z* factor for the conjugated substituent is used when either the substituent or the double bond is further conjugated with other groups.

Metal and Enamel

Above: these pieces demonstrate sgraffito and stencilling techniques, part of the enamel course.

Edited by Linda Fox

Published by Marshall Cavendish Books Limited
58 Old Compton Street
London W1V 5PA

This material was first published by Marshall Cavendish Limited in the publication *Encyclopedia of Crafts*

First printed 1978

Printed in Great Britain

ISBN 0 85685 265 1

Introduction

Metal and Enamel is a complete introductory course in two absorbing crafts which can be combined in many ways. Both can be practised in a kitchen or any suitable working area in the home with a minimal outlay on equipment, and both can be used to produce exciting and imaginative decorations for the home, beautiful jewelry, and original gifts of all kinds.

Learning the techniques is easy following the simple step-by-step instructions given here, accompanied throughout by dozens of colour photographs and diagrams. You can progress at your own pace, acquiring tools and materials as you need them. And there are projects to demonstrate each skill, so you start making things straight away. For example, you will learn how to handle copper and silver wire by making earrings, bracelets, rings and pendants: and there is a copper wall plaque to make, a pewter mirror mount, metal sculptures and beautiful soldered silver rings. Projects in the enamel course include a wall panel, a lovely champlevé dish, and lots of jewelry. Enamel colours are beautiful, and even early efforts can have a really professional look.

Combine your metal and enamel skills to cut your own metal blanks for enamel jewelry or to build up whole panels from enamelled metal shapes. Any one of the dozens of skills explained and demonstrated so clearly in this book can be developed into a fine art. You may want to try everything! Or you may want to develop just one particular skill – if so, there are plenty of tips on creating your own designs. Whatever your aims, *Metal and Enamel* is packed with facts, hints, ideas and projects to suit every taste, talent and level of skill.

Contents

Metal

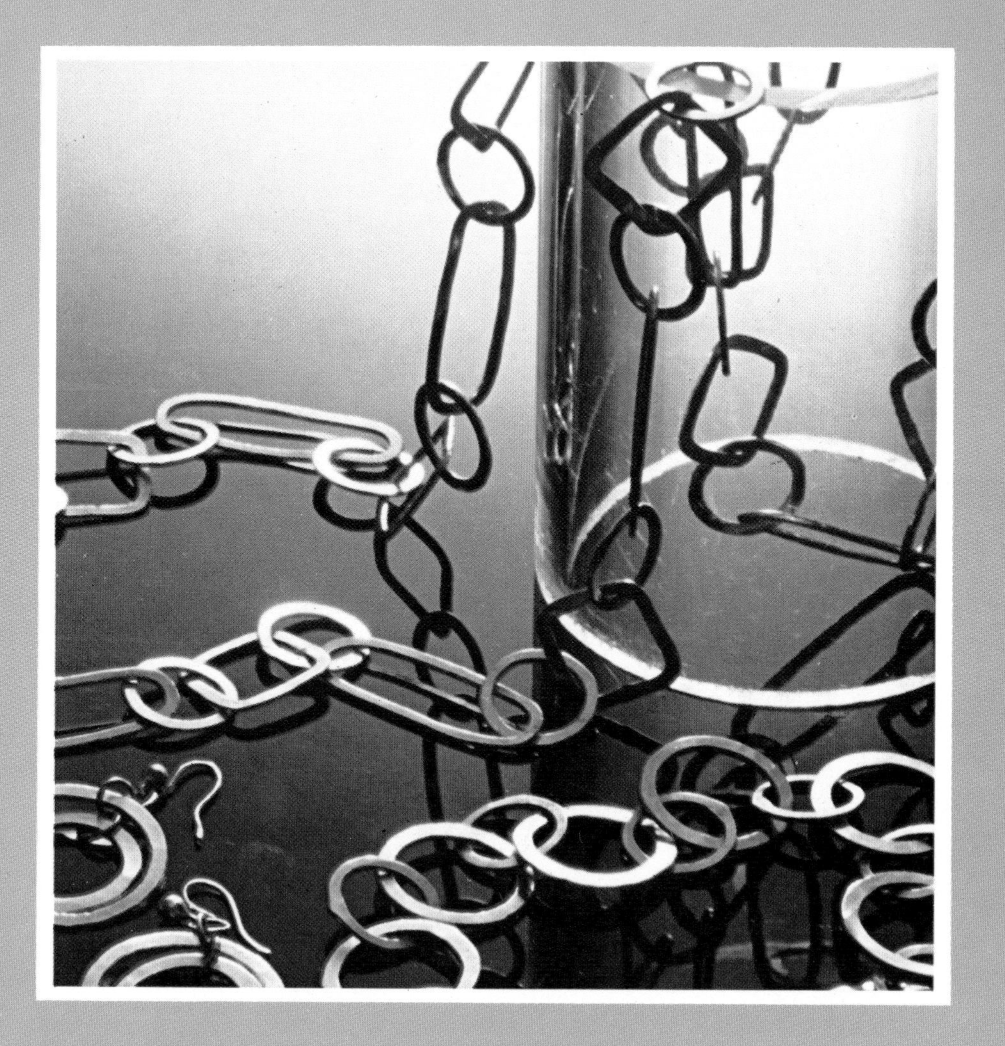

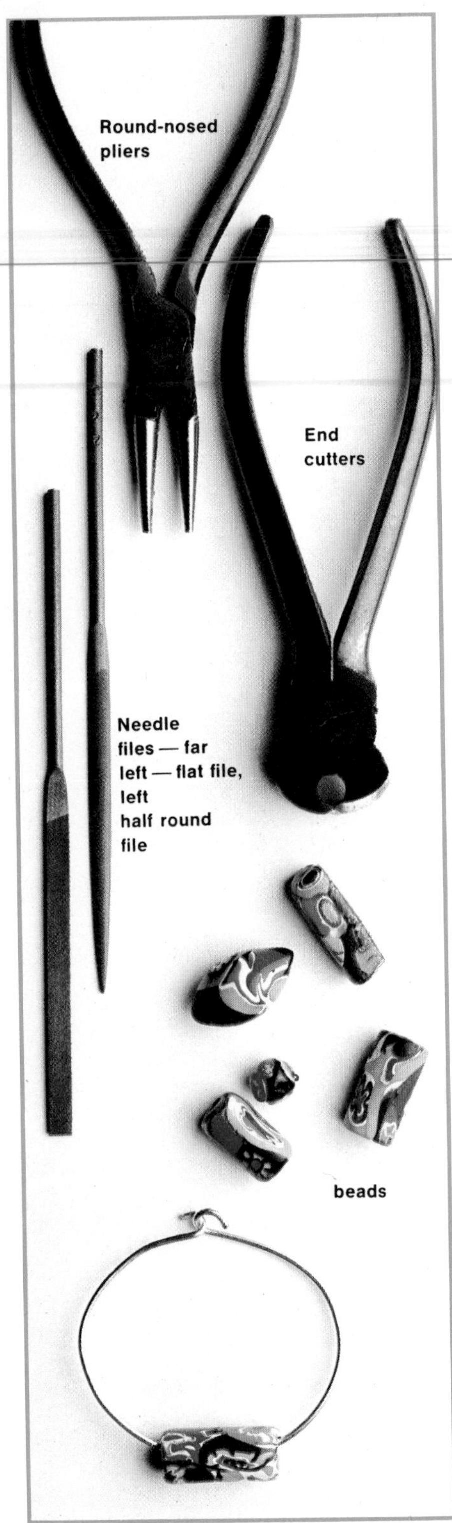

Begin with wire jewelry

Because most people tend to think that working with metal involves using heavy machinery, it is rather neglected as a craft material. But primitive man did marvellous things with the minimum of tools and now modern, small-scale aids make it possible for anyone to take up a wide variety of metalworking techniques at home. And since metal is so versatile the creative design possibilities are endless, once the basic techniques are mastered. This course is arranged in a progressive order of techniques so you can acquire tools gradually and advance at your own pace.

Wire

Many early pieces of jewelry were based on patterns created with wire which had been made by hand. As an introduction to the possibilities of working with metal, interesting jewelry can be made from various types of wire. The techniques can be developed to form very intricate pieces and wire combines well with other materials, like leather and all kinds of stones. Economically, it is a good idea to start with copper, brass or silver-plated wire. Once you have acquired the knack of bending the wire in various ways you can graduate to fine silver wire.

Keep stocks of wire labelled with their thickness and gauge number until you have learnt to recognize the various sizes. Wrap the wire up in plastic bags to prevent discolouration when you are not working with it. Always keep wire as straight as possible. If the wire has been kept in a large coil it can be straightened out by lightly smoothing with the fingers. Completed pieces should be cleaned by immersing them in a liquid silver cleaner. Wash in soapy luke-warm water and dry. To prevent future discolouration, spray with a metal or clear varnish.

Note how joined pieces hang. The pieces do not stay flat but alternate – one flat and the other at right angles to it. Designs must follow the pattern otherwise the piece of jewelry will not hang in the correct way. A piece of work will never appear the same when lying on the work bench as when it is suspended, so always hold your work up to look at it. It is difficult to correct or straighten a piece of wire that has gone wrong without it looking untidy and overworked so, if a mistake is made, it is better to start again with a new piece of wire. You can always use odd pieces of discarded wire for making jump rings.

Tools

Round-nosed pliers need not be expensive but check that the jaws make contact along their entire length. The inside of the jaw should be smooth otherwise it will damage the surface of the wire. If they are not smoooth, cover the jaws with adhesive tape, although this does make it slightly more difficult to grip heavy wire firmly. All-in-one pliers that cut and bend round and square lines are only suitable in the early stages. Once more progressive techniques are reached a higher quality of tool is required. The size of the curl made in the wire depends on how you position it in the jaws of the pliers. A small curl is made with the wire at the front of the pliers and a large one with it at the back. The end of the wire in the jaws should not stick out beyond the pliers. Grip the pliers firmly and turn in a clockwise direction using the thumb of the other hand to apply pressure to the wire close to the jaws. You can work in an anti-clockwise [counter-clockwise] direction if you find it easier, but whichever way you do, make it a habit to work in the same direction. Try to complete a curl without having to regrip the wire.
Diagonal wire cutters or end cutters It is easier to cut jump rings with diagonal wire cutters, but end cutters will do.
Metal file You can use a medium sized metal file or you can buy a selection of Swiss or needle files. You will not need all the files to make the pieces shown here, but it is economical to buy a set of about six as they will be useful later. It is sensible to spend as much as you can afford on good quality tools.

Bracelet with bead

File ends of wire and thread the bead. Shape with fingers and make a curl at either end to hook the bracelet together. The bracelet is illustrated opposite, along with the tools required to make it.

Earrings

Make two jump rings by placing the end of the wire halfway down the jaws and working it around until you have two complete circles (fig.1). Line up the end cutters with the end of the wire and cut through both coils (fig.2). File the ends of the rings flat and smooth. Jump rings are used to attach and assemble all kinds of jewelry so it is worth making a large number. You can coil the wire around any cylindrical shape of the right size (fig.3).

Bracelet

You will need:
18cm (7in) of 1.5mm (gauge 15) silver-plated wire
One bead that will fit on to wire
Round-nosed pliers
Metal file

Earrings

You will need:
31cm (12in) of 1.5mm (gauge 15) silver-plated wire
A pair of earring findings [backings], clip-on or screw-type, available from craft shops or jewellers. The findings must have a hook or eye to which a jump ring can be attached
Round-nosed pliers
Metal file
Diagonal wire cutters or end cutters

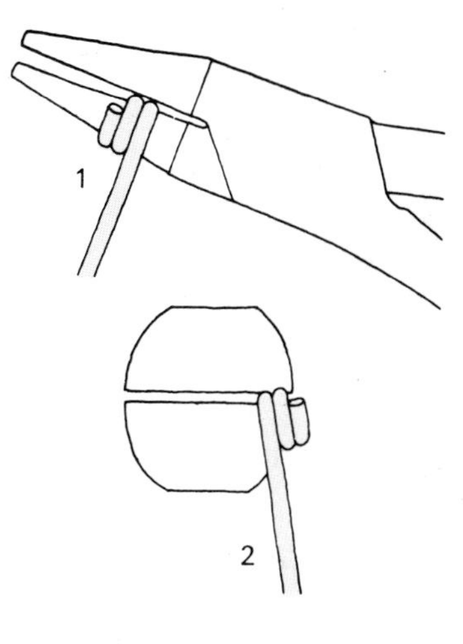

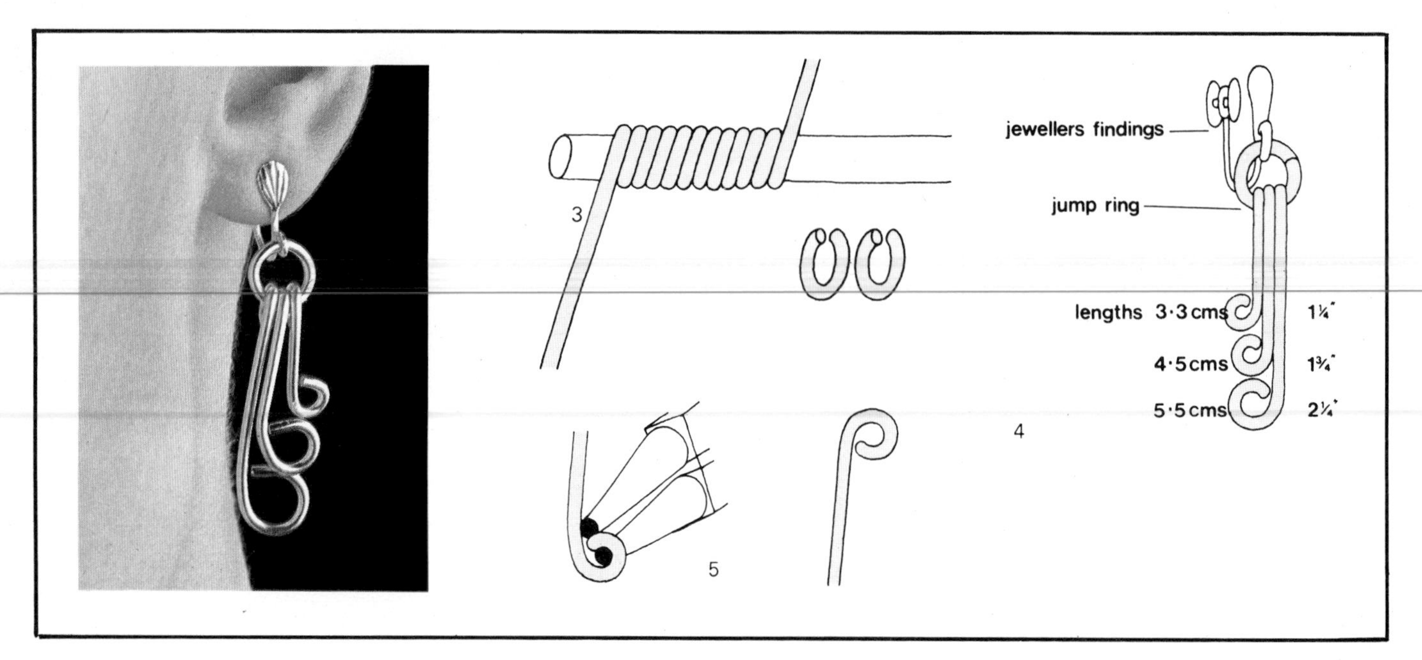

Above: The earrings are assembled and attached to jump rings.

To make the hanging drops, cut two of each of the lengths shown in fig.4 and file ends. Make a small loop at one end of each piece of wire (fig.5). These will thread on to the jump rings. The curls at the other end must be made at right angles to the loop. You must curve three pieces in a clockwise direction and three in the opposite direction. To do this, divide the lengths of wire into two equal groups so that you have the same number of equal lengths in each pile. Working from one pile, hold the loop to curve away from yourself and make a curl at the other end so that it is in a vertical position when the loop is horizontal. Make a small curl for the short piece of wire and larger curls for the other two pieces, the longest piece having the largest curl.
Repeat with the other three pieces but hold the loop so that it curves towards you. Assemble the pieces with the jump rings and findings [backings] as illustrated. Close the jump rings with the pliers, holding the earring carefully to avoid bending any pieces.

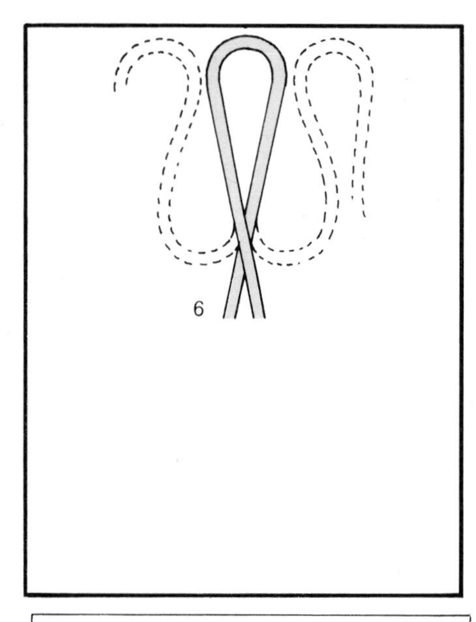

Curvy bracelet
You will need: 38cm (15in) of 1.5mm (gauge 15) silver-plated wire Tools as for the earrings

The curvy bracelet

At the centre of the wire make a large curl with the pliers to let the wire cross about 2.5cm (1in) from the curve. Bend as shown in fig.6. Make the next curve where the wire crosses and continue to form the pattern illustrated. Bend the wire with your fingers so that it fits round the wrist. The bracelet is not circular but elliptical. Bend each end into a U-shape so that they hook into one another. The size of the bracelet can be varied by opening or closing the angles of the curves.

The pendant

The six suspended pieces on the inside of the horseshoe (fig.7) are made in the same way as the six earring pieces. Make the horseshoe with 16cm (7in) of wire. File the ends and fold the wire around a cylindrical object of 3cm (1¼in) diameter. Do this by holding the halfway point of the wire to the cylinder with one hand using the thumb to keep the wire firmly in position. With the other hand work the wire around the cylinder, starting by applying pressure to the wire closest to the thumb of the other hand (fig.8).

To make the ten suspended pieces at the bottom of the horseshoe cut two each of the following lengths: 2.5cm (1in), 3.3cm (1¼in) 4.5cm (1¾in), 5.5cm (2¼in), 6.5cm (2¾in). File the ends and make a small loop at one end of each piece. Divide the pieces into two groups and proceed as for the earrings. The three shorter pieces from one pile are made with small curls, the next piece is slightly larger and the longest piece has the largest curve.

Thread the pieces on to the horseshoe to form the pattern illustrated. Make a large curl at either end of the horseshoe but do not close them completely. Thread the six pieces and close the open ends. Make any adjustments to the pieces to even out the pattern.

Pendant

You will need:
91cm (37in) of 1.5mm (gauge 15) silver-plated wire
45cm (18in) of 0.6mm (gauge 23) silver-plated wire for the thong
Leather thong 56cm (22in) long, about 6mm (¼in) in diameter
Tools as for the earrings

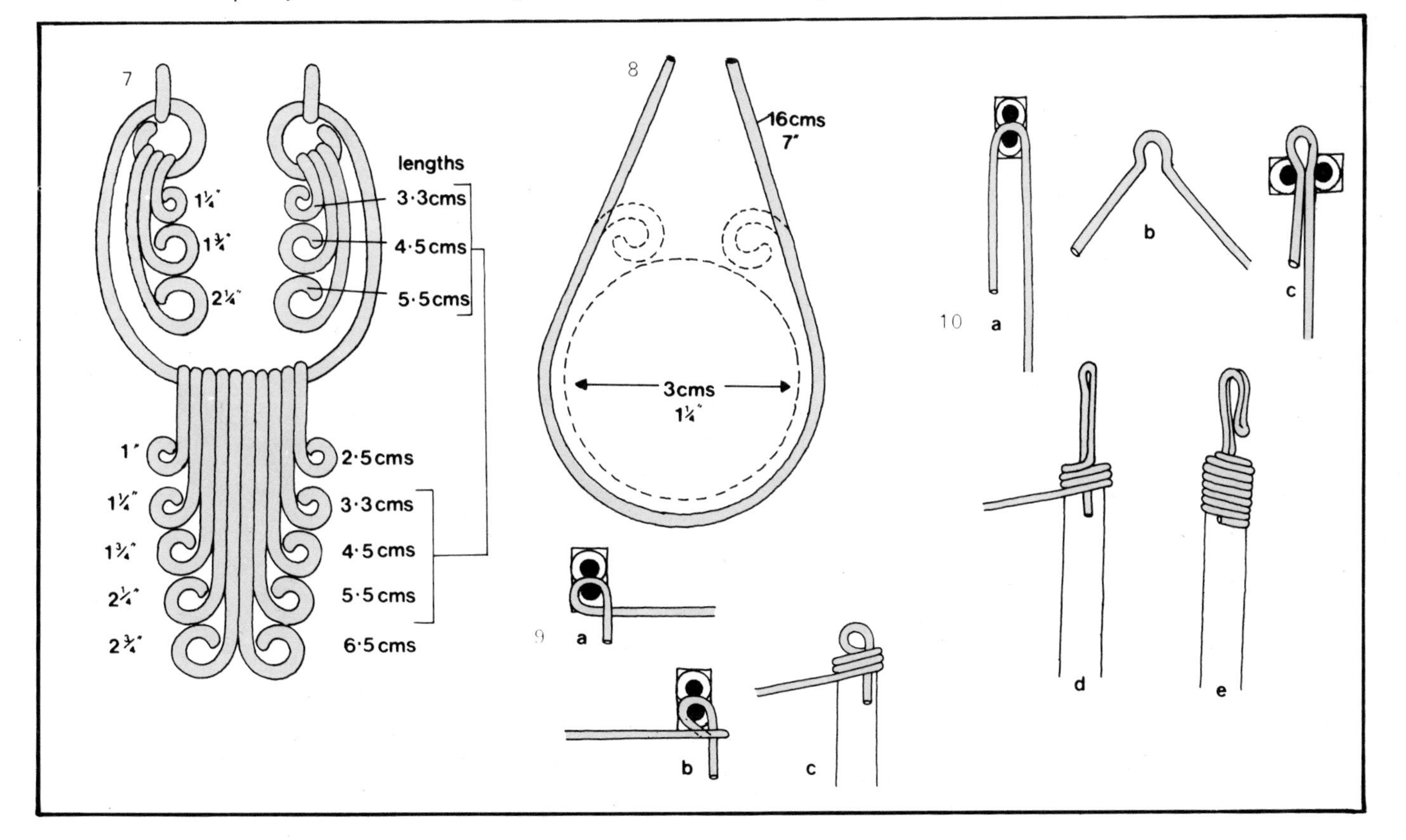

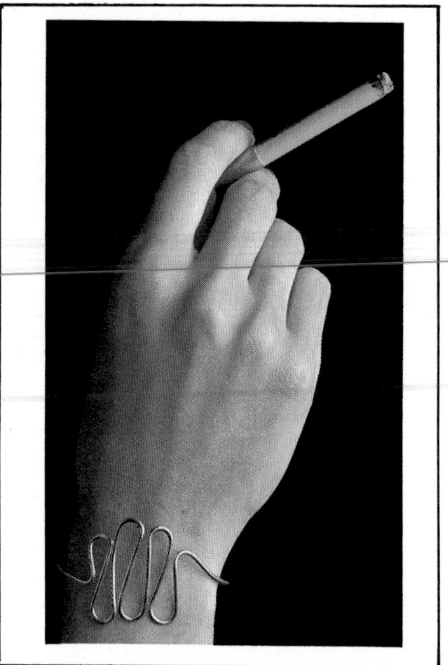

Above: A series of curls forms the curvy bracelet. The pendant (right) develops into a large necklace (centre). Hang the completed necklace over an upturned mixing bowl and make adjustments so that it hangs evenly.

Hook two jump rings through the two curls on the horseshoe. Close the jump rings and thread the leather thong through them.
Hook and eye closure For the thong use the 0.6mm (gauge 23) wire For the eye use 20cm (8in) of wire and make a loop 3cm ($1\frac{1}{2}$in) from the end of the wire (fig.9). Wind the long end of the wire around the thong four times. You may find it easier to do this if you start a short distance from the end of the thong and trim it when you have finished. Cut off the short end of the wire and continue winding the wire to the end. Press the end firmly into the thong with the pliers, without damaging the coils.
To make the hook use the remaining wire and make a U-shape 5cm (2in) from the end (fig.10). Start winding the wire around the thong 2.5cm (1in) from the end of the U-shape and proceed as before. Bend the end around to form a hook and make any adjustment necessary so that it fits the eye exactly.

Preparing designs

Once you have a working knowledge of different wires you can design your own jewelry. Always start by drawing a full-scale, linear diagram with a felt-tipped pen to the same thickness as the wire you intend to use. Make your motifs according to this diagram or a prototype made from it. It is difficult to make every section identical, but by using the diagram you can retain the scale, which gives an overall unity to the design.

Silver-plated rings

A great advantage of plating is that you can use a cheap wire and then, if the results of your work are successful, you can have it silver-plated. Silver-plating is not easy to do at home but most silversmiths or jewellers will do it for you at a reasonable price. This obviously makes the possibilities of working with wire more interesting as it expands the scope of the material. It lends itself particularly well to jewelry as you can use any wire such as copper, brass or tinned copper to make brooches, earrings, pendants and bracelets. Once the object has been finished satisfactorily, perhaps after a number of attempts to work out your own designs, you can have it silver-plated to enhance the appearance and to give your rings and other jewelry a very professional finish.

Silver, when new or polished, has a reflective surface but, if left for a period of time, it will oxidize naturally and will need to be polished to remove the oxidization and restore the shiny surface. You can use metal varnish to avoid this.

Oxidization

Oxidization is a process which can be speeded up with an oxidizer if you particularly want an antique finish. It can be very effective

These three rings are made from copper wire which was silver-plated and oxidized after the rings were completed.

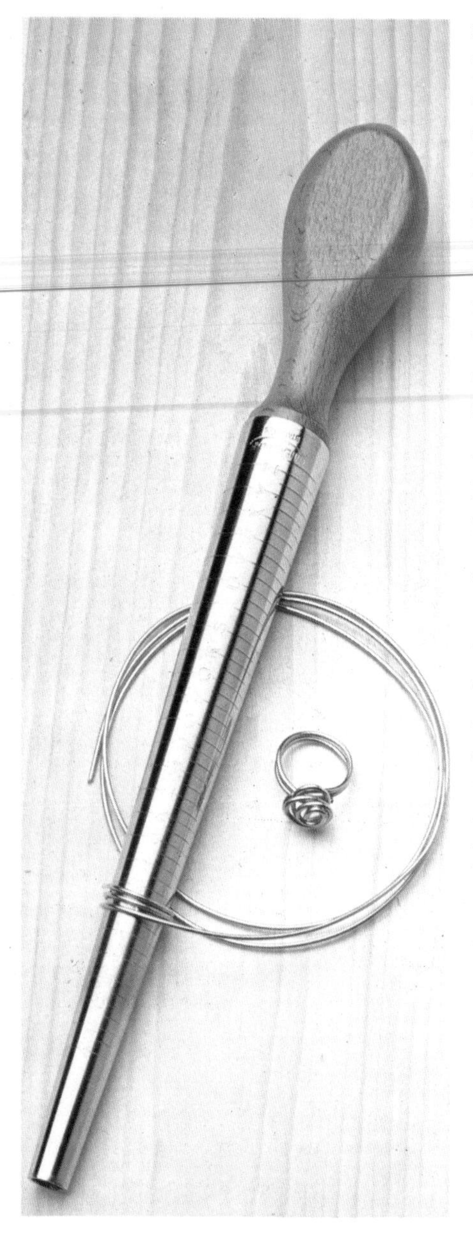

A ring stick makes it easy to repeat rings of a particular size.

as the more prominent outer surfaces of the silver can be lightly polished to shine the surface, leaving the recesses darker. An oxidizer is applied with a sable-haired brush or by immersing the object in the oxidizer. The oxidizer can be bought or mixed at home by diluting potassium sulphide with water. If you mix your own, follow the instructions supplied with the potassium sulphide. If the mixture is too strong the oxidization might be patchy or flaky so, if you are in any doubt, make a weak mixture and immerse object and check the results—you can always strengthen the mixture. Before oxidizing anything make sure it is clean and free of grease by washing and scrubbing it in hot soapy water and drying it off carefully with a clean cloth.

Filing

Plating also means that any imperfections, like indentations caused by the pliers, can be removed by filing down. But while the plating will hide the filing, it will not hide indentations as it is too thin. You can also use a file to make the shape of the wire more interesting by giving it a flat surface, varying the final appearance by filing down different parts of the design. To the layman all files might appear to be similar. They are not: some files give a flat surface, others give a groove, or a series of grooves. A file must only be used for doing what it was made to do. So do not confuse files used in carpentry with metal files. All you will need for the rings is a hand file 15cm (6in) long. Needle files can used be for finer work and grooves but they are not essential. The file only removes metal on the forward stroke or when you are pushing the file away from yourself, not on the backward stroke.

Ring sticks

A ring stick is a tapering cylinder with various sizes for rings marked on it. If you are making a large number of rings it is useful as you will be able to make a ring the exact size you require and duplicate it any number of times. Simply select the size you want on the ring stick and work the wire around it once or twice as required. As an alternative you may be able to find a piece of dowel—or even something like the handle of a kitchen tool—which is the right thickness.

To make silver-plated rings

The rings illustrated are made of copper wire. The surfaces were filed down before silver-plating and the plated rings were then oxidized. The stones were glued to the finished rings.

If you are going to use stones in the rings then make up the designs

to accommodate the stones. That is, bend the wire in such a way that the stone appears to fit naturally in its setting once the ring is complete. Make any adjustments to the ring to hold the stone comfortably before plating. Don't try to make adjustments once the ring has been plated as you might damage the plated surface.

It is economical to make jewelry like this as you can always discard any items that are spoilt and have only the successful ones plated. You can experiment with various designs and, as you become more accomplished, you will find ways of starting the rings with the design and then making the coils to fit the finger afterwards. But this is only recommended once you have gained some expertise and want to make more intricate jewelry.

The rings illustrated here are suitable for men and women. You can design your own shapes and sizes, although generally the more intricate designs look better on women.

Cut the required length of wire, allowing slightly more than you expect to use, since you can always trim off the excess. Holding one end of the cylinder, work the wire around it once or twice as required. Bend the wire far enough so that the end does not stick out as shown. Using the round-nosed pliers continue bending the wire to form the design you have in mind. To complete the design always bend the end of the wire underneath a previous curl or let it end where its line will flow naturally into the overall design. Cut off any excess wire. Using the hand file, file this end so that it will not cut into your finger or catch on clothing when it is being worn. Fit the ring on to the finger. If it is slightly too big, use the round-nosed pliers and make a small kink in the coil that goes around the finger, just below or next to the design, where it will not show. File the surface down until you get the desired effect. Also file down any imperfections before silver-plating. The plating will not alter the size of the ring. Send the ring to your jeweller to be silver-plated once it is completed to your satisfaction. The ring can then be oxidized and polished as already described.

Silver-plated rings

You will need:
Copper wire 1.8mm (gauge 13)
Stone, if required, and suitable glue
Round-nosed pliers
Diagonal wire cutters or end cutters
15cm (6in) hand file
Ring stick or any cylindrical object the same size as the ring you wish to make. The length of wire will depend on the design you make and also on whether you make a single or double coil to go round the finger. For a single coil ring you will need about 5.5cm ($2\frac{1}{4}$in) and twice that for a double coil, plus the length of wire required for the design itself.

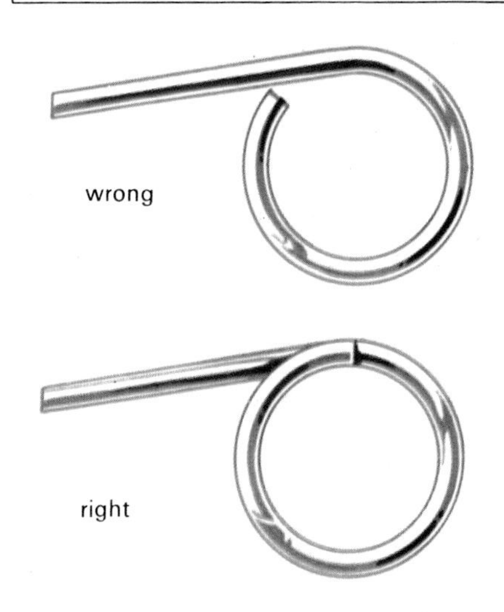

Above left: When you have made a few rings you can try gluing pretty stones to them for a different effect.
Above: The wire must be curled around to form a complete circle.

Silver wire jewelry

Making jewelry from silver wire requires plenty of practice. It is best to familiarize yourself with the tools and techniques first using copper, brass or silver-plated wire, since silver wire is expensive to use for a trial sample. If you do spoil a piece of silver wire don't throw it away—you can use it for making jump rings

Tools required include a hammer, anvil, and flat and round-nosed pliers. The wire used here is made from silver but copper can be substituted to get some practice before using silver.

and findings. The silver can be cut, sawn or hammered to shape it, depending on its form. However, the more silver is worked, or 'shaped', the harder it becomes, and eventually it will be too brittle to work, so try to work slowly and carefully.

Silver

Fine silver Pure silver is referred to as fine silver. It is extremely malleable and ductile, and lends itself to being worked by hand. Fine silver can be used as a base for enamelling because its lustre and reflective qualities are ideal for showing transparent enamels at their best. Fine silver is extremely soft and will not retain its shape unless it is handled carefully.
Sterling silver Copper is normally added to silver to form a strong alloy. It strengthens the silver without reducing its malleability. To qualify as sterling silver an alloy must consist of at least 95% fine silver with no more than 5% copper added for strength.
Silver can be bought from silversmiths in various qualities. Since fine silver is so soft, sterling silver is more suitable for most jewelry purposes. It is available in sheets, strips, and as wire in various sizes, shapes and thicknesses. Explain your requirements to the silversmith who will cut to size and then weigh the silver. In the imperial measuring system 'troy weight' is used for measuring precious metals, and in the metric system grams are used. The terms used are:
24 grains = 1 pennyweight (dwt)
20dwt = 1 ounce troy
12 ounces troy = 1 pound troy
(1 ounce troy = 1.1 ounce Avoir = 31.1 grams)

Techniques

Making designs using silver wire is no more difficult than working with other types of wire. But as it is more expensive a bit more care should be taken so as not to waste it. Working with silver wire has the advantage that it can be hammered to change its shape. Copper wire can also be hammered, but it is more difficult since it is harder, and silver-plated wire should never be hammered as the plating will be destroyed. The hammering will also weaken the silver after a while, so if you hammer the wire do so with firm blows without making the wire too thin in any one place. The hammering will change the appearance of the silver—it will lose its lustre and become matt and covered with fine scratches. For this reason try to work on a smooth surface and use tools that have good, unscratched surfaces. Polishing will restore some of the lustre to the silver once the item is completed.

To soften overworked silver it is heated and then plunged into cold water. This process will restore the silver to its original state and is called annealing. It is not an easy process as strict temperature controls are required. So do not work it more than is absolutely necessary to complete the design.

Tools

You will need all the tools described for simple wire jewelry. You will also need an anvil, a hammer and a pair of flat-nosed pliers, all of which are available from most hardware stores and jeweller's suppliers. The anvil does not have to be large as long as it has a flat surface of about 8cm (3in) square. You could make do with any small, flat steel surface. A Warrington or cross pein hammer is ideal for beating metal. Hammers are sold by the weight of the head and for planishing (metal beating) a 227 gram (8 ounce) hammer head is ideal. The head should be polished and the handle made of ash wood. A wooden handle has a natural 'spring' to it and is therefore more satisfying to work with than a metal handle which has no spring. There are many other types of hammer, each with its own specific function. No doubt you already have some form of hammer and if it's not too heavy, and it has one flat surface, it will do, providing the surface is smooth and you keep it just for beating metal and for no other purpose which might damage the surface.
Flat-nosed pliers are used in the same way as the round-nosed pliers but they make a square corner or bend rather than the curves made by round-nosed pliers. The insides of the jaws of the pliers must be quite smooth or they will damage the silver surface. Scratches can be rubbed down with a metal file.

Square pattern pendant

You will need:
1.93m (6ft 4in) sterling silver wire
1mm (gauge 18-19) thick
Round-nosed pliers
Metal file
Diagonal wire cutters or end cutters
Anvil
Hammer
Flat-nosed pliers

Square pattern pendant

The pendant is approximately 7.5cm (3in) at its widest point and 11.5cm ($4\frac{1}{2}$in) long from the bottom tip to the neckband. You can make it smaller by only making one square instead of three, or you could make two squares side by side. Depending on the number of squares the length of wire will vary. For each square you need 46cm (18in) of wire, which includes the straight length which is attached to the neckband. For an average sized neck you need 41cm (16in) of wire which includes 5cm (2in) for the hook and eye to secure the ends. Always allow a little more wire than you expect to use and make any necessary adjustments to the length later on.
It is tricky to make the interior part of the design. Start with a small circular curl and then make a square (fig.1). Practise this with a piece of copper wire (or any cheap wire) before using the silver wire. You might find it easier to use the flat nosed pliers as shown in

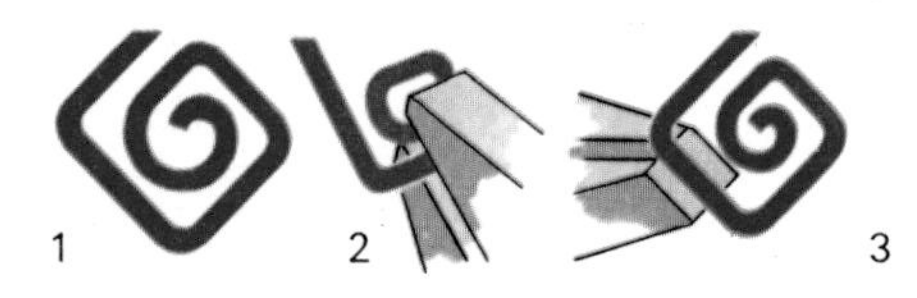

1 *Starting the square*
2 *Working with flat-nosed pliers*
3 *Bending the square corners*

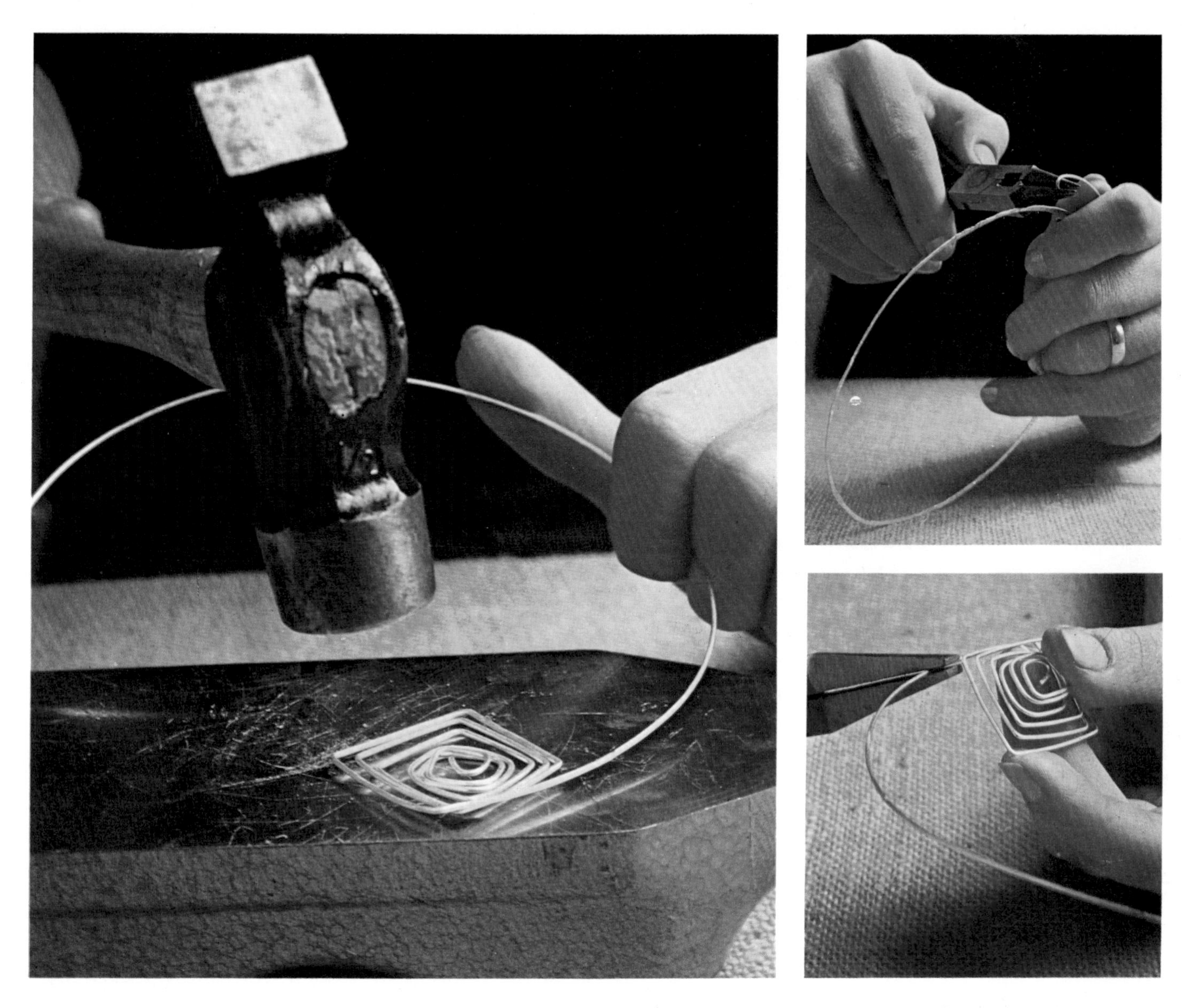

fig.2. Once the work is larger you can use the pliers as shown in fig.3. Make sure to bend the wire to an angle of exactly 90° to keep the geometric shape. When you feel confident to start on the pendant itself, cut three lengths of wire 45cm (18in) long for the three squares. Work the design from the centre and make the square so that each side is made up of nine strands of wire, counting from the centre. Hammer the square patterns to flatten them but don't overwork them or they will lose their straight lines. Just flatten the surfaces slightly to give the wire an angular appearance. Assemble the three squares, arranging 'stems' of wire as shown,

Above: The flat-nosed pliers are used to bend the wire at right angles.
Top: Make a hook and eye at the ends of the wire to which the pendant is attached.
Left: The completed square is hammered to flatten it before assembly.

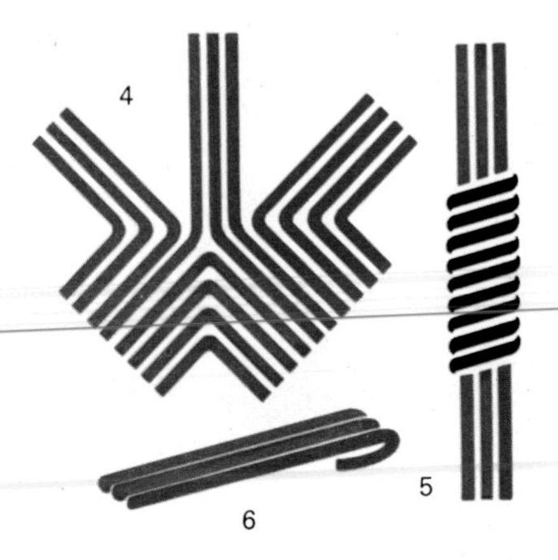

4 The square patterns are arranged together before binding.
5 The wire is held securely by winding a piece of wire around it. in a tight spiral.
6 The stem is folded back to make a hook which is attached to the neckband.

Curly pendant
You will need: 7 pieces of wire, each 16cm (6½in) long plus 16cm (6¼in) for the binding and 41cm (16in) for the neckband Tools as for square pattern pendant

running parallel to each other (fig.4). Cut 41cm (16in) of wire for the neckband. Work it round with the fingers and make a hook at one end and an eye at the other end. Make any adjustment necessary to hook the ends together comfortably. Use the remaining wire and wind it around the three 'stems'. It will help if you rest the three squares on the edge of a flat surface with the ends jutting out. Start from the back, and end at the back, bending the wire as tightly as possible (fig.5). Place the pendant on the anvil, right side up. Hammer the binding around the stems. This will help to secure the pieces of wire. Hammer the wire pieces evenly so that they all spread to the same extent. Cut the three ends to the same length.

Hold the pendant to the neckband, allowing 12mm (½in) for a hook, and check the pendant does not hang too low. Shorten the stems if necessary. Bend back the end of the stem (fig.6) to form a hook. Insert the neckband and close the hook around it.

Curly pendant

The curly pendant is about 5.5cm (2¼in) wide and 8.5cm (3¼in) long. Make the curls by starting from the inside of the pattern, using round-nosed pliers. Make the long centre piece and hammer the top end flat before running it to the front to form a hook for the neckband.

Complete the individual pieces. To get them all to match, lay them down on the design (fig.7) and adjust accordingly. You can hammer the curls slightly to flatten the wire. Concentrate the hammering on the lines that will form the outer part of the design. To assemble the pieces of wire, bind them with wire starting from the back and ending at the back. Hammer evenly to secure the pieces. Attach the neckband by bending the centre piece forwards. Make a neckband as for square pendant to finish.

Alternative designs

The designs given here are based on a central piece of wire attached to a neckband, and a number of wires on either side, both sides having the same number of wires. In the square pattern pendant only one end of the wire is used to form the design, but in the curly pendant both ends are used. You can use any combination but it is always advisable to make a full-scale diagram. Draw one side and then fold the design in half along the centre line and duplicate the other half of the pattern. This will also give you some indication of how much wire you will need. Simply lay lengths of thread along the design, measure and add them up, allowing a small extra piece for any adjustments.

7 The different colours on the pattern show the separate pieces of wire used to make up the design. You can vary the design by leaving out any pair.
Left: The round-nosed pliers are used to make curls for the pendant.

Cleaning and polishing

If you have handled the wire carefully the surface should not require too much work. Any obvious indentations can be removed with a small metal file. Do not remove more silver than is absolutely essential. A slower and safer way to smooth the surface is to use a fine grade wet and dry emery paper [sandpaper]. Tear small strips from the paper, wet it and rub the surface until it is smooth. Wash in luke-warm soapy water and thoroughly dry the article. Cleaning and polishing materials are available from jewellers' suppliers and some hardware stores.

Pumice stone in powdered form will also smooth silver. Simply mix the powdered pumice stone with a little water and apply it to the silver with a soft cloth. Rub it well for a good finish. French chalk and whiting (a variety of chalk) are finer than pumice and will polish the surface. Use in the same way as pumice. You can also polish the silver with a metal polish or immerse it in a silver cleaning liquid. An application of metal varnish to prevent future discolouration is optional.

Cleaning and polishing are time-consuming operations, but will make all the difference to finished articles.

Tin can punching

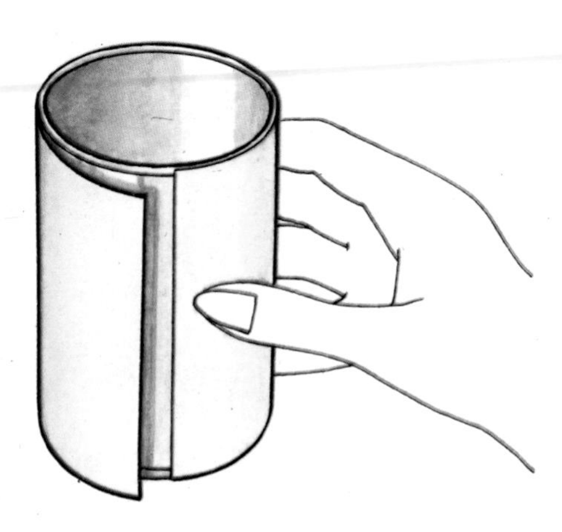

1 *Cut paper to fit around can.*

Apart from paper, tin cans are probably the greatest throw-away items in any kitchen. They come in a variety of shapes and sizes, from ordinary round cans to the more unusual shapes often used for patés. Some cans incorporate a corrugated design in the curved surface. The catering trade buys fruit juices in large tin cans which invariably end up on the rubbish heap in just the same way as the cans which we buy and take home.

The natural resources of the earth are dwindling away and, although it is impossible to replace them, everyone can certainly attempt to conserve and make better use of them. A tin can, once empty, has served its obvious purpose but, instead of throwing it away, why not spend a little time punching a design in the tin and then using it as a candle-holder or a receptacle for pot plants? You can attach pieces of wire to the tin cans, fill them with trailing plants and suspend them from the ceiling. If you use them as candle holders, the holes in the tin will form an intricate pattern of light spots around the room, a cheap and effective form of party lighting.

Designs

Try to work out a design rather than punching holes at random. To do this, cut a piece of plain paper to fit around a tin can (fig.1). Open the piece of paper to a flat surface and divide it into a number of sections of equal size. You can measure the sections to make them exactly the same or you can judge the size by eye (fig.2). The number of sections does not matter, but the larger the can the more sections you should make. In each section draw circles to the size you want to punch the holes and repeat pattern in each section (fig.3). Or you can reverse them in alternate sections (fig.4). Use a ruler or tape measure to help you with straight lines and a cup or saucer for curved lines. You can make very intricate designs, varying the shape of the holes by using different implements as your punching tools. Some suggested tools are shown overleaf.

Prepare the tin cans by cleaning them and removing the labels. Flatten any sharp edges around the lip with a hammer.

Punched tin cans

You will need:
Hammer
Any or all of the following tools:
Nail-punch or large nail for round holes.
An old screwdriver for elongated holes. Can opener (of the kind used for soft drinks or beer cans) for triangular shaped holes
A piece of wood or dowel large enough to fit neatly inside the cans
Empty tin cans of your choice

2 Divide paper into sections.

3 Draw pattern on to the paper indicating the size and shape of the holes you wish to punch.

4 The designs are easy to change. Any one design can be varied by reversing the pattern in alternate sections.

Punched tin cans make attractive holders for plants or candles.

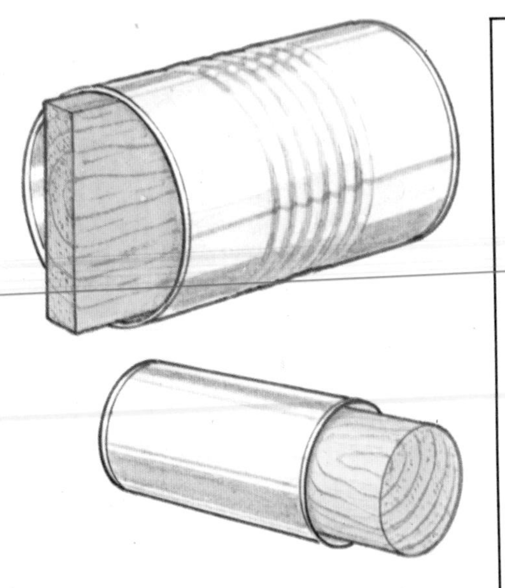

5, 6 *To prevent can from buckling while being punched, insert a piece of wood to support it. Right: Tools required for punching, indicating the various shapes they will produce.*

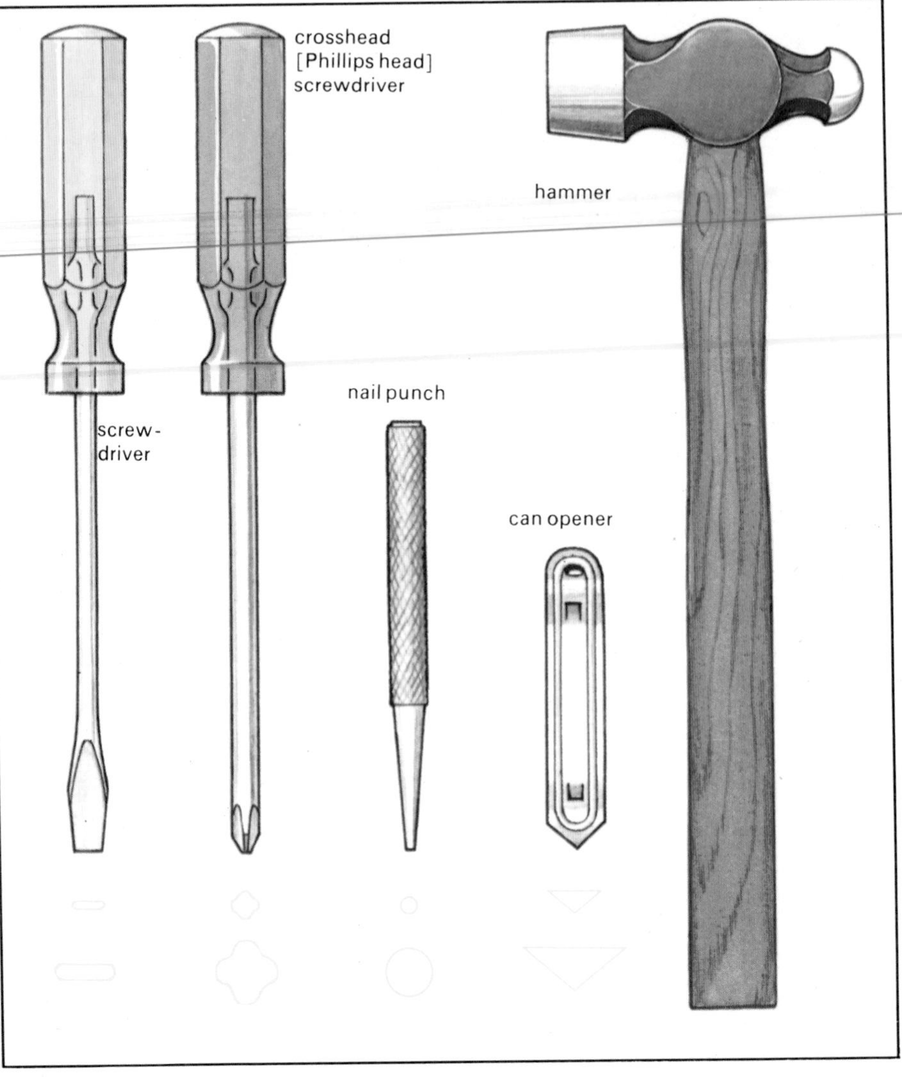

Position for punching cans

If you punch holes from the bottom towards the open end, the can will dent and lose its shape. To avoid this the can should be supported from the inside. Any piece of wood or dowel can be used for this purpose. Insert a block inside the tin can where you are about to punch holes (fig.5). Or slide a length of thick dowel into the can (fig.6). Wrap the design around the tin can and secure by taping the ends together. Use the nail-punch and hammer to make holes at the required spots. If you want to re-use the design make very light indents in the tin can and remove the paper design before actually punching the holes in the can. If you use the tin cans for plant pots, paint or varnish them to prevent them from rusting and to enhance the shine of the metal.

Cutting metal decoratively

Various sheet metals, such as copper, aluminium and silver, can be used extensively around the house if they are cut into decorative shapes. Silver is mostly suited to jewelry but copper and aluminium can be used to make name plates, number plates for the front door and also to make decorations around key holes or hinges. The tools used allow you to cut complex and delicate shapes, the surfaces of which can be textured, engraved, painted or polished. So choose something with an interesting outline for your first design.

Sheet metal

Sheet metal is sold by weight and is available in various thicknesses. Copper and aluminium are ideal to start with – copper shapes are useful as a base for enamelling. Once you are familiar with the techniques you can use silver, which is available from silversmiths and once again it is sold by weight and is available in varying thicknesses. Fine silver is too soft so use sterling silver. For small items you can buy off-cuts, but if you have a number of designs a larger piece is more appropriate as you can fit the designs on to the sheet in a more economical way. The thickness of the metal you will need depends largely on its function. For jewelry and most other small projects, .9mm (gauge 20) is quite sufficient. Larger items that are mounted on to another surface such as wood can also be made from this particular thickness. For example, you can make decorations to put around drawer handles or hinges on an old chest of drawers or sideboard. Don't forget that tin cans are a free source of thin metal: cut off the tops and bottoms, cut down one side and flatten the tin sheet with a mallet.

Tools

A jeweller's piercing saw frame and blades are used to cut the metal and are available from jeweller's suppliers and some hardware stores. To insert or replace a blade, clamp one end of the blade in the screw at the end of the frame – make sure that the teeth slope towards the handle – and at the same time tighten and tension the other end of the blade in position. The metal with the design on it

Tools for decorative metal cutting. A piercing saw makes it possible to cut very fine and intricate designs. The surface can be decorated by engraving, punching or drilling.

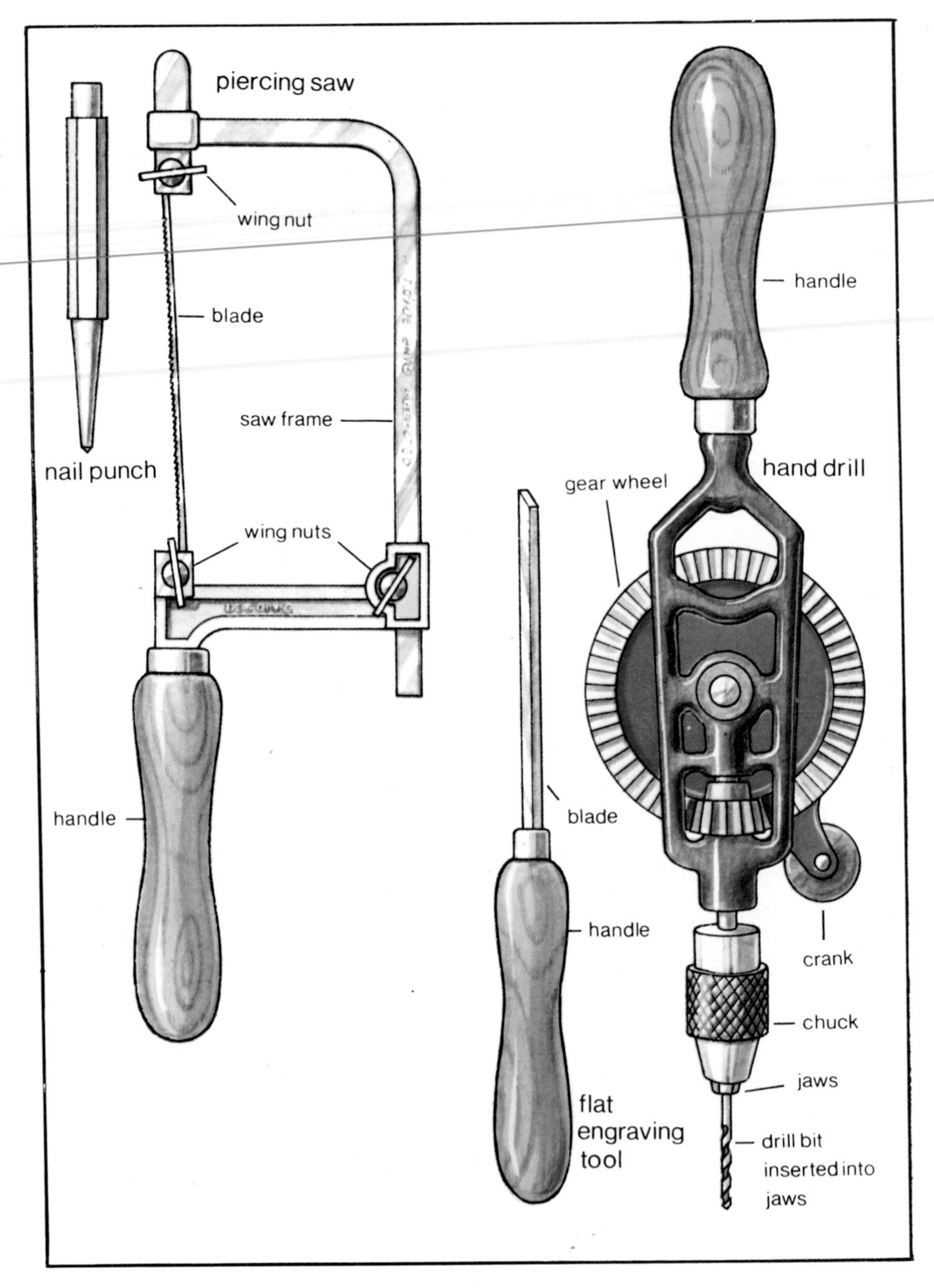

is placed on the bench peg which is clamped securely. Hold the metal firmly with one hand and work the saw with the other hand. If you find it difficult to saw, increase the tension of the blade slightly until you can control the saw easily with a rhythmic movement. The saw cuts on the downward stroke. Each downward stroke removes a small piece of metal, allowing you to follow an exact line. It is quite normal to break blades at first so do not worry – it takes a bit of practice to adapt to the rhythm of sawing. Do not handle the saw roughly. Blades for the saw are available in varying sizes. The thinner the metal the finer the blade must be. To get the correct blade to cut a particular thickness of metal refer to the instructions supplied with the piercing saw. A finer blade must also be used for scroll work, ie curved shapes. If the blade breaks when cutting a curve the chances are that the blade is too large to cut the curve, so use a finer blade. Practise on scrap metal before actually cutting out a design and use various blades to cut curves. You will find that the finer the blade, the finer the curve can be cut. A hand drill is used with a metal bit to drill holes in the metal. You will also need a bench peg [board pin] to place the metal on when cutting. The bench peg is secured with a G-clamp [C-clamp] on the edge of the working surface – a table edge will do. A bench peg can be purchased from a jewellers' suppliers and is inexpensive. Or you can make your own by cutting a V-shape into the thin end of a wedge-shaped piece of wood as illustrated right.

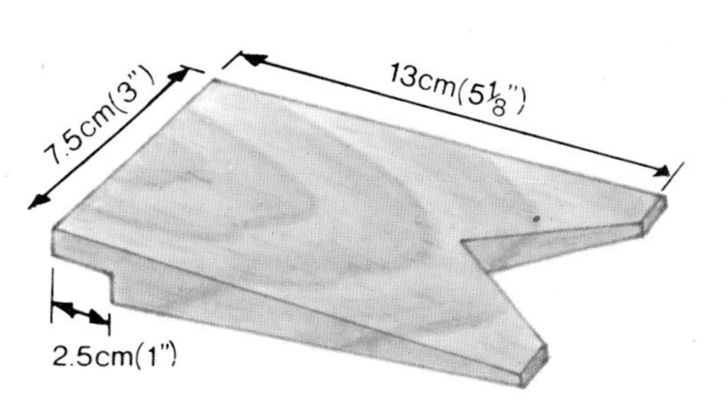

The bench peg [board pin] is cut from a wedge shaped piece of wood and clamped to the working surface with a G-clamp [C-clamp]. The cutting is done within the V-shape.

Cutting the metal

The metal is placed on the bench peg for support while cutting and the cutting is done within the V-shape of the bench peg so that the metal is supported around the area being cut. To turn a corner continue with the sawing up-and-down action, but with less pressure, and ease the saw gently round so that the cut starts to follow the line of the curve indicated. Continue sawing as before. It is easier to work in stages. If the saw has to be held in an awkward position to cut then place the metal in such a position that the saw can be held comfortably.

To cut the inside of a design drill a small hole using a hand drill and a metal bit – large enough to let the blade pass through it. Open one end of the saw, pass the blade through the hole and tighten and tension the blade to the saw frame as before. Cut out the required shape so that it drops out, then open and remove the saw blade from the metal. Don't be discouraged if your early efforts are not very accurate. Practice will improve your skill very quickly.

Tools for decorating

Engraving is done with a flat engraving tool. First mark the line to be engraved on the metal with a pencil. Using the engraving tool place the point on the line and with a wrist movement wiggle the engraver from side to side, 'walking' it along the metal surface. Engraving tools can be bought from jewellers' suppliers. They are available in various sizes and shapes. Alternatively, a small jeweller's screwdriver is inexpensive and also makes a suitable engraving tool. You can file it down to sharpen it if necessary but you will not be able to use it as a screwdriver again. You can also make do by flattening and then shaping a nail with metal files.

Punching The surface can be textured with a centre punch (nail punch). Hold the punch in position and tap it with a hammer to indent the metal surface.

Drilling Clusters of holes made with a hand drill and a metal bit can form part of the design. Or use the drill to make indentations by drilling only part of the way through the metal.

Polishing Use metal polish to shine the surface. If a satin finish is required rub the metal in one direction with fine steel wool. The use of metal varnish is optional.

Designs

Inspiration can be drawn by observing and copying shapes and patterns from man-made and natural objects such as spiky flowers, frost patterns on windows, intricate machine parts and even letters

Below: Trace pattern for the brooch made up of two separate pieces
Bottom left: The metal with the design on it is cut with tin snips [metal cutters].
Right: Hole drilled on waste side of design to hold saw blade

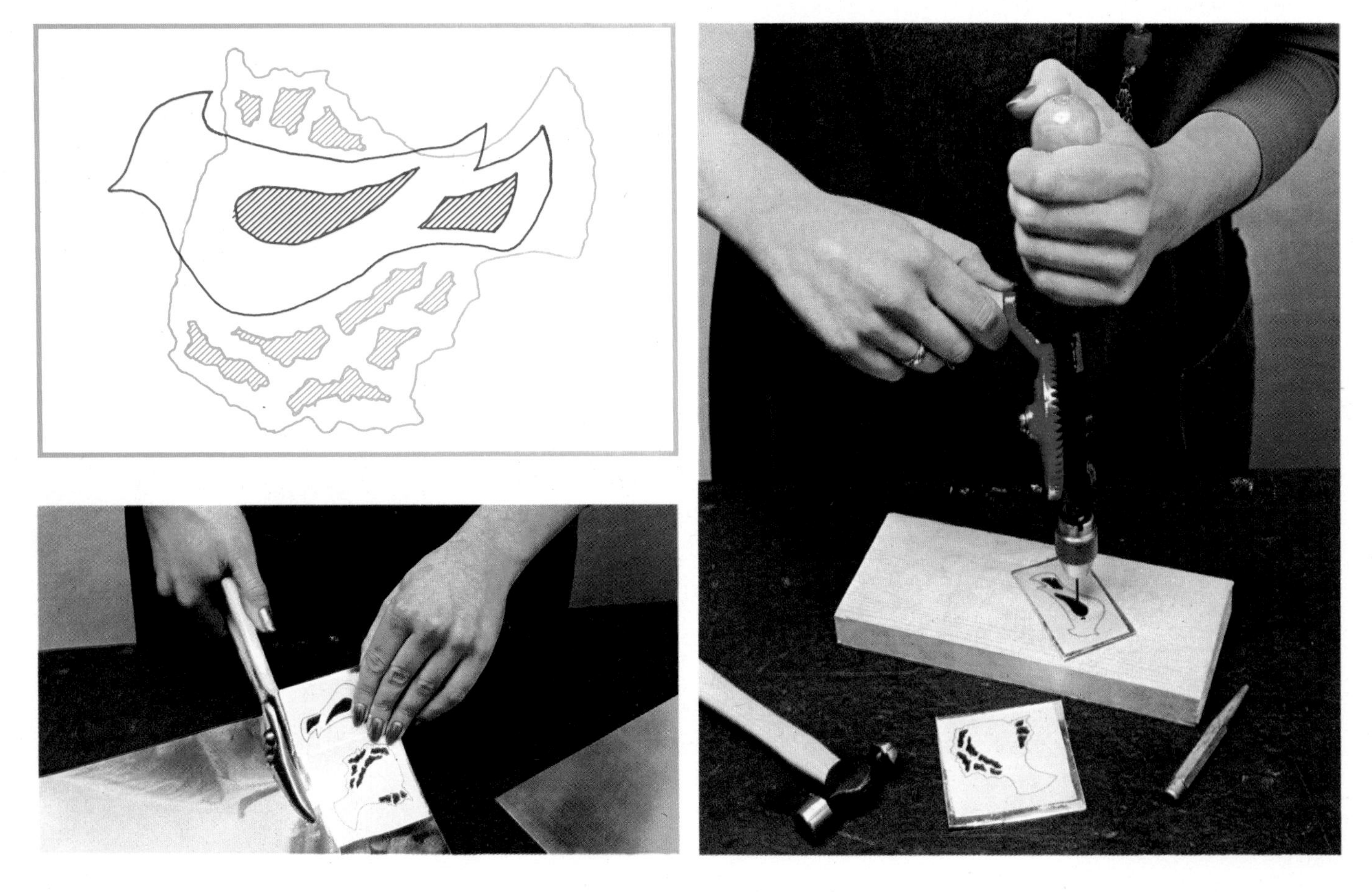

of the alphabet. Draw your ideas on paper so that you can see the exact shape, form and size clearly. Try combining various shapes until you are satisfied with the results. You can further enhance the design by texturing the surface. You can drill holes or texture parts of the surface by using a centre punch or a nail punch, or you can engrave and paint the surface.

Transferring a design on to metal can be done in various ways. The point is to transfer the design in such a way that it will not rub off while the metal is being handled. Draw the design on a piece of paper (tracing paper will do) with a ball-point pen. Cut out the design slightly larger than the outline and glue it to the metal with any household glue. Leave to dry, then shade in the parts of the design that are to be removed. This will prevent cutting the wrong pieces from an intricate design. Another method is to paint the metal with poster [watercolor] paint. Transfer the design on to it using carbon paper. Use a ball-point pen or felt tipped pen to shade in parts of the design to be removed. Surprisingly intricate and beautiful designs can be produced in this way.

Left: The saw is used vertically and moved up and down.
Below: The cutting completed, the edges are smoothed with needle files. Use a triangular file to reach into corners.

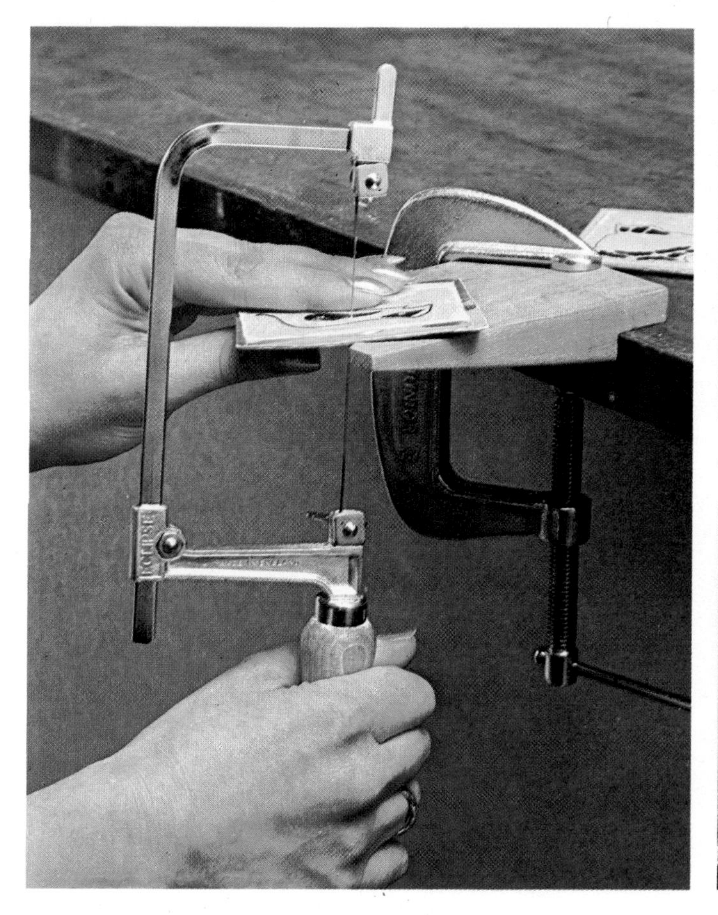

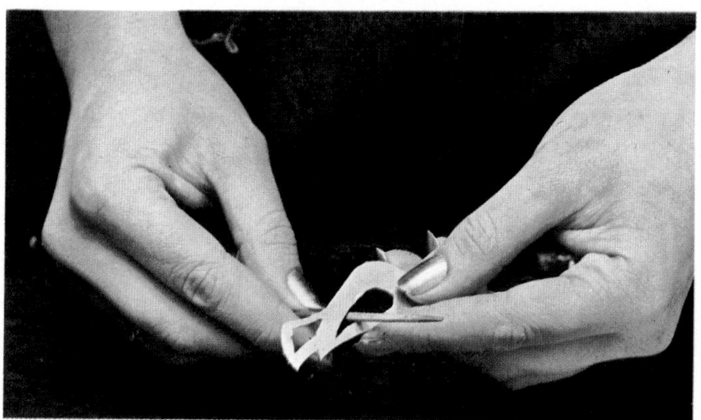

The surface is decorated with an engraving tool.

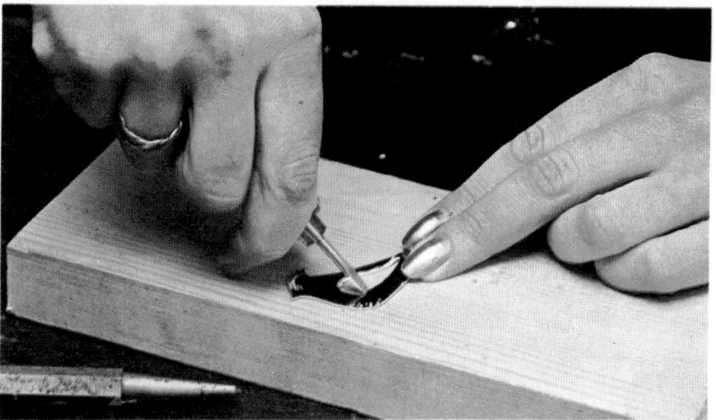

Bird brooch
You will need: Jeweller's piercing saw frame and blades Bench peg [board pin] G-clamp [C-clamp] Tin snips [Metal cutters] Mallet—a plastic one or a wooden one with a hide head Needle files Hand drill with metal bit and centre punch or nail punch Flat engraving tool (optional) .9mm (gauge 20) 10mm × 7.5mm (4in × 3in) copper, silver or aluminium sheet Brooch pin, epoxy resin adhesive and metal polish

Bird brooch

Transfer the design on to the metal by one of the methods described. Using the tin snips [metal cutters], cut the design from the metal, allowing for a waste area around the design. On a solid surface gently flatten the metal with the mallet. Use the centre punch to make marks on the waste side of the metal for each cut (the shaded part of the design). Drill a hole through each mark. The hole must be just large enough to let the blade pass through it. Using the piercing saw cut away the waste areas starting with the parts inside the design. Cut out the outlines last. Wash off the poster [water-color] paint or tracing paper.

Use the needle files to smooth all the edges. Decorate the metal using the engraving tools. Glue the bird shape on to the other piece of metal to form the design illustrated. Polish the brooch and glue the pin to the back with epoxy resin adhesive.

Other decorating ideas

If you use two or more pieces of metal superimposed, the cut out parts of the design will form recessed areas, and these can be filled with clear-cast resin with added colouring pigment. The most beautiful effects of all will be achieved with fired enamels, and cutting metal well is a must for the skilled enameller.

You are not limited to flat surfaces. If you have cut out a flower shape for example, you can use a pair of round-nosed pliers to curl the individual petals. Do this after any engraving or punching has been completed, but before decorating.

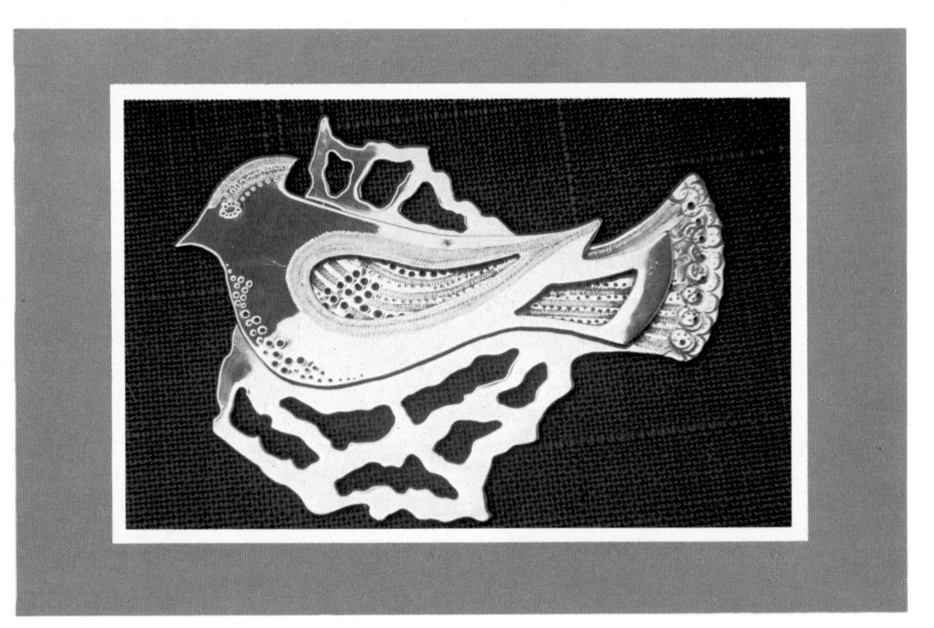

Bird brooch with punched and engraved surface

How to model sheet copper

Modelling thin copper sheet is an ancient oriental craft. For centuries copper has been used to decorate surfaces – even covering whole walls. Pictures, wall panels, fire screens and jewel boxes can be decorated with copper. The completed items look very solid but in fact a very thin sheet of copper is used and then mounted on to a base board. The effects can be varied. For example you may cover a cigar box and oxidize the copper to give it an antique finish or, for a wall panel, you can leave it fairly bright. Another idea is to use metal blanks (like the blanks used for enamelling) and cover them with the modelled copper to form jewelry. You can make labels for decanters similarly. The technique does not involve hammers at all. The design is traced on to the face of the copper and then modelled from the back. The indentations are then filled in and the facing surface is polished to complete the object. No anvils are required and it can be done on any firm flat surface – the kitchen table will do. Copper sheeting and modelling tools can be obtained from craft stores.

This fire screen is made from a thin sheet of copper which has been tooled and mounted on to a hardboard base.

Trace pattern for making one of the soldiers illustrated (top right). The background is textured and the finished panel is mounted on a wooden base board and framed.

Tools and materials

A dead ball point pen or a knitting needle, with not too sharp a point, is used to trace the outline of the design on to the face of the copper. You will need a soft surface over the work table such as a folded towel, an old bath mat, or a piece of foam rubber of similar thickness to give the copper the right kind of support.

The modelling is done from the back of the copper with various spatulas or modelling tools. Here you can improvise; the handle of a teaspoon or knife can be used, but the disadvantage is that the grip is not comfortable unless you wrap the tool with a piece of cloth and masking tape so that it fits comfortably in your hand – it is held like a pencil but requires more pressure.

A filler is used in the indentations so that the copper will not dent once the design has been completed. There is a variety of suitable fillers. Use whatever you have to hand or what is economical; candle wax, beeswax, wall filler and clear-cast embedding resin with hardener are all suitable fillers. Any of these can be used for flat surfaces but if a slightly curved surface, such as a label for a decanter, is being made, use beeswax as it will not crack or break when curved. It is melted and poured into the indentations before the metal is curved. Once the wax has set the metal can be curved gently with the fingers.

The copper can be given an antique appearance by applying an oxidizing agent sometimes called copper patina. You can buy it from craft stores or mix your own by adding six parts of water to one part of potassium sulphide from chemists [drugstores].

Technique

Practise on a piece of scrap copper before actually making up a design. On a small piece of copper draw a circle with a dead ball point pen. Turn the copper over and draw another line as close as possible to, and within, the circle. Place the copper on a towel or old bath mat with the second line you drew facing upwards. Use a modelling tool and push it gently backwards and forwards, using short strokes within the circle. Turn the copper over and check your results. Do it evenly so that the surface does not have bumps or mounds, which are impossible to remove. You can continue until the copper becomes thin. You will soon get the 'feel' of it, and if the copper does tear it will give you a guide as to how much you can force the copper out and will prevent you from doing it when making up a design. Any picture will do as the basis of a design, from vintage cars to abstract motifs of your own choice. Draw full size designs on to paper (or trace them) and mark in any necessary detail. Simple designs are best to begin with.

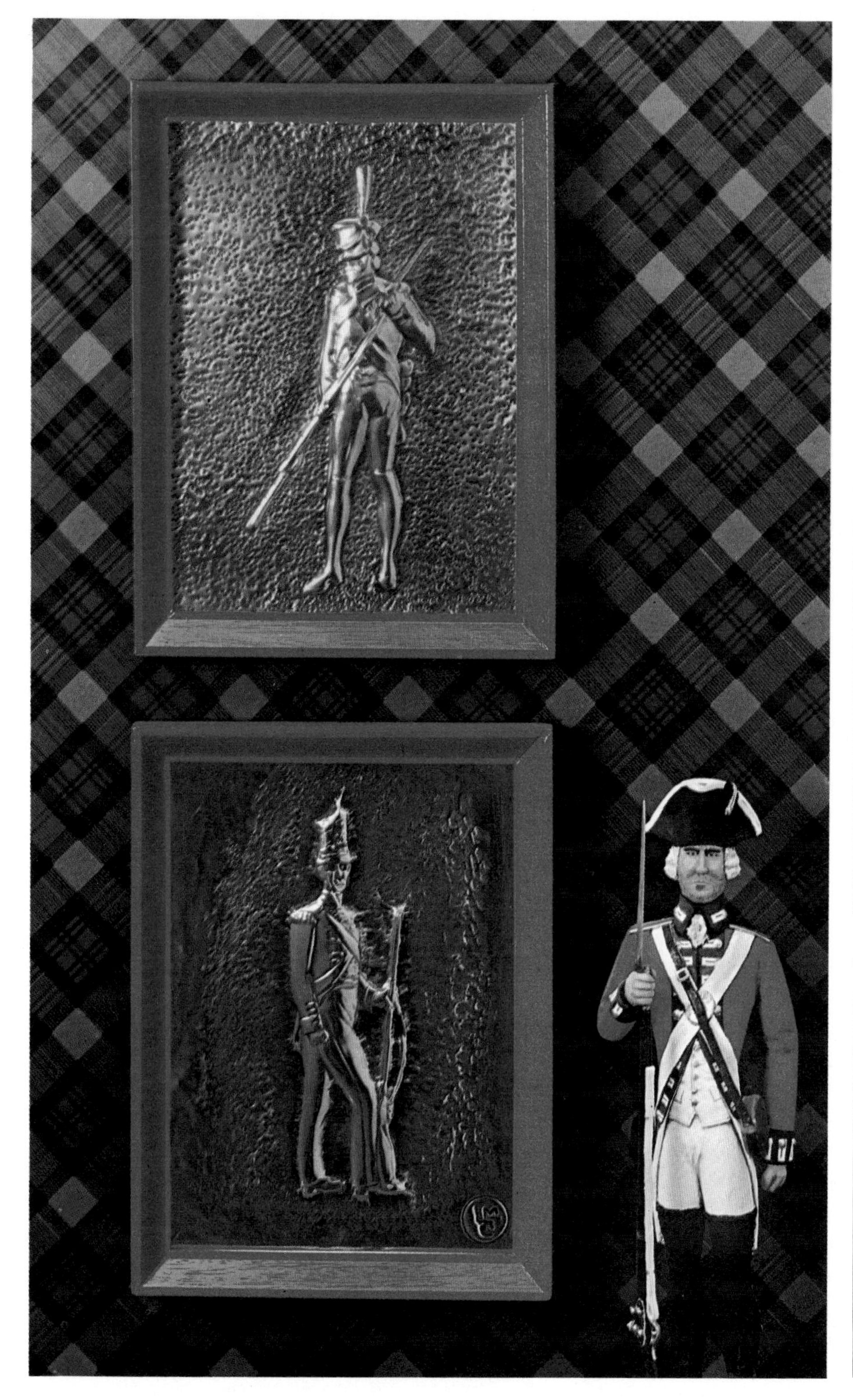

Copper wall panel

You will need:
Suitable design on tracing paper
Dead ball point pen or knitting needle
Modelling tools
Copper sheet .13mm (gauge 35), 22cm × 17cm (8¾in × 6¾in)—any copper around this thickness will do. Craft stores often refer to it as .0056in.
Masking tape or equivalent
Filler—any of the materials already mentioned
Base board such as chipboard [particle board] or plywood, 20cm × 15cm (8in × 6in), 6mm (¼in) thick. The copper is larger than the base board so that the sides can be folded around the baseboard to hide it.
Potassium sulphide—optional—to give the copper an antique appearance
Methylated spirit or acetone (optional)
All-purpose adhesive
Lacquer or clear varnish—optional—to seal the copper surface and prevent it from discolouring. Metal polish. Piece of glass or old tile to work on

Transferring the design

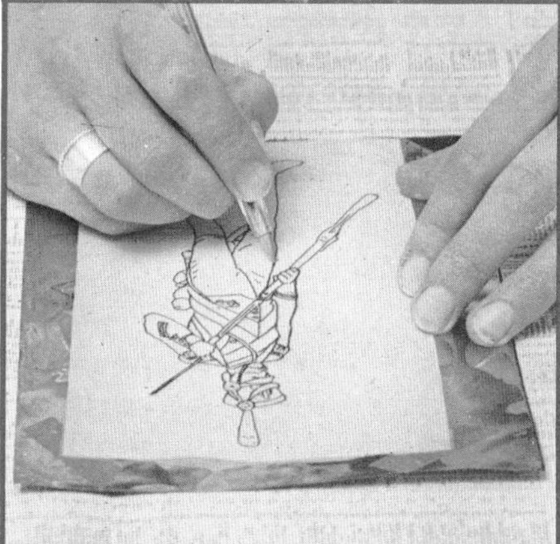

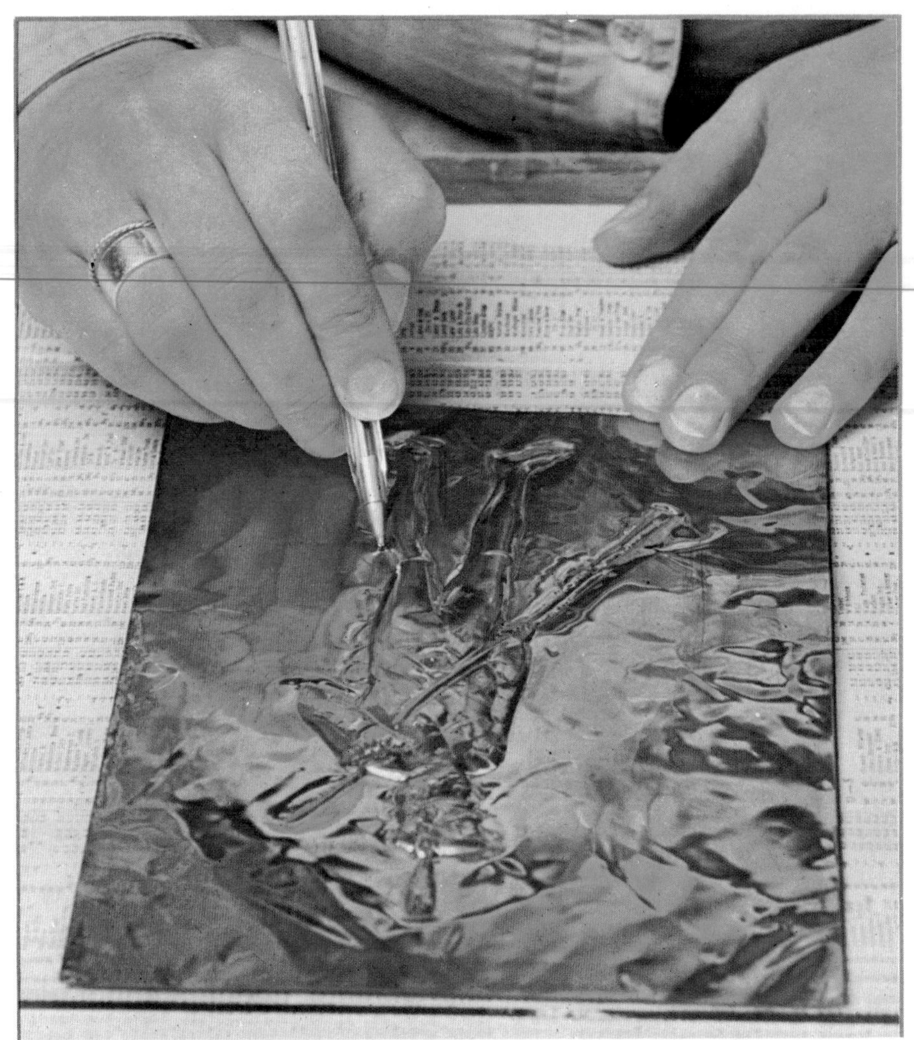

The outline is repeated on the wrong side of the copper just inside the original design (top picture), and the modelling is done from the back with a spatula (above). Right: The outline is repeated as the copper is shaped to prevent it from buckling.

Copper wall panel

The panel measures 20cm × 15cm (18in × 6in). The frame is optional, but if you do frame the panel then the copper sheet will not need to overlap the base board at the sides.

Cover the work surface with paper. Place the more attractive surface of the copper facing upwards. Use masking tape to hold the design on the copper in position. Using a dead ball point pen or knitting needle trace over the design. Keep the pressure even and make sure to go over all the lines making up the design. Lift the copper and look at the back of it to see that the design is complete and that the lines are even. Remove the design from the copper and lay the copper face downwards. Now draw the design on the wrong side just inside the lines of the design, but around the main outline

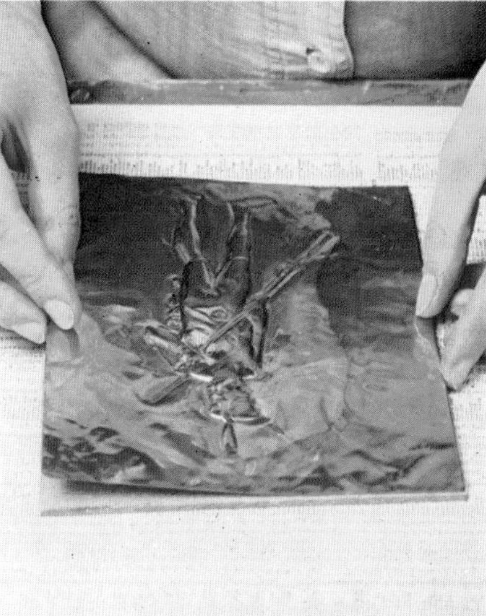

only. Look at the facing side of the copper and you will see that the design is now slightly raised.

Place the copper, face downwards, on a soft surface, such as a folded towel or a bath mat. Use a modelling tool and start to press the copper down with firm strokes, moving the tool backwards and forwards. Work evenly all over the design but within the sections of the pattern. Some parts need to be pushed out more than others. Turn the copper over and check your results. Continue until you get the desired effect. You will develop a 'feel' for the copper and know when it is getting too thin and is likely to tear. Once the copper has been pushed out it is impossible to push it back, so try not to make any mistakes. Work in stages and check your results as you progress.

Turn the completed design face upwards on your working surface without the towel. Go over the exact outline firmly with the ball point pen. (This will prevent the metal from buckling and keep the design flat.) Do this carefully, turning the copper as required so that you can reach the outline without putting any pressure on the shaped copper. Make sure the surface on which you are working is quite flat – a piece of glass or old tile is ideal. Add more detail if necessary by working on the towel again. Once you have finished go over the outline as before. This is important as it keeps the work absolutely flat which is necessary to mount it on the base.

Once the design is complete the indentations must be filled to protect them from possible dents. Use a filler of your choice. If you are using wax melt it and then carefully pour it into the recesses. Once it is has hardened you can scrape off any excess or

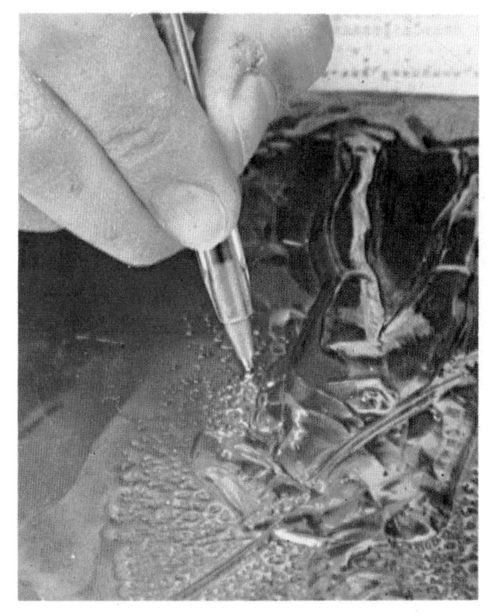

Left: To prevent the shaped copper from being damaged a filler is used to fill the recessed areas at the back. Once the filler is dry the entire back of copper is glued and placed on the base (top picture). The background is then decorated and textured by a series of random dots around the design.

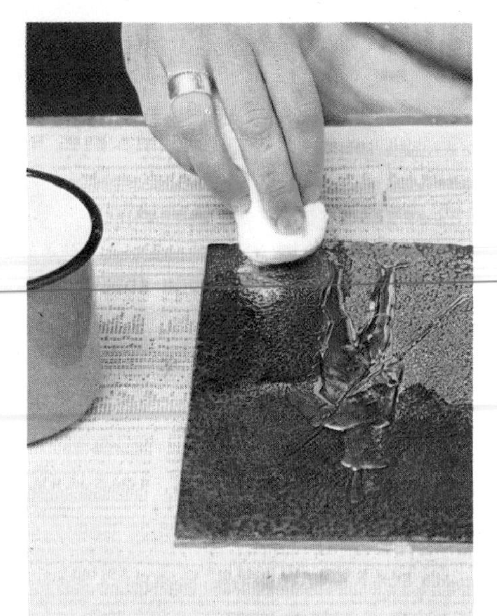

Oxidizing agent blackens the copper. Metal polish is then applied to highlight raised areas.

spilt wax with a knife. If you are using another type of filler, work neatly and make sure there is no filler spilt on the copper outside the design as it will show on the completed outside surface. Resin takes a while to set hard but the results are excellent – so place it in an airing cupboard and be patient until it is absolutely hard. The tacky resin surface is also a good adhesive when you mount the copper. Once the filler is dry spread glue over the entire back surface of the copper – including the filler. Apply glue to one side of the base board. Place the copper on to the glued surface of the base board. Make sure it is in the correct position and gently push the copper down. Work from the area around the design towards the outer edges of the copper. Go over the outline again to make sure it is quite flat. Using scissors cut the corners of the copper where it overhangs the base board and fold the copper down as close to the edge of the base board as possible. If the picture is to be framed then it is not necessary for the copper to overlap the base board.

You can leave the background as it is or use a metal knitting needle to make indentations, at random, all over, or starting from the design and decreasing the texturing as you work away from it. Another method of decorating the background is to use the round head of a hammer and to lightly beat the metal. Be very careful when doing this so that you do not damage any of the raised surfaces. Clean the surface with metal polish. Do this a number of times if necessary until the metal is very shiny.

To darken the copper for an antique appearance, use the potassium sulphide. Mix one teaspoon of potassium sulphide with six tablespoons of water. This mixture is hard on the hands so protect them with rubber gloves. The copper surface must be free of grease to make this successful so wipe the surface with methylated spirits or acetone before applying the patina. Apply the mixture to the copper with a piece of cotton wool [cotton ball]. Do it evenly until the copper is completely black. Wipe off any excess liquid. Using a small amount of metal polish start polishing the surface, concentrating on the pushed out parts of the copper. This will highlight the design. The copper will start returning to its original colour. Continue this polishing until only the recessed parts of the design are left black. If you are not happy with your result you can darken the surface again and start over until the right effect is achieved and you feel satisfied with the results.

To protect the surface and to prevent it from discolouring apply a coat of lacquer or varnish to the design. If you do not do this the copper will need repolishing. The design can now be framed, or hung as it is if the sides have been neatly folded back.

Modelling and tooling pewter

Pewter is a gray alloy of lead and tin. It is very soft and lends itself to tooling or modelling. The pewter used for modelling is softer than copper and therefore more detail can be added. The techniques for copper and pewter tooling are very similar. More tools

Mirror frame modelled in pewter. The centre is cut out and can be re-used if desired.

The tools used to model the pewter vary in shape to work small or large sections of a design.

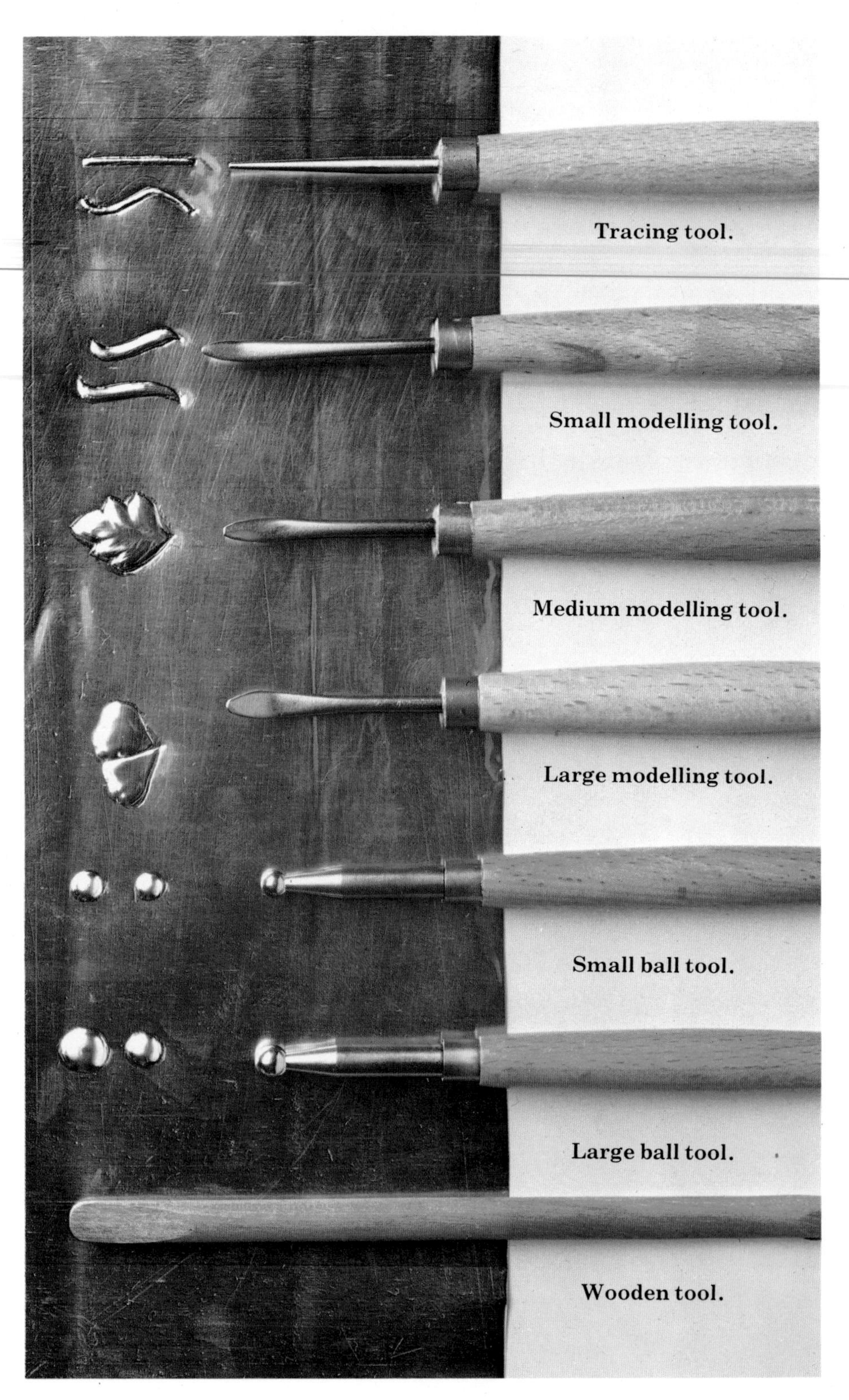

are required for pewter because it is possible to do much finer work than on copper. Pewter can be used for wall panels, pictures, covering boxes, jewelry and frames around mirrors. You can set stones and beads in it to add colour.

Tools and materials

All the tools and materials needed are normally available from craft stores. The pewter is bought in sheet form in varying sizes. Buy .13mm (gauge 35) pewter, which is often referred to as .0056in or .006in. This is the most suitable thickness for modelling.
You will also need a metal tracer for drawing the design on to the pewter. You can improvise with a metal knitting needle. Modelling tools are used for the actual modelling – you will need a small and a large one. Ball tools are useful for making spherical shapes and are available in metal or a very strong glass. A wooden or bone tool will keep the pewter flat without damaging the facing surface, but you can make do with a bone knitting needle. A pair of curved nail scissors is used to cut the metal. In addition to these basic tools you will need a soft cloth or piece of thick felt on which to do the modelling. Fold the cloth so that you have two layers of it on which to press. Pewter patina is used to oxidize the metal in the same way as copper patina. Methylated spirit is used to clean the

Note for U.S. readers:
Lead has been forbidden by law in the composition of American pewter for about forty years. Modern American pewter is an alloy of tin and antimony with a little copper added to give malleability. In appearance it is brighter and shinier than the British pewter shown here, and does not oxidize with age. Consequently it does not require the application of pewter patina or polish. It is available in sheets ranging in thickness from 30 gauge to 2 gauge. It is still a soft metal, but not as easily worked as British pewter, so ask your supplier for advice when buying.

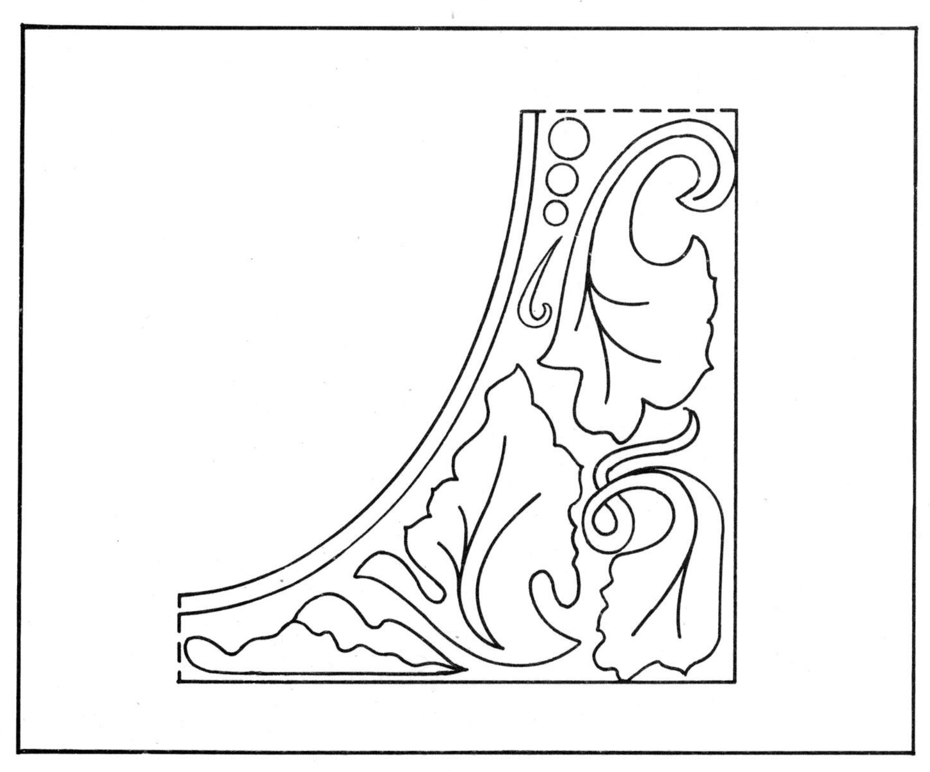

Trace pattern for one corner of the frame. Repeat it four times to form the frame illustrated.

metal surface before applying the patina as it will not take if the metal is greasy. Ordinary household scouring powder will do the same thing. An all-purpose adhesive is needed to glue the pewter to the base or backing board, and a filler to fill the tooled or modelled parts of the design (see the previous chapter). The design is worked out on tracing paper and held in place with adhesive tape while the design is transferred to the pewter.

The pewter is always mounted on to a base of some sort. For a wall panel or picture use 6mm ($\frac{1}{4}$in) plywood and for jewelry you can buy metal blanks from craft stores or cut them yourself. If you are covering an existing surface, such as a cigar box, the pewter is mounted straight on to this surface. Metal polish can be used to shine and highlight the pewter, and metal varnish to prevent discolouration is optional.

The tools are not expensive, and each has a modelling tip at either end of a different size. Smaller modelling tools are used for fine work and the larger ones for pushing out larger areas. Use each tool in turn and draw lines on a small test piece of pewter to see exactly what each one does. Use the tools on a folded cloth to push out the pewter to find out how much you can work it before the pewter tears. Always use the same pressure and repeat the backwards-and-forwards motion to increase the depth of the pewter. The ball tools are used in a small circular motion.

Pewter mirror mounts

You will need:
Pewter 30cm × 30cm (12in × 12in). This is enough to leave you a piece on which to practise
Pewter patina
Metal polish
Filler
1 mirror tile 15cm (6in) square
The design on tracing paper
A hard surface such as a tile or a piece of glass
Tools as previously given

Pewter mirror mounts

Trace the pattern on to tracing paper and repeat it three times so that it forms a square with an oval in the middle. Place the pewter with its brighter side facing upwards on the working surface. This will be the front or 'right' side of the pewter. Secure the design on the pewter with adhesive tape. You can make the frame in one piece and then cut out the centre, or make four corner sections and use them individually, or combine them on the mirrored tile. Go over the design with the tracing tool. Check the back of the pewter to make sure that you have not missed any part of the design, or lift part of the tracing paper and check the results. Make sure that the design is complete and that all the lines meet. There must be no gaps between the lines forming the outline. Remove the tracing paper, checking the tracing carefully as you go.

Place the pewter with right side facing upwards on a single thickness of cloth and using the tracer go over the design with enough pressure to leave an imprint on the metal. This must be done carefully. Once there is a mark on the pewter it cannot be removed easily and must therefore become part of the finished object. Work slowly and follow the exact line of the design. Turn the pewter as

Opposite: Various items covered with modelled pewter. The paper weight is made from clear-cast embedding resin with a pewter base. Note the different ways of finishing backgrounds – parallel lines, spotted for a 'beaten' effect, or left plain.

necessary so that you work from left to right – or right to left – whichever you find easier. Do not work in a direction where you find it difficult to handle the tracer (this applies to all the modelling tools) since you will not be able to work accurately.

Turn the pewter over and then work on the back surface. Draw a line just on the inside of the design using the tracing tool. Concentrate on the main outline of the design and any lines that outline a section to be modelled. Exclude fine lines which detail interior parts of a section. These are added later. This line on the inside of the design forms a slight edge. If you now turn the pewter the right way up you will see the design slightly lifted.

Place the pewter wrong side upwards on two thicknesses of the cloth. Use a modelling tool or a ball tool and start pushing the pewter out by moving the tool backwards and forwards. Work only on those parts of the design which are to be raised. Do one section at a time using a tool suited to the size of the section being worked. A small section will need a small tool that fits within its lines and a larger section will need a larger tool to push out evenly. The ball tools are best used for circular shapes where a ball-like effect is required, and are used with a circular motion.

Place the pewter right side up on a hard flat surface such as an old tile and press the area around the design flat using the wooden or bone tool. This keeps the work flat and prevents it from buckling. Part of the design might buckle when you flatten the surrounding pewter so it might be necessary to do a bit more tooling to restore the section. Carry on like this until the work is flat and the design intact. Add any small detail lines to the design still working from the back and with the pewter on the cloth. If you add lines from the front you must work very lightly so that you don't push the raised sections down.

Once the design is complete and the surrounding area is flat fill the recesses at the back of the pewter. Do this carefully and do not let the filler spill on to any part of the design which has not been raised. Allow the filler to harden. Clean the pewter well with scouring powder and make sure the surface is not greasy by wiping it with methylated spirit. Apply the pewter patina to the entire design and leave it to dry for a few minutes. The pewter must be a dull, dark, almost black colour. If any bright pewter shows the metal must be polished again and more patina added. To highlight the design use a little metal polish on a cloth and go over the pewter lightly. Do it slowly and check the pewter at intervals until you achieve the effect you want. The more metal polish you use the lighter and more metallic the pewter becomes. The pewter takes a lot of polishing. Keep at it and don't get impatient – good

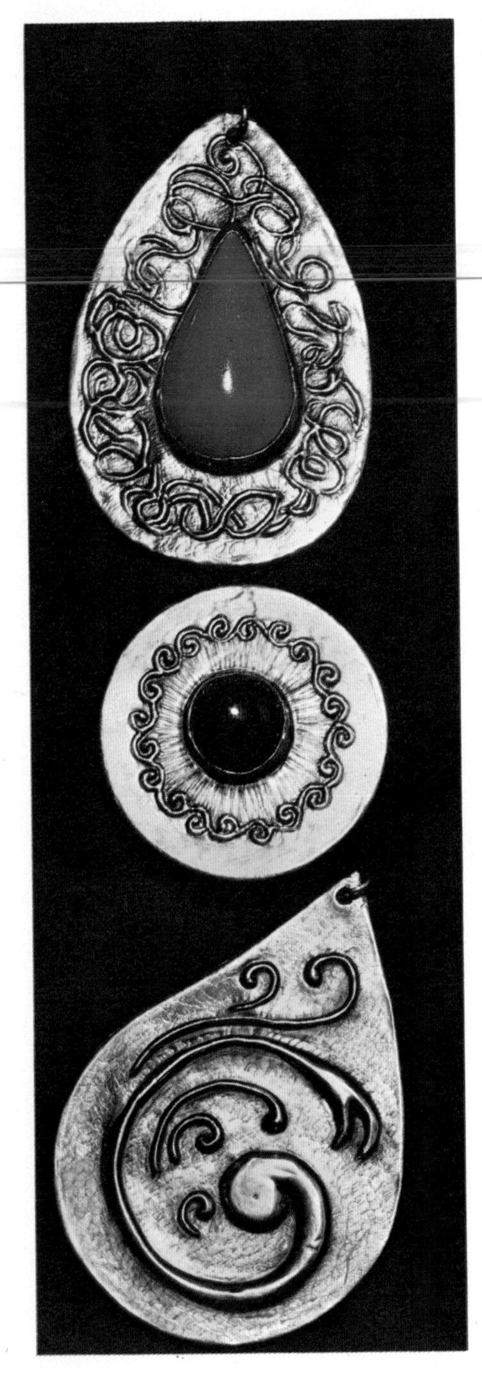

Jewelry made from pewter. The modelled pewter is mounted on to metal blanks which you can buy or cut yourself.

results will require some effort. Apply metal varnish if you wish. Using the curved scissors carefully cut out the design. Apply all-purpose adhesive to the back of the pewter and position it on the mirror. Once it is dry check around the edges to see that they fit against the mirror. If there is a slight gap use a modelling tool and gently work the side of the pewter until it closes.

Covering boxes

If you are covering a flat surface, such as the lid of a box, you will not cut out only the design but the area around it to fit the lid. This surrounding area must be kept flat. Use a bone or wooden tool or wrap a bit of tissue paper around a modelling tool to prevent it from scratching the pewter surface. Use the flatter part of the tool and, working on a hard surface, gently push any uneven parts of the pewter down the sides of the lid.

Background texture

You can decorate the background of a design once the design is complete but before polishing it. Using a tracer you can 'spot' the entire background or concentrate around the outline of the design and decrease the spots as you move away from the design.

You can also use a ball tool in the same way for a softer effect. Lines can be scrolled freehand to cover the background, but some designs lend themselves to straight lines.

Mounting and finishing jewelry

The design in pewter must fit on to a metal blank. Model and fill the pewter as before and set a stone if you want to. Cut out the design from the pewter slightly larger than the outline. Cut little V-shapes into the excess pewter so that the point of each cut stops just short of the outline. Glue the design to the blank and fold the excess area to the back of the mount. Cut out another piece of pewter to fit the back of the blank and glue it in position for a neat finish. Glue jeweller's findings [backings] to the back if necessary to complete the jewelry.

To set a stone it must be incorporated as part of the design. Draw the stone's outline at the required position on the design. Leaving this part flat model the rest of the design, then fill and polish it. Draw a line just inside the outline of the stone (fig.1). Using the curved scissors cut out the flat part where the stone is to be set (fig.2). Place the pewter on the stone and gently work it over the stone. Place the stone and pewter on to the mounting and glue them in position (fig.3). Clean the pewter carefully when the glue is dry, and apply pewter patina and polish if desired.

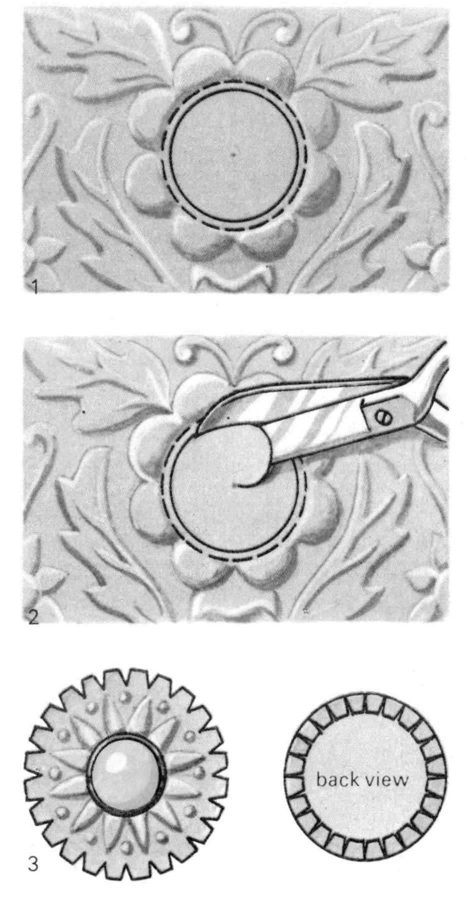

1 First stage to set a stone. The dotted line indicates the outline of the stone to be set.
2 Inside cut out with small scissors
3 Outline cut and folded to the back

Soft soldering techniques

Soldering is a simple method of joining two pieces of metal together by means of a molten metal filler. There is more than one kind of soldering, though the method is always similar. The difference between them is that some solders are harder and stronger than others and melt at different temperatures. This chapter introduces the tools and materials, and the following chapter will show working techniques and projects for soft soldering.

The three essentials are solder, flux and heat. There are different types of solders and fluxes, and the heat is provided by a soldering iron or a blowtorch. These must be combined in the correct way to suit the metal to be soldered together (see chart). Don't become confused with different metals, fluxes and solders by thinking of all the things you want to make. Instead, think in terms of a particular application: for example, solder tin cans together as shown in this chapter to make a tool tidy [caddy] or a plant hanger, using a copper bit, mild flux and a 60-40 solder. All these terms are explained and the chart shows you which solder and flux to use with different types of metal. If you have never done any soldering before you may be surprised at how simple it is, but you will need practice to perfect your technique.

A blowtorch works by the direct heat of a naked flame, the blue inner flame being the hottest part. A copper bit can be fitted to a blowtorch, making it more versatile.

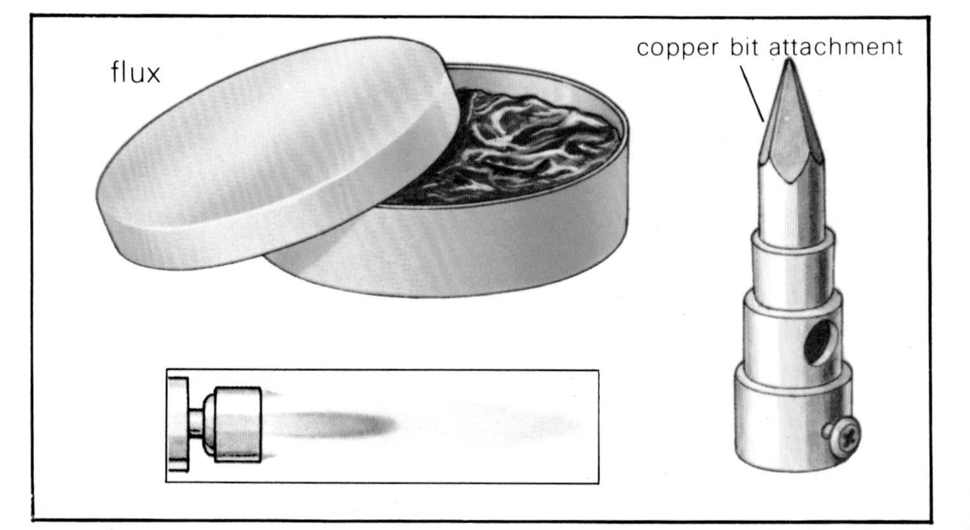

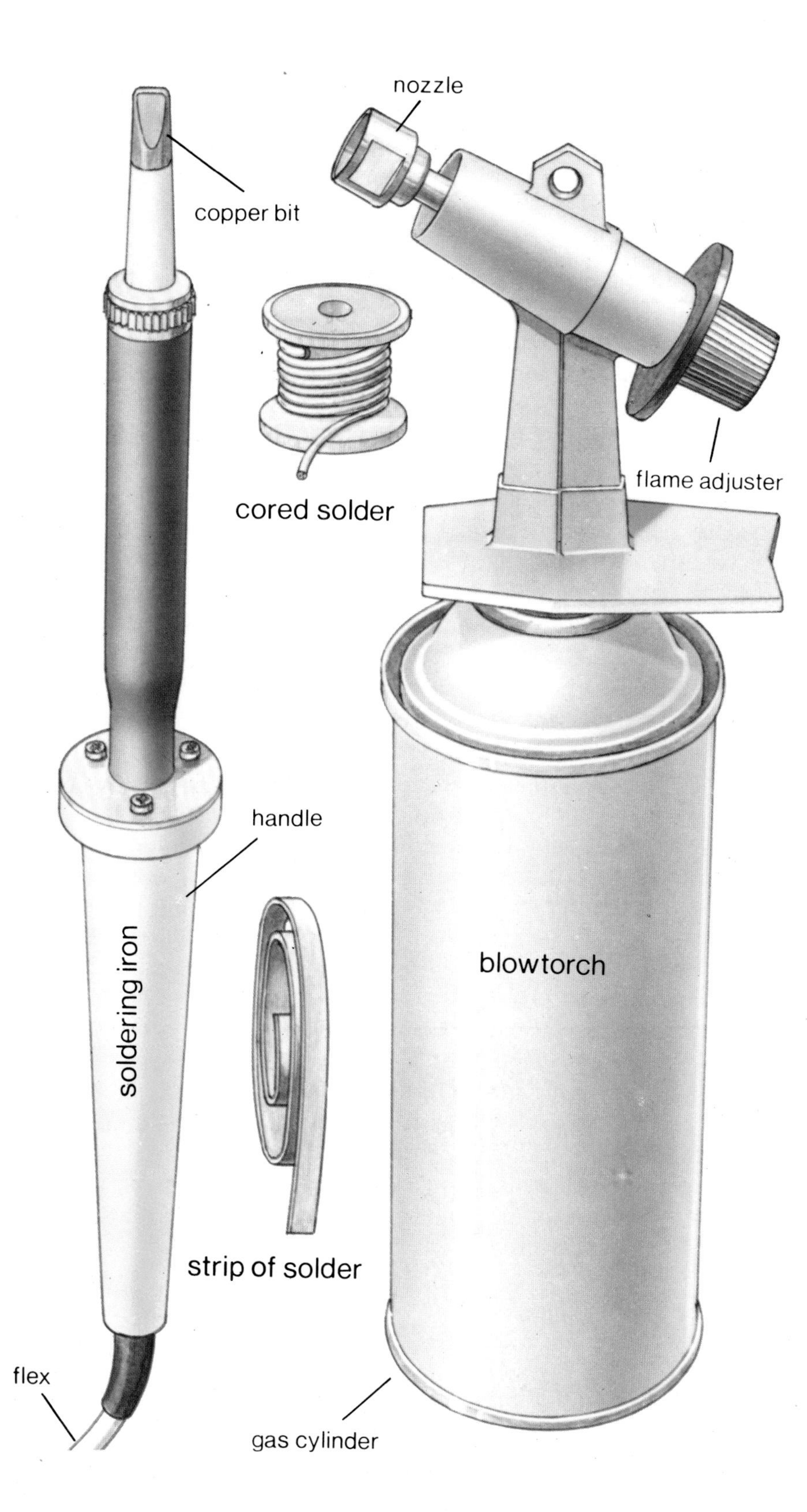

A soldering iron has a copper bit which is usually electrically heated. The blowtorch illustrated on the right works off a portable gas cylinder.

Soft solder and flux selection guide			
Metal to be joined	Solder % Tin	% Lead	Flux
Brass or Copper	50	50	Acid or Mild
Tin	60	40	Mild
Lead	50	50	Mild
Iron or Steel	50	50	Acid
Gold or Silver	70	30	Acid
Nickel	50	50	Acid

Heating equipment

Most people have at some stage or other seen the basic soldering tools – a blowtorch or soldering iron – being used. The blowtorch is reminiscent of a flame thrower and looks a ferocious piece of equipment but in fact it is a very useful one. It is not difficult or dangerous to use providing the necessary safety precautions are taken. A soldering iron does the same thing as a blowtorch but it is quite different in appearance and not as versatile. It resembles a screwdriver with a pyramid-shaped copper tip which is either heated over a naked flame or has an element which is electrically heated. It is used for very fine work such as joining the intricate wires of a transistor radio and for making fine jewelry. Despite the difference in the appearance of the two pieces of equipment, both have the same function and the same technique is used for both. They both supply the necessary heat to solder or join two pieces of metal together. If you have the equipment already you probably have some idea what you can do with it but, if you are buying it, you must decide which you prefer – a blowtorch or a soldering iron – how much you want to spend, and how useful either is going to be. If you decide on a blowtorch it is worthwhile getting one to which a copper tip – also known as a bit – can be added.

Blowtorch The modern blowtorch is a very versatile item. It is available in various sizes, from something not much bigger than a cigarette lighter to heavy industrial equipment. The easiest to use around the house is a fairly small one with portable gas cylinder which is not much larger than a can of hairspray. A very small blowtorch will supply the necessary heat for soldering but will not lend itself to much more. The slightly larger ones are far more versatile as they can be fitted with various attachments and are also suitable for hard soldering. With this type of blowtorch the flame can be adjusted to do small or large joins. A copper bit can be added so that it works exactly like a soldering iron which means you can work with mild and acid fluxes.

The flame on a blowtorch is easily adjusted by turning a knob. The inner part of the flame, which is blue, is the hottest and the length of this inner flame indicates the amount of available heat. For very fine intricate work, as when making jewelry, the inner flame should be about 6mm (¼in). This will only heat a very small part of the metal and is easily controlled. It also means that if you are soldering near to another soldered joint the heat will not spread and undo this joint if you work fast. For less intricate work, but where the join is still reasonably small, a 12mm (½in) inner flame setting is needed. For larger jobs, where the heat has to be spread over large areas, set the inner flame to 25mm to 30mm (1in to 1¼in).

Soldering iron Electrically heated soldering irons are available in various sizes. A 15 watt iron is for very small intricate work and a 100 watt is suitable for light sheet-metal work. It is easy to work with and light to hold. Make sure that the soldering iron is correctly wired and it must be earthed. The iron runs off any power socket. The copper bit or tip of the soldering iron (or blowtorch attachment) must be clean before you start soldering. Remove all traces of dirt or solder with a metal file. Heat the bit and dip it into the flux. Contact between the hot bit and the flux will give off fumes – avoid breathing them. Then apply the solder; let the solder melt and cover the working top of the soldering iron. The iron is now ready for use.

Suitable metals for soft soldering

Most metals can be soft-soldered. Aluminium is the exception and zinc is difficult but this still leaves you with readily available metals such as brass, copper, tin, lead, iron, steel, nickel and silver. Gold is also suitable but it is of course expensive and not readily available. The various metals are all soldered in the same way but some need a different type of solder and flux for a particular application.

Solder

Solder is the 'metal' which is heated to form a molten filler to join metal together. It must have a lower melting temperature than that of the metal being joined together. It is an alloy made from tin and lead and may also contain small amounts of other metal such as copper. It is identified by its composition, ie 60-40 solder is 60% tin and 40% lead – the tin content is always indicated first. For the best results use the solder, indicated on the chart, with the different types of metal.

Soft solder melts below a temperature of 450°C (842°F). It is sold in strip or rod form in various sizes. A 6mm ($\frac{1}{4}$in) thick strip is suitable for most general soldering. It is referred to as tinman's solder. Soft solder in the form of wire normally includes its own flux. It is known as radio quality solder or cored solder and is meant for electrical work.

Hard solders melt at temperatures above 450°C (842°F) and are mainly used for structural purposes or where a very strong joint is required. These are dealt with in a later chapter.

Flux

Flux is a chemical compound that prepares the metal surfaces for joining. It also helps the molten solder to flow smoothly. The usual type of flux is in the form of paste and is sold in tins. A

Three tin cans of the same size are soldered together to make this hanging plant holder. The size can be varied to adapt the design to a tool container.

metal surface might look bright and clean but there is always a certain amount of oxide present on the surface which increases when the metal is heated. The flux dissolves these oxides that form as the metal is heated. The molten solder flows freely where flux is applied so it is essential to apply it neatly and only on that part of the metal where solder is required. There are two types of flux, mild and acid.

Mild fluxes are non-corrosive. They are especially suitable where it is difficult or impractical to clean a join after soldering is complete. *They must only be used in conjunction with a copper soldering bit as they are inflammable and will burn if exposed to the hotter temperature of a blowtorch flame.*

Acid fluxes are corrosive. They are available in liquid or paste form and as they contain acid you should clean any of it that you might spill. Avoid contact with clothing, skin and eyes. Wash off any spilt flux with plenty of soapy water. Acid fluxes have better solvent properties and are therefore generally more effective. *They leave a corrosive residue which must be removed after soldering is completed.* This is done with soapy water. Acid fluxes can be used with a blowtorch flame or soldering iron.

The fluxes shown on the chart are the recommended ones. They are meant as a guide and not as a rule. Acid fluxes can usually be used where it is possible to wash off the residue after soldering is completed. A mild flux must be used when it is impossible to clean the join after soldering as the corrosive properties of an acid flux will damage the metal eventually. If in doubt ask your dealer to recommend solder and flux for specific jobs.

Soft soldering using tin cans

Soldering has various applications, but start by getting familiar with the equipment and techniques that have already been described before embarking on a project. The first bit of soldering you do will probably not be very neat so soldering together a few tin cans makes an ideal first project before progressing to more intricate work such as jewelry. Or you can make an abstract metal collage or sculpture. It will save you disappointment and expense if you have a bit of practice first.

To ensure the best possible results the following conditions are necessary. The solder itself and the metal surfaces to be joined together must be at the same temperature, and this temperature must be just above the melting point of the solder. If the solder does not flow the metal is not hot enough. The surfaces must be clean and free from dirt and corrosion. The right kind of solder must be used for the metals being joined together (see chart).

Working area

You do not need a large working area (depending of course on what you are making) but keep it tidy and uncluttered. The equipment is not dangerous if you use it sensibly – it is important to remember that it produces heat and that you should handle it with the same degree of care as you do a hot iron. If you do not have a workshop you can use the kitchen. Work on a heat resistant surface, preferably an asbestos mat. Always rest the soldering iron on a stand when it is hot so that the copper bit is not in contact with anything. Bend a wire coat-hanger to make a stand for the hot soldering iron (fig.1). For an electric soldering iron you will need to work near a power point [outlet]. If you are working with a blowtorch do not point

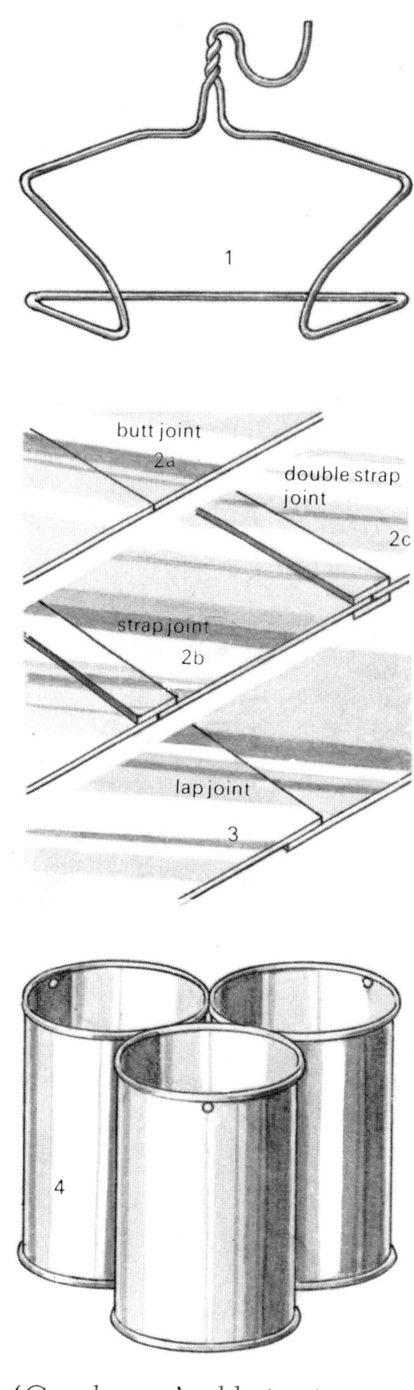

1 *'Coat-hanger' soldering iron stand.*
2-3 *Various joints*
4 *Cans positioned for soldering*

the flame towards yourself, nor towards the wall. Keep the flame as small as possible and only turn it up when you are soldering. If it makes you feel safer keep a small fire extinguisher near your working area, but accidents are unlikely with care.

Joints

The solder used to join two pieces of metal together is usually weaker than the metal. To make the joint as strong as possible the area of contact between the two pieces of metal should be as large as possible to increase the 'hold' of the solder on the metal. A butt joint is the weakest type of joint (fig.2a). It will hold the metal together if that is all that is required. Strengthen a butt join by adding one or two strips of metal to make a strap joint (fig.2b and 2c). Lap joints (fig.3) are stronger than butt joints. The bigger the overlap is, the stronger the join. If the overlap is very big the surfaces making contact must first be tinned, ie covered with solder. They are then put together and heat is applied to 'sweat' them together. Practise the joints on odd pieces of metal.

After the areas of contact are fluxed, heat the tin cans with the hot iron for soldering.

Technique

The area of metal to be soldered must be clean and free from grease. Rub the metal down with steel wool, glass-paper [sand-paper], emery cloth or, if the metal is rusty, use a metal file to clean it. Remove any traces of grease with a scouring powder.
While you are preparing the metal, warm the soldering iron or light the blowtorch and adjust it to a small flame. Apply flux with a small paint brush or wooden spatula where the solder is required. Do it neatly as the solder flows wherever the flux is. The copper bit or tip of the soldering iron, or the copper bit attachment of the blowtorch, must be tinned before use. Make sure the tip is free from rust. Dip the heated tip into the flux and then touch solder to the tip to cover the working surface.
Heat the metal evenly where it is to be joined by putting the bit or tip on the metal. The solder will flow from the bit once the metal is hot enough. If more solder is required touch the solder to the tip. Once the metal is hot enough the solder can be touched directly to it so that the solder flows to fill gaps. If you are using a blowtorch without an attachment, heat the metal where it is to be joined with the inner blue flame. Touch the solder to the metal. If it does not flow the metal is not hot enough to melt the solder so apply more heat.
Leave the solder to cool. If you have used an acid flux, thoroughly wash the join with soapy water. If the join is untidy, clean it carefully with a metal file until it is reasonably smooth.

Wire, washers and nails

Soldering allows you to work with a wide variety of metals, and most metals, apart from aluminium, are suitable for practising your soldering techniques. It is a good idea to start a collection of scraps, since most stores tend to sell metal in sheets and wire in coils – rather large quantities for the home craftsman or woman. Do not let this deter you – there is still a vast quantity of available materials which need not be bought in bulk. Nails, screws, washers, horseshoe nails and so on can be used creatively whether for jewelry or abstract metal sculpture. Spokes from old bicycle wheels and wire coat-hangers are especially useful. Collect any bits of scrap metal you might find lying around. Scrapyards are another good source for materials. They are often the best place to look for a specific piece of metal as you do not have to buy in large quantities. Do not be deterred by the colour of the metal. It does not matter if it is tarnished as this is easy to clean, but avoid metal which is flaked with rust. Test a piece of metal by cleaning it with steel wool and metal polish to see if the surface is suitable for whatever you have in mind.

Soldering jewelry

Many of the jewelry projects already covered lend themselves to soldering, which in many cases will produce a more professional finish. For example, if the design requires jump rings or a chain, the links can be soldered together. Solder the two ends of a jump ring together and when the solder is cool use a small metal file to tidy the join. Insert another jump ring into the soldered link and once again solder the two ends together and tidy the join with a file. Continue like this until the chain is the length you require. Small jump rings are best soldered with a small soldering iron. If you use a blowtorch adjust the inner flame to the smallest possible size and only heat the jump rings around the open end. Chains made with jump rings are fairly chunky and can be combined with pieces of metal to make heavy jewelry such as pendants – horseshoe nails are especially suitable.
Attractive jewelry can be made using horseshoe nails and wire, but

Trembling flowers

You will need:
Horseshoe nails—the number depends on how they are arranged on the washer. They can be the same size or you can alternate larger and smaller sizes
2 washers with 18mm-25mm ($\frac{3}{4}$in-1in) diameter
Jeweller's finding [pin backing] for a a brooch, or jump ring for a pendant, or length of wire on which to mount the flower for the sculpture. The wire can be cut from a wire coat hanger.
Wooden base for sculpture—the size depends on the number of flowers you wish to assemble on it
Glass beads (optional) and glue
Soldering equipment

instead of holding the nails together with wire, they can be soldered. If you are using a blowtorch the nails can be heated with the flame to make the bending and curling easier. Washers can be combined with nails and lengths of wire to make a variety of designs. Use a washer as a centre piece and add horseshoe nails (screws or ordinary nails will also do) to form 'petals' for a flower-like design (fig.1). This can be adapted as a brooch by attaching a jeweller's finding [pin backing] to the back, or as a pendant, or several can be mounted on to lengths of wire to form a metal sculpture. The loose ends of the wire are inserted into a wooden base to keep the flowers upright. Any strong base is suitable providing that it is heavy enough to support the assembled sculpture.

When working with wire and nails always prepare a full-sized drawing and bend the nails and wire as required. Assemble the pieces of metal on the design and make sure that the necessary contact is made to solder them together.

Trembling flowers

The metal flower can be adapted to make a brooch or pendant, or a number of flowers can be used to make the metal sculpture. The sculpture illustrated is assembled and brazed on to steel washers (see page 54). However, the flowers can equally well be mounted on a wooden base. To do this, drill small holes in the base to house the wire or hit a small nail into the wood and then remove it to make the necessary holes. If the base is fairly deep the wires will be held firmly in place.

Assemble the nails on a washer in the pattern desired. Clean the

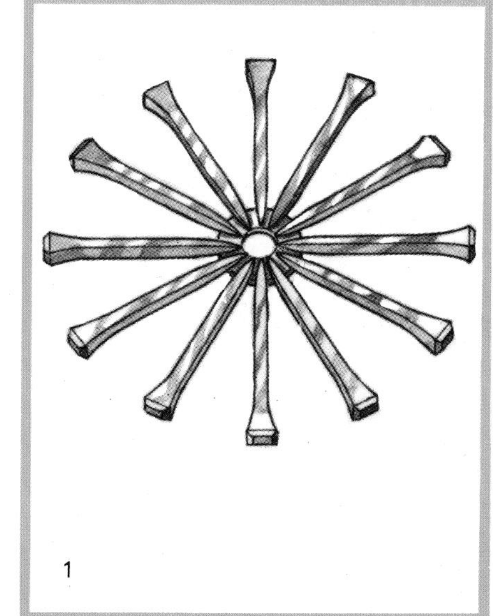

Horseshoe nails, washers and wire are combined to make the sculpture illustrated left. The diagram shows how each flower is assembled. The number of horseshoe nails used can be varied.

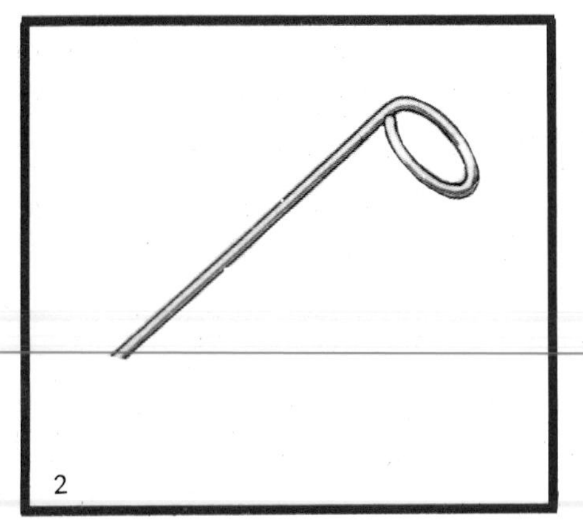

2 Strip of wire cut from wire coat-hanger shaped to form the stem of the flower

washer and nails to remove all traces of grease, rust and dirt from the parts to be soldered. Tin one side of the washer by applying flux to the metal surface and then covering with molten solder. Sweat the assembly together by applying heat. For this part of the project a blowtorch is more suitable than a soldering iron. Use more solder if necessary. Leave to cool.

Turn the washer over and solder a brooch pin to the back or solder a jump ring to the end of a nail to make a pendant. For the sculpture you need to solder a length of coat hanger to the washer. Soldering the back of the washer can cause difficulties. Heat applied to the back of the washer will cause the solder on the other side to melt which means that the nails will fall away from the washer if you are not careful. This is overcome by placing the nails face downwards on a flat surface so that if the solder does melt, the nails will not move and will remain in position once the solder on both sides cools. Prop the assembly on a metal washer, if necessary, to keep it flat. If the washer sticks to the nails after soldering leave it intact on the facing surface. The soldering iron (or blowtorch with copper bit attachment) is more suitable for this as it spreads less heat. Every time solder is melted and allowed to cool the temperature required to re-melt it is slightly higher. This means that once the front of the washer has been soldered it will need a higher temperature to re-melt it which also helps to prevent the front from falling away while the back is being soldered.

To solder a piece of wire to the back of the washer, bend the end of the wire into a loop (fig.2). Solder the loop to the washer and push the other end of the wire into a hole in the wooden base of the sculpture after the solder has dried and the joints have been cleaned and tidied up if necessary. Cleaning is very important and will take as much time, if not more, than the actual soldering process. For a professional appearance always remove any excess solder (with a file if necessary). Rub surfaces with fine steel wool and use metal polish for a shiny finish.

Napkin ring

Cut a strip of tin from a suitable tin can. Remove the label and clean the surface. Shape the strip of tin with the pliers into a circle or hexagon and solder the two ends together. Then solder lengths of wire to the strip, creating a random effect or according to the pattern illustrated, whichever you prefer. When soldering the wire around the soldered ends of the tin, hold the assembly together with a pair of pliers until it is cool otherwise the tin strip will come apart along its seam. File any rough edges and rub surfaces with fine steel wool for a smooth finish.

Napkin ring

You will need:
Strips of wire of various lengths cut from wire coat hangers
Tin can
Household pliers
Tin snips [Metal cutters]
Soldering equipment

Above: Stages in soldering the horseshoe nail flower. Below: napkin ring.

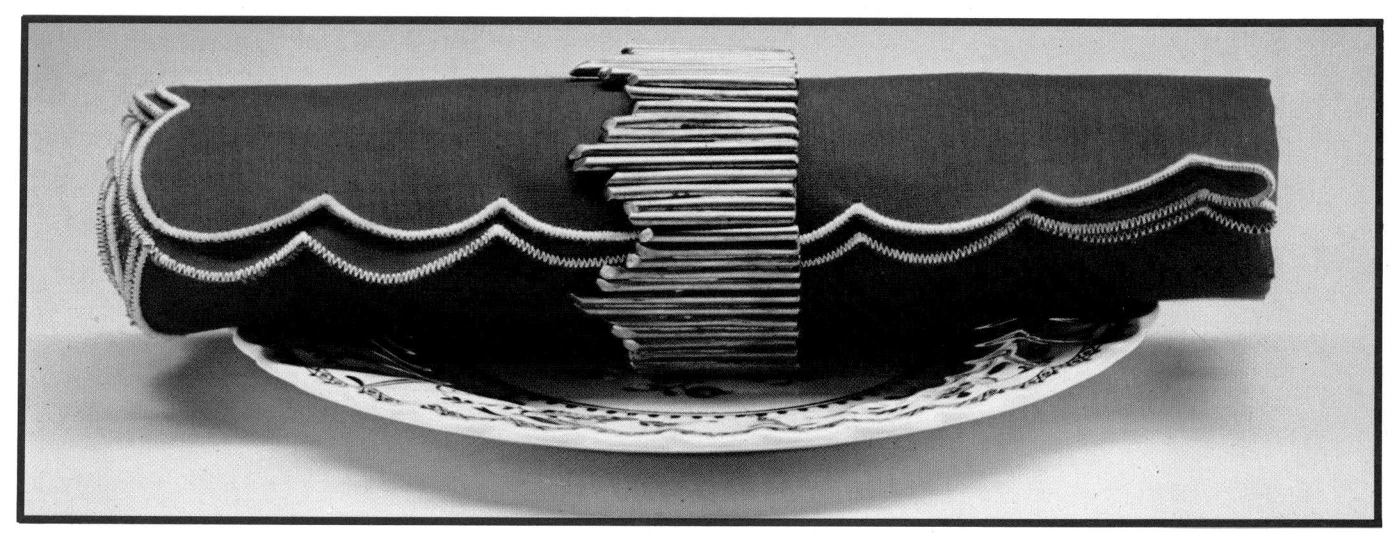

Pictures from nails

Nail panel

The nail panel is 30.5cm (12in) square and is so designed that you can put any number of them together and retain the flowing lines of the design.

You will need:
Hammer
Bradawl, or biro
Pair of household pliers—to remove any bent nails or nails that are out of line
Wooden base—chipboard [particleboard] or plywood will do–30.5cm (12in) square
Tracing paper
Carbon paper
Brass dome-head upholstery nails, 25
Brass cone-head upholstery nails, 30
Copper nails or rivets, 30, 31mm ($1\frac{1}{4}$in) long
Clout head galvanized nails, 227g (8oz), 25mm (1in) long
Clout head galvanized nails, 910g (2lb), 44mm ($1\frac{3}{4}$in) long
Wire nails with small heads, 455g (1lb), 44mm ($1\frac{3}{4}$in) long
Wire nails with large heads, 227g (8oz), 25mm (1in) long
Wire nails with large heads, 227g (8oz), 19mm ($\frac{3}{4}$in) long
Blue tacks, 113g (4oz), 12mm ($\frac{1}{2}$in) long
Blue tacks, 455g (1lb), 25mm (1in) long

Nails are taken for granted in most households. They tend to accumulate as more and more are purchased for various odd jobs. In themselves they are useful yet decoratively uninteresting; however, if several sizes of nails are taken and hammered into a wooden board to make a pattern they are transformed. The individual nails become part of an ornamental design consisting of ordered lines of metal dots. Whether the nails form geometric or abstract patterns is immaterial. The effect is like a drawing that is made up of a series of broken pencil strokes or spots of colour.

The process of producing these effects is so simple that you do not need to be a trained artist to achieve some spectacular results. Designs can be drawn with a pair of compasses, a ruler and a pencil; or you can improvise with some crockery for the curved lines.

The working drawing should be prepared on a piece of stiff tracing paper the intended size of the picture. The design is then transferred to the surface of the wood base. Nails of varying lengths and types are then hammered into the sections formed by the pattern. Some nails are left protruding more than others. The size of the nail heads and their colour can also be used to make the pattern more interesting.

You are not limited to a wood finish, but you should start with a plain wood surface to familiarize yourself with the materials and the technique. Once you have done this you can progress to variations. A mirrored polyester film can be glued to the wood and the design created on it, leaving some of the polyester film exposed to form part of the decoration. A thin sheet of copper can also be used to cover the wood base. Nail it in position, polish and varnish it before proceeding with a design. The tools and materials are easily available. A selection of nails from a hardware store is probably all you will have to buy if you do not have a nail box already.

The components of a design should not be too small as they will tend to lose their shape and break up the flow of the lines unless you can fill sections with small nails.

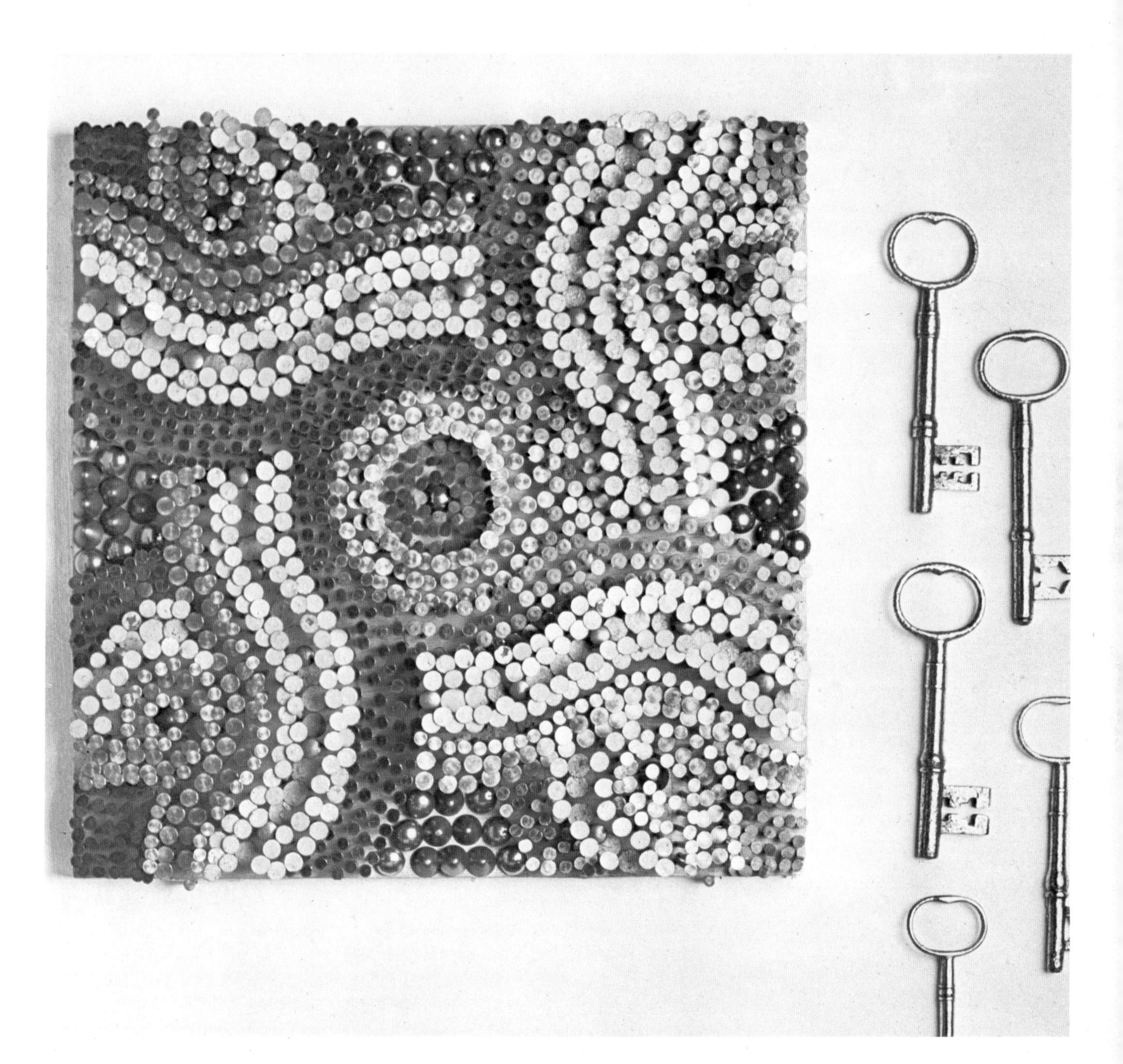

A good selection of nails is essential to make this attractive wall panel.

Nail panel

Trace the pattern of solid lines from the trace pattern four times to make up the complete design.

Place the trace pattern on the wooden base with carbon paper underneath it. Tack it in position and, using the bradawl or

1 The complete trace pattern is placed over the wooden base.
2 The design is traced on to the base using carbon paper.
3 The nails are hammered in, working outwards from the centre.
4 The circular centre completed.
5 Starting the surrounding area.
6 Nail heads are left at different levels as the pattern is built up.

biro, draw the line on to the base. Start filling the sections with the nails. A brass dome-head nail is hammered into the centre of section A and then surrounded by a circle of 12mm (1½in) blue tacks. Section B is filled with 19mm (¾in) wire nails and completed with an inner circle of 25mm (1in) wire nails overlapping the other nails. The four C sections are filled with 25mm (1in) blue tacks.

The four D sections are made up from a double row of 44mm (1¾in) galvanized nails adjacent to the C sections. Down the centre of D sections, five 44mm (1¾in) copper rivets are equally spaced to the edges. These are interspersed with groups of three 25mm (1in) galvanized nails. The D sections are then completed with another double row of 44mm (1¾in) galvanized nails on the edges adjacent to the four E sections.

The four E sections are made up of a double row of 25mm (1in) wire nails hammered into place next to the double row on the edge of section D. Section H is made up at all four corners of 12mm (½in) blue tacks hammered half their length into the wood. Section F consists of a central triangle of five brass cone-head upholstery nails.

On either side of this triangle, and adjacent to section K and the bottom edge of section C, three brass dome-head upholstery nails are hammered on either side of the triangle.

To complete sections K, J and I, hammer over the same pattern of nails found in the opposite edge of sections C, D and E so that two panels next to each other will match. Note section G is left blank.

The solid lines form one quarter of the trace design.

Silver soldering sheet metal

Silver soldering is a form of hard soldering used on small items such as jewelry. The technique is not only used on silver, it is suitable for other metals as well (see chart). It is an alternative to brazing and strong joints are easily achieved. If you have been using a blowtorch you will not need any additional equipment and you do not need a special working area either. Remember to take the usual precautions. Work on an asbestos mat placed on a heat-resistant surface such as fire bricks, turn down the blowtorch flame when you are not actually soldering, and point the flame away from yourself.

Materials

Whether you are soft soldering, brazing or silver soldering, apart from a heat source you will need some form of solder and flux. The solder and the flux will differ according to the temperatures being worked at and the metals being used.

Silver Solder is an alloy of copper, zinc and silver and, sometimes, cadmium. It is available in three types: easy, medium and hard – the last named contains the most silver and consequently has the highest melting point. The more silver there is in the solder of course, the more expensive it is.

Flux Borax is used as a flux and sometimes boric acid (see chart). Several proprietary fluxes can be obtained for the different solders but they are not essential as borax is suitable for most purposes and will generally prove less expensive.

Technique

The same preparations must be made for silver soldering as for previous soldering techniques. The metal around the joint must be cleaned with a file or an abrasive paper. The flux must be applied on to the parts that are to be joined. For good results try to hold or grip the metal firmly together in some manner during the soldering process so that you have both hands free.

Copper tea caddy [canister] with a cork lid

Use tin snips [metal cutters] in order to cut the sheet of the required size. The outline of the container can be marked on the copper with a felt-tipped pen before you start cutting.

Hard solder and flux selection guide

Metal to be joined	Solder	Flux
Brass or copper	Silver solder	Borax
Tin, zinc or lead	Not possible	—
Gold or silver	Gold solder Silver solder	Boric acid
Nickel	Silver solder	Special flux
Iron or steel	Brazing solder Silver solder	Borax
Cast iron	Brazing solder	Cuprous oxide

Silver solder is often cut into very small pieces and placed along the joint as opposed to adding it to the joint when the metal is hot as described in previous chapters. The advantage of this method is that it avoids wasting expensive solder and also reduces the amount of excess solder to be cleaned later. If you use this method remember to apply the flux before placing the pieces of solder in position. The flame must be used gently at first so that the pieces of solder are not blown off. After a short while the solder will adhere to the heated flux and the flame can be then increased. The silver solder melts and flows easily, the whole operation taking no time at all, so there is no stopping halfway through. If you work carefully and follow the above instructions there is little that can go wrong except with the amount of solder used and that is easy to estimate once you have done a practice joint or two. If you use too much solder it can be filed away.

Multiple soldering It is quite usual for several silver solders of different hardnesses (or melting points) to be used on one piece of work where several joins are being made. This becomes necessary if a piece of metal is being soldered near to a previously soldered joint as the solder used for the new joint should have a lower melting point than that used in the other joint. This will prevent the first joint from melting when the new joint is heated. In other words a hard solder is used for the first joint, a medium solder for the next, and then an easy solder. It is a particularly useful technique when doing fairly intricate work where several pieces of metal have to be joined in close proximity.

Copper container

The container shown is 10.5cm (4¼in) in diameter and 14cm (5½in) high. A lid can be made to fit over the top or you can use a cork lid, or cut one from a piece of wood. The inside of the container is varnished to seal the metal. This prevents the copper from oxidizing. The container can be used as a tea caddy [canister], and the size can be varied to suit your requirements. Copper is readily available and inexpensive, and making the container will give you some experience in silver soldering. By the time you have finished it you should be able to make a reasonably neat joint. Also the same technique is used when working with silver so this is a good foundation before embarking on more ambitious and expensive items in silver.

Cut the copper to the width required – 14cm (5½in) in this case – and the length about three times the required diameter – 31.5cm (12¾in). The smaller the container, the easier it is to make, so you may prefer to make a smaller container to begin with. The copper should now be annealed. This is a process used to soften metal: the metal is heated and then plunged into cold water, which makes it softer and more malleable. To anneal the copper, heat it with the blowtorch until the metal is a dull red. Pick up the copper with a pair of pliers and immediately plunge it into cold water. This will make the copper soft enough to be bent with the fingers.

Shape the copper roughly to a cylinder so that the edges butt. Flatten along the joint faces by placing the copper between the pieces of waste wood and using a hammer or mallet – the wood protects the copper from indentations from the blows of the

Copper container

You will need:
- Copper sheet .9mm (gauge 19-20)—a fraction either way is immaterial. To make the size shown here you will need a piece of copper 30.5cm × 46cm (12in × 18in). This is sufficient for the lid as well as the container.
- Binding wire—thin galvanized wire, brass picture hanging wire, or similar
- Strips of waste wood
- Piece of pipe around which to shape the cylinder
- Metal file and tin snips [metal cutters]
- Hammer or wooden mallet, and pliers
- Wet and dry emery paper—medium and fine
- Fine steel wool and metal polish
- Blowtorch
- Medium and easy silver solder
- Flux—borax

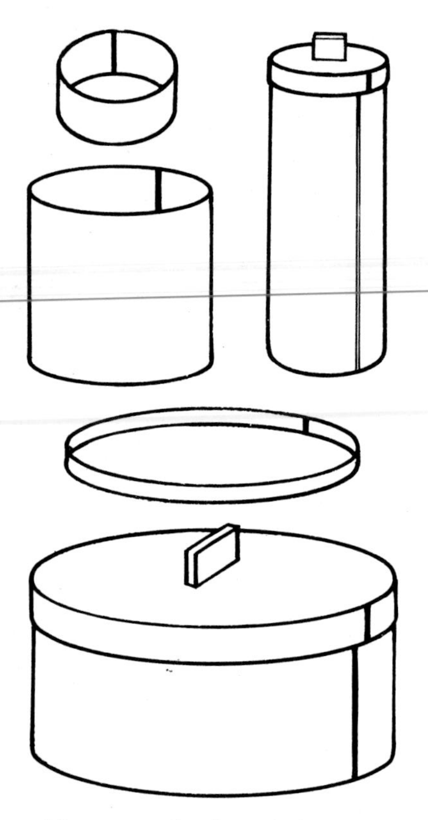

Various cylindrical shapes with copper lids. The handles on the lids are secured with solder.

hammer (fig.1). Run the joint faces along the file (fig.2). It will be easier if you clamp the file in a vice or with a G-clamp [C-clamp]. Butt the edges together again and repeat the process of flattening and filing until the joint is perfect.

Make a small loop in a piece of wire and tie the wire around the cylinder twisting the ends together. Push a pencil through the loop and turn it to tighten the wire around the copper so that the edges meet. Twist the wire until it pulls the copper tight along the joint (fig.3). Do not worry too much about how circular the cylinder is at this stage as long as the joint fits neatly. Mix borax with a little water and apply it to the joint along the inside. Place small pieces of medium solder along the joint and heat until it flows. Leave the copper to cool. If you have used too much solder and the joint looks untidy you can smooth it down carefully with a fine file.

The cylinder can now be made more circular by placing it over a suitable cylindrical 'former'. This can be a piece of thick dowel or a length of metal pipe (fig.4). If the copper is trapped between the pipe and a hammer or mallet it will stretch just in that area. By gradually working along the cylinder and rotating it so that every part has been worked, this local stretching will form a perfect cylinder. Do not rush this part of the operation, since the final appearance depends on this careful and even hammering.

The bottom end of the cylinder must be made perfectly flat. To do this rub the bottom on a piece of medium wet and dry emery paper placed on a flat surface. Cut a rough circle for the base of the container – at least 6mm ($\frac{1}{4}$in) larger all around than the bottom of the cylinder. If the joint between the base and the cylinder is not a perfect fit, soften (anneal) the base as described previously and flatten it between two pieces of wood. Stand the cylinder on the base and flux the joint. Add pieces of solder (easy) to the outside of the joint and heat until the solder flows. Leave to cool. Cut off surplus metal around the base with tin snips or metal shears. File down to exact size of the cylinder.

Clean the copper with medium and then with fine wet and dry paper. Finish with fine steel wool and metal polish. A lid can be made for the container. A circle of wood can be cut or a cork top is also suitable. To make a copper lid, repeat the soldering process already described. The lid must be made slightly larger to fit around the container. The base (now the top) is put on as before. The copper surface can be textured by using an engraving tool, a punch, or a hammer. Alternatively the hammered effect made while shaping the cylinder can be accentuated to form a decoration. If a perfectly plain surface is required special care must be taken to form a perfect cylinder before soldering.

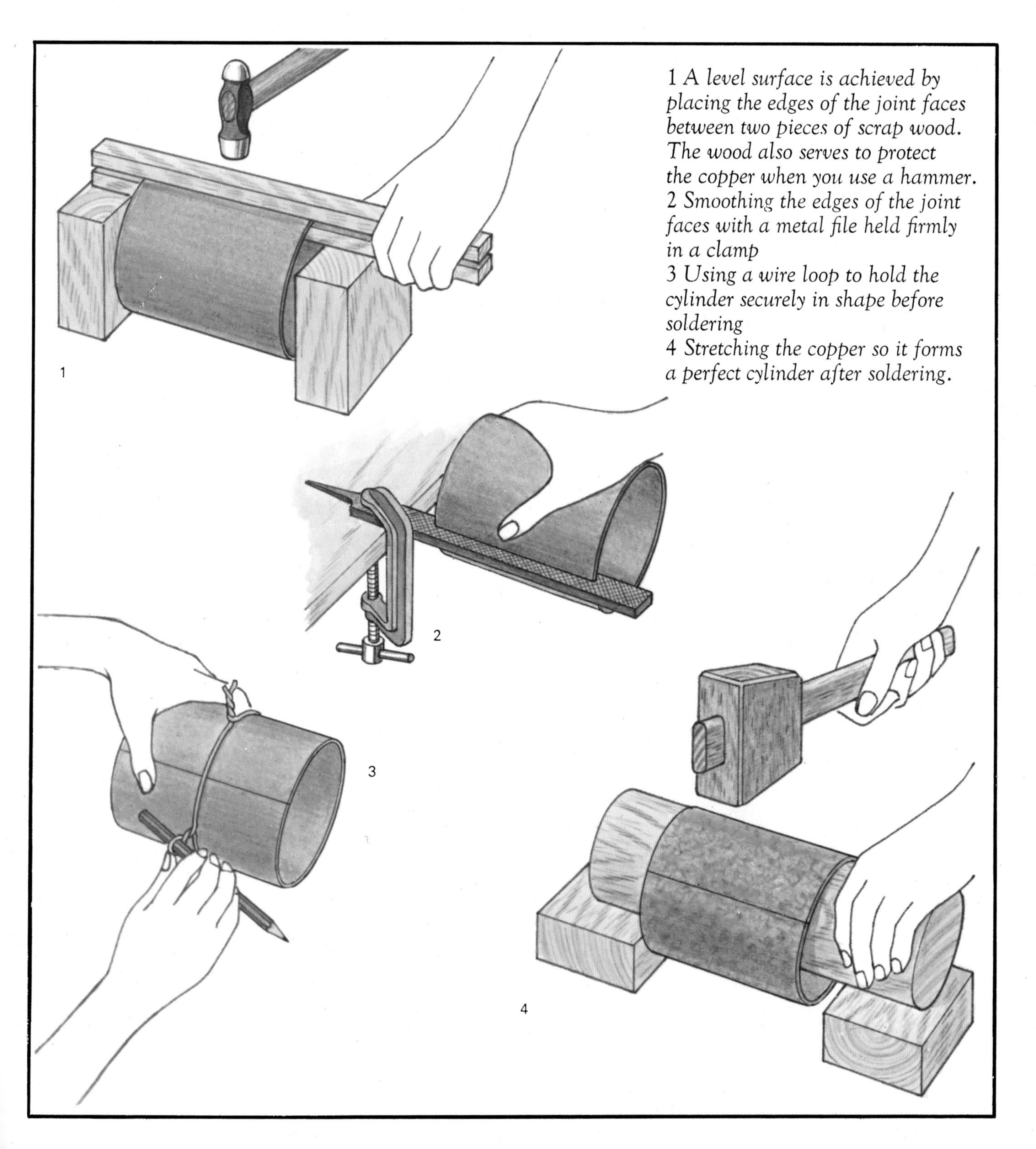

1 A level surface is achieved by placing the edges of the joint faces between two pieces of scrap wood. The wood also serves to protect the copper when you use a hammer.
2 Smoothing the edges of the joint faces with a metal file held firmly in a clamp
3 Using a wire loop to hold the cylinder securely in shape before soldering
4 Stretching the copper so it forms a perfect cylinder after soldering.

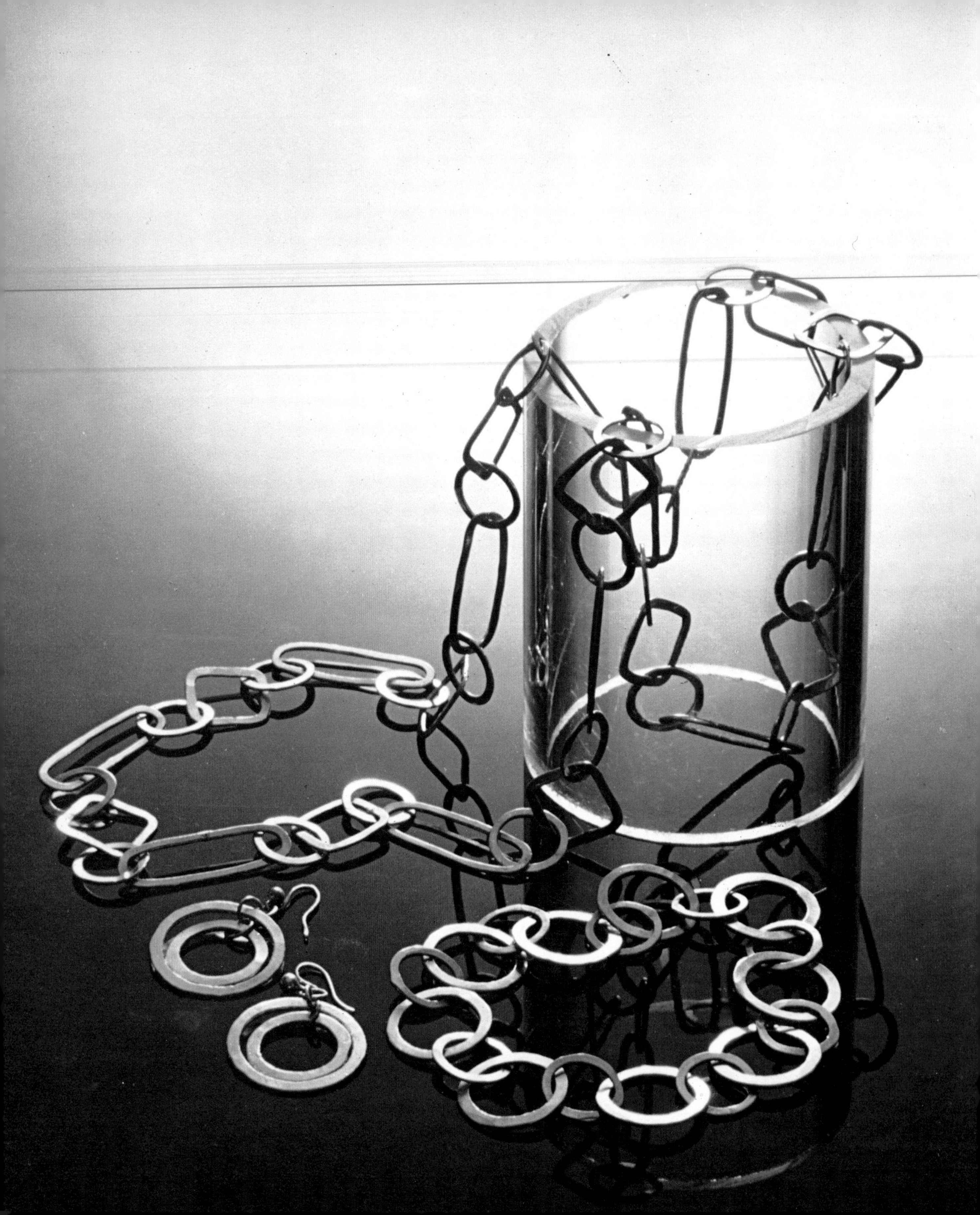

Soldered silver wire jewelry

Gold and silver can be joined by soldering and once you have familiarized yourself with the soldering techniques already described you can try your hand at simple silver jewelry. These articles are made from silver wire, cut and shaped into links which are soldered together. Care must be taken during the actual soldering to ensure that the joins are not too noticeable. Gold and silver wire are obtainable from silver and goldsmiths and are sold by weight, length and thickness (see page 69). Although round wire is used for the jewelry described here, square wire is also available. It is more expensive as it is heavier. Start working with silver wire before you progress to more expensive gold wire.

Opposite: Make this silver necklace, bracelet and earrings from soldered silver links for a beautiful matching set of jewelery.

Silver necklace

Annealing or tempering makes the wire soft and pliable so that it can be bent easily. Wind the wire into a coil about 5cm (3in) in diameter. Tuck the ends in securely to stop it unwinding. With the blowtorch heat the wire all over, taking care to move the flame steadily around the circle of wire as too much heat in one place might melt or distort the metal. When the wire is a dull red colour, remove the flame and cool the wire in water.

To make the links take 60cm (24in) of the tempered wire and wind it tightly around the piece of square dowel as near to the end of the wood as you can. With a piercing saw cut through the wire, pushing the coil to the end as each link is separated (fig.1). Repeat this with 80cm (32in) of wire wound around the round wooden dowel and with 75cm (30in) around the ruler. You will need 12 square, 12 oval and 24 round links.

The square and oval links must now be soldered separately to make them whole. The round links are not soldered till later. With the metal file, file the cut edges of the link flat so that they fit together snugly. Solder will not 'jump' or fill gaps caused by rough edges. Cut the medium solder into very small pieces by cutting down the strip with tin snips [metal cutters] then cutting across as finely as possible (fig.2). Place several links on the asbestos and heat gently with the blowtorch. Using a fine brush, drop a little

Silver necklace

You will need:
2.15m (7ft 2in) of 1.6mm (gauge 14-16) silver wire
A slightly domed hammer and a steel block to use as an anvil
15cm (6in) length of round wood dowel, 10mm ($\frac{3}{8}$in) in diameter
15cm (6in) length of 10mm ($\frac{3}{8}$in) square wood dowel or beading
Wooden ruler about 2.5cm (1in) wide
Piercing saw, fine metal file and tin snips [metal cutters]
Asbestos mat
Tweezers and fine brush
Alum
Medium solder and an acid flux such as borax
Fine emery paper, tripoli polish, jeweller's rouge and a fine cloth
Household detergent and an old toothbrush
Soldering tools and equipment

borax flux on each join. The metal should be just hot enough to make the borax splutter. Pick up a piece of solder with the tweezers and place it on the join. Repeat the procedure with the other links. Heat with the blowtorch until it is red hot and the solder melts. Remove the flame immediately and move on to the next link, soldering all the links in the same way.

Make up an alum mixture of one level teaspoon of alum (bought from a chemist [drugstore]) and half a pint of water. Drop the soldered links into the mixture to remove the flux. This is called a 'pickle' and is used to remove the hardened borax from silver jewelry after soldering. Finally remove any excess solder with a fine file.

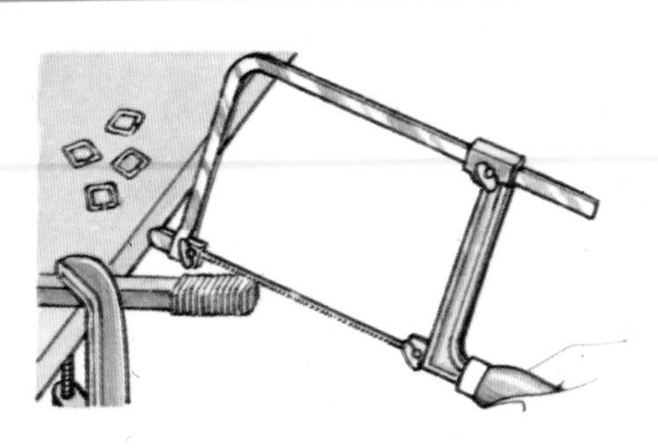

1 *Cutting links from the coiled wire with a piercing saw*

The links now need to be hammered. Place one of the links on a steel block – you can use a large, flat pebble or the underside of an old iron held in a vice – and hammer gently, moving it around so that it thins evenly. Turn the link over and hammer the other side, just enough to give a beaten surface texture. Do not heat too hard as this will make the link too thin. Repeat the procedure with all the oval and square links.

The round links are now attached. They go between each square and oval link and great care must be taken when soldering these to ensure that they do not get soldered to the others. Thread a square and an oval link on to a round link. Hold together the ones not to be soldered and place on the asbestos mat. Keep the round link join as far to the opposite side of the other two as possible, and solder (fig.3). Join all the links in this way until the chain is complete. The round links must now be hammered to match the others. This must be done carefully by holding the square and oval ones on either side in your fingers and extending half the round one over the edge of the flat surface, hammering and moving it round continuously. The chain is then ready for polishing. If you want a very high polish, start by rubbing the article with a fine emery paper, then polish with Tripoli on a soft cloth. Finish off with jeweller's rouge. Should you want to retain the beaten surface appearance of the silver, use only the Tripoli and jeweller's rouge. If, however, parts of the link which have been filed look scratched then first smooth out the scratches with emery paper.

Wash the whole chain in detergent, scrubbing gently with an old toothbrush to clean off excess polish. Dry on a towel.

Bracelet

The bracelet is made in exactly the same way as the chain but without the square links. Use round dowelling of 10mm ($\frac{3}{8}$in) and 12mm ($\frac{1}{2}$in) diameter to shape the links. You need 83cm (33in) of 1.6mm (gauge 14-16) silver wire to make nine large links and nine small. Anneal, solder, hammer and polish as with the necklace so that the links are joined in alternate sizes and make a complete circle. Like the necklace, the bracelet has no fastening catch as it is just big enough to slip over the hand.

Earrings

The earrings are made with similar links to the bracelet but a larger dowel, 16mm ($\frac{5}{8}$in) diameter, is used with 12mm ($\frac{1}{2}$in) diameter dowel. Four main links are made from 30cm (12in) or 1.6mm

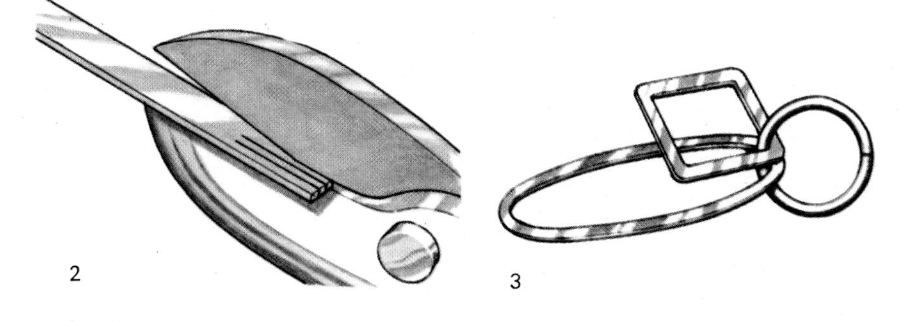

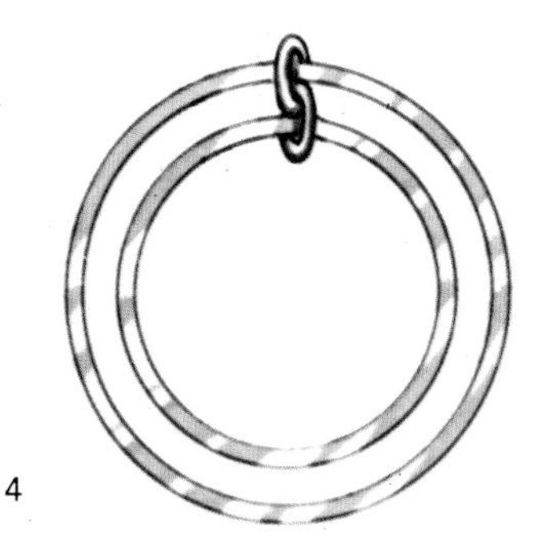

2 Using tin snips [metal cutters] to cut pieces of solder
3 Soldering the round links
4 Linking the earrings with jump rings

(gauge 14-16) silver wire. These are joined with jump rings made from about 10cm (4in) of .9mm (gauge 19-20) wire. Alternatively, jump rings can be bought from jeweller's supply shops, but it is much cheaper to make them yourself.

Anneal the wire and make two links of each size, solder and hammer separately until flat. You will need six jump rings. To make these wind the thin wire around a No.9 [US 5] knitting needle and cut to form a link. Do not hammer these. Put one jump ring around each link and link them together so that the smaller of the hammered links sits inside the larger (fig.4).

Now solder the jump rings. Great care must be taken when soldering as the joins are so close to the other links. Use very small pieces of solder, keeping in mind that as the jump rings are so much smaller than the other links, they will melt sooner. Remove the flame as soon as the solder flows or the rings will melt. Put another jump ring on the outside of the earring and through the top jump ring, then solder.

Polish both earrings and decide on which fittings you want. You can use either hooks or wire, or wires with butterfly backs, which are suitable for pierced ears. If screw fastenings are used, they will have to be linked to the jump ring before the jump ring is soldered. These accessories are available from jewelry supply shops. As you perfect new techniques your jewelry will become more and more professional-looking, so use high quality findings.

As your technique improves, more and more intricate designs become possible using silver. This symbolic charm necklace, designed by Pat de Menzes, incorporates motifs cut from sheet silver, with silver wire used for surface detail.

Strips of plain or patterned silver can be used to make bracelets as well as rings. These bracelets are made from sheet silver and thick, round silver wire. Silver balls are soldered to the ends of the wire.

Soldering silver rings

Besides using silver wire to make jewelry, flat strips of patterned silver can be used to make rings. Ring bands can also be made from strips of sheet silver. The band ring is simple to make and has no adornments attached. It is possible to glue a stone direct on a flat ring mount, using a strong epoxy resin adhesive, but if you want to mount a stone properly a claw setting is more attractive and the stone is more secure. This chapter explains the techniques for both the simple band ring and the claw setting.

When you have mastered these techniques you will be able to make a variety of simple silver rings, which make beautiful presents, for a fraction of the buying price.

Buying silver

Silver for making jewelry can be purchased in different shapes, sizes and thicknesses. The thickness of silver (and other metals) is measured in terms of its 'gauge'. There are three different types of gauge system in use at the moment: the Brown and Sharpe – B & S – used in the United States, the Standard Wire Gauge – SWG – used in most countries, and the Birmingham Metal Gauge – BMG – which is still used in the United Kingdom.

The BMG system is usually applied to silver and gold but, in order to avoid confusion, specify which gauge you are referring to when ordering. If you know the thickness (usually in millimetres) of the silver, mention it as well when ordering. Silver wire, either square or round, ranges from about 25mm (1in) thick to 3mm (0.12in). Silver tubing is available in a wide range of square or round seamless sections. The thickness of the silver used for the tubes varies. Sheet silver ranges between 12mm ($\frac{1}{2}$in) thick to 3mm (0.12in) and silversmiths will cut pieces to the required dimensions. Flat strips of patterned silver, called gallery silver, are available in a wide range of different widths and patterns. A selection of types of silver is illustrated on page 71. This shows all the types mentioned here, including patterned bracelet strips and gallery silver. The selection will vary according to your stockist and area, but suppliers will usually advise you when purchasing silver.

Silver band ring

You will need:
A length of thin string and tin snips [metal cutters]
Gallery silver—buy enough to go round the finger for which you intend the ring
Metal file
Ring stick or triblet
Wood or leather mallet
Asbestos block and some small pieces of asbestos or charcoal
Medium silver solder
Blowtorch, fine brush and borax flux
Pickle solution
Jeweller's rouge
Soft cloth and household detergent —for finishing and polishing

Silver band ring

This ring of gallery silver is easily made and is a good introduction to the fine soldering required in silver jewelry work. With a length of thin string measure round the finger for which the ring is intended. This is the length of gallery you require. Cut the gallery to the exact length and file the ends so that they are square with the top and bottom edges. This is to ensure a good join.

Place one end of the gallery about half way down the ring stick and hammer gently to bend it round (fig.1). Do the same with the other end and then work towards the middle of the gallery until you have a circle. Bring the ends together so that there is no gap between the solid strips in the pattern (fig.2) and place on the asbestos block. Prop the ring up with small pieces of asbestos or charcoal so that the join is on top. Place a small piece of medium silver solder where each solid strip is to be joined. Warm the ring gently with the blowtorch and brush borax flux on to the join. Heat the ring, moving the flame steadily around. When the ring is a dull red the solder will melt and form the join. Make sure you do not over-heat the ring as the silver is thin and may move or distort. Remove the flame as soon as you see the solder begin to run.

Drop the ring into the pickle and leave until any excess borax has disappeared, then rinse in cold water. You will proably find that the ring is not quite circular. If so, put it back on the ring stick as far down as possible without it sticking, and hammer gently turning the ring stick all the time. As the ring stick is slightly conical in shape the hammering tends to stretch the ring at the lower end. After going around the ring once, slide it off, reverse it on the ring stick and hammer around again.

The ring is now ready for polishing. This is done with jeweller's rouge applied with a soft cloth. Wash the ring with a detergent to get rid of any excess polish and polish again using a clean cloth.

Pictured opposite are some of the types of silver available. These include gallery silver in various patterns (top left), patterned bracelet strips (four types are illustrated here), square and round tubing, wire in various thicknesses, and sheet silver.

Ring with claw setting

You will need:
6mm (¼in) wide strip of sheet silver, .8mm thick (gauge 20-21)—enough to encircle the finger for which you intend the ring.
22mm × 30mm (⅞in × ⅛in) rectangle of sheet silver, .8mm thick (gauge 20-21) for the setting.
A flat-bottomed, cabochon shaped stone, 20mm × 15mm (¾in × ⅝in).
Pair of dividers [calipers], jeweller's piercing saw, and square edged pliers.
Easy and medium silver solder
Other tools and materials as for the band ring

Ring with claw setting

The method described here is the easiest way to set a stone on a ring. The ring is made from a 6mm (¼in) wide band of sheet silver long enough to fit round your finger. The support for the stone is 22mm × 30mm (⅞in × 1⅛in) sheet silver.

Choose a cabochon shaped stone that is flat on the bottom. In the illustration an oval stone has been chosen, 20mm × 15mm (¾in × ⅝in). If a larger stone is used, you must make the band around your finger slightly wider to balance the overall proportions. The pattern for the setting is easily adapted to suit either an oval or round stone. Make sure that the bottom of the stone and the setting are flat so that there are no unsightly gaps when the stone is set.

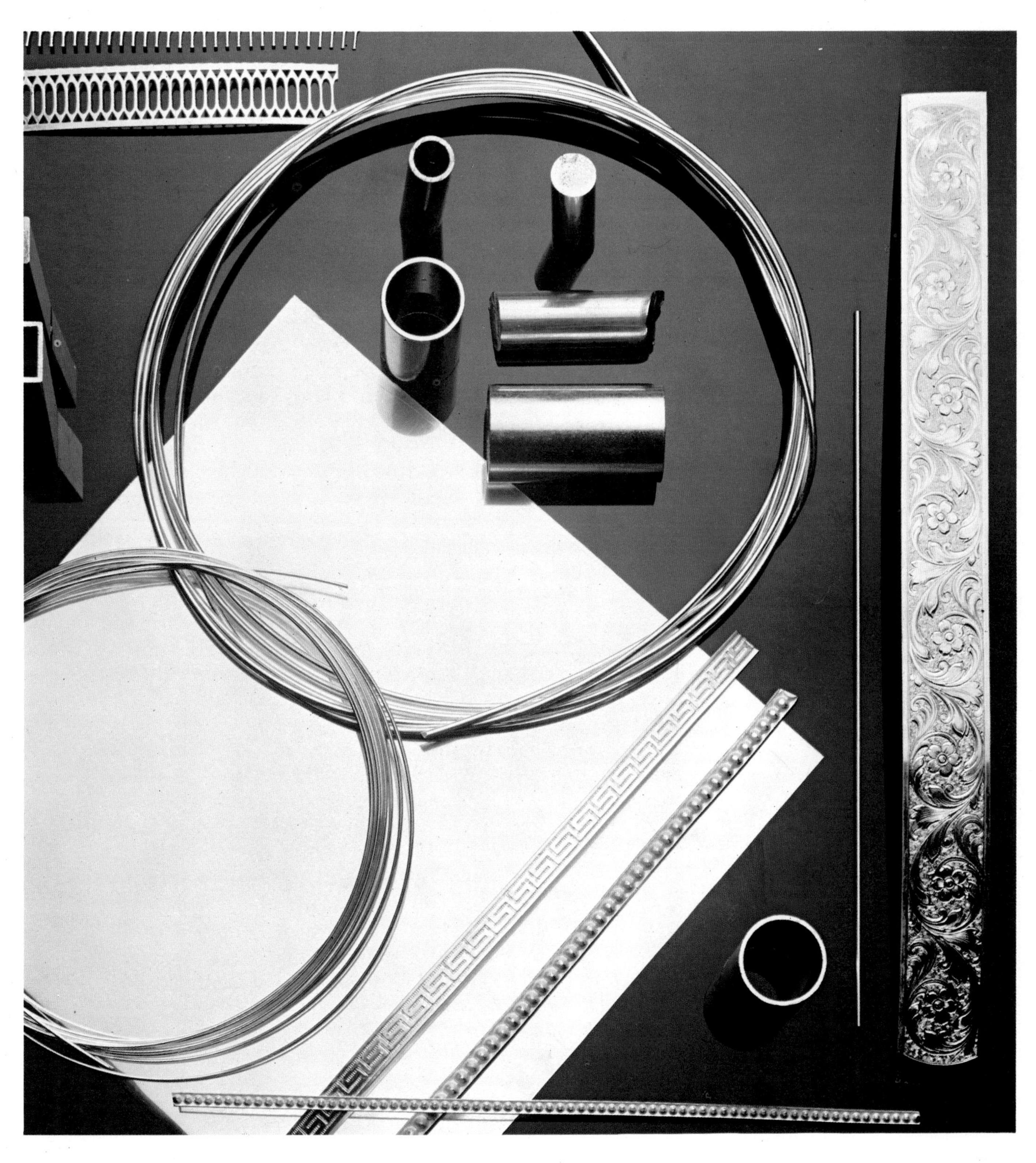

Cut the silver to be used for the band. If the edge of the silver sheet is straight, simply measure off the correct length with a pair of dividers [calipers] and cut with tin snips [metal cutters] or a saw. Hammer each end on the ring stick as for the band ring and make a circle with the ends joining.

Lay the ring on the asbestos mat. It is not necessary to prop it up as only one piece of solder is used which will run down the length of the join. Warm the silver slightly and brush borax on to the join. Put a small piece of medium solder on the join (fig.3). Heat the ring gently, moving the flame around all the time to get an even heat. When the solder begins to run, remove the flame. Allow to cool and place in the pickle. Rinse in clean water and polish.

The base and claws of the setting are cut from the rectangle of sheet silver in one piece. Place the stone in the middle of the silver piece so that there is at least 3mm ($\frac{1}{8}$in) to spare around it. Draw lightly around the stone with the point of the pair of dividers or a very sharp pencil. Mark off 3mm ($\frac{1}{8}$in) around the outside of this line (fig.4). Divide the oval into six roughly equal segments (fig.5). Draw the triangles for the claws.

The band ring and the ring with claw setting

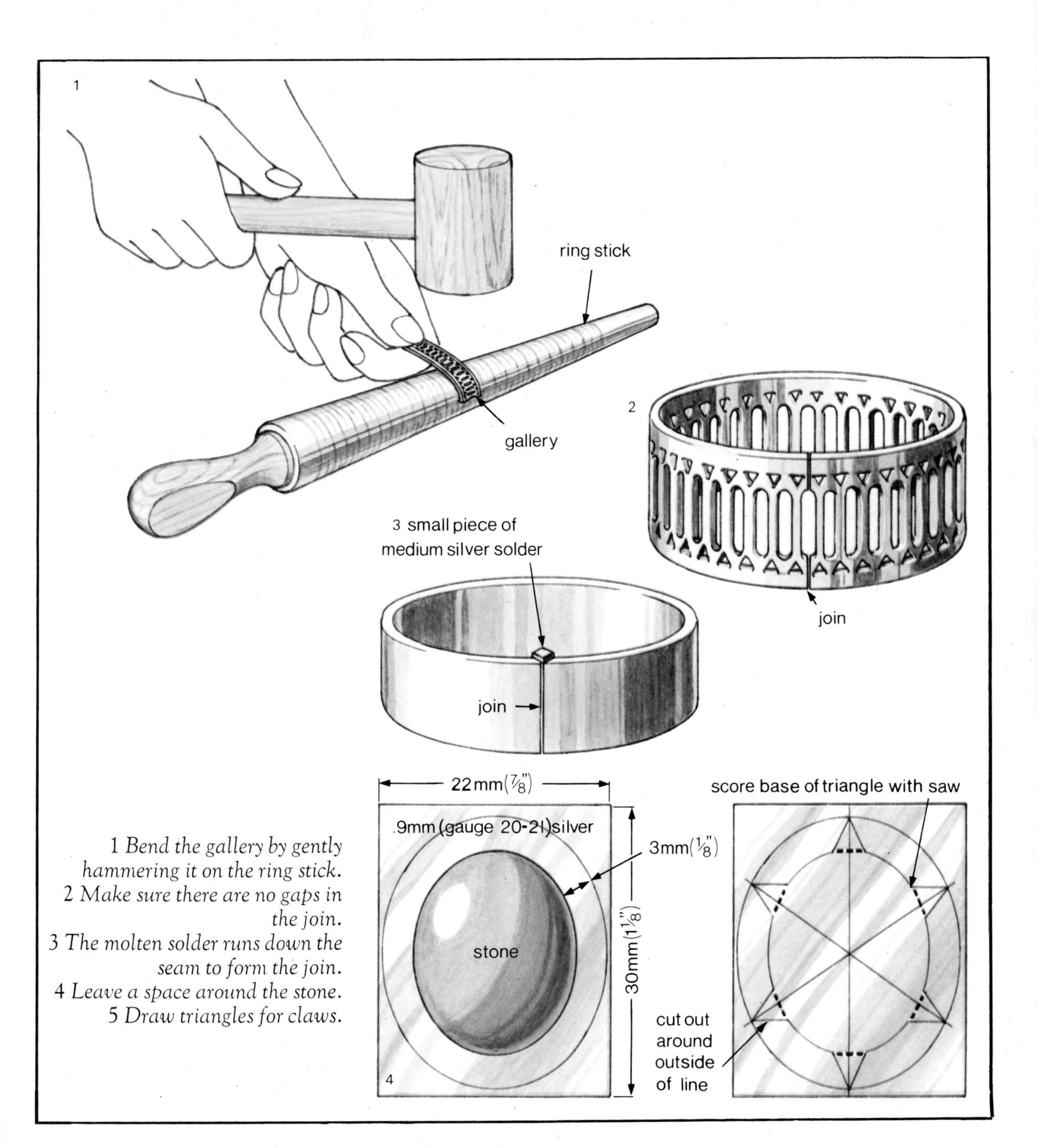

1 Bend the gallery by gently hammering it on the ring stick.
2 Make sure there are no gaps in the join.
3 The molten solder runs down the seam to form the join.
4 Leave a space around the stone.
5 Draw triangles for claws.

Using the jeweller's piercing saw cut out the setting. Cut on the outside of the line as any excess can be filed off later. With the saw, carefully cut a shallow crease across the bottom of each claw at the point where it will bend up to hold the stone (see fig.5). This helps to make the angle of the fold a right angle, but do not cut too deeply as it will weaken the claw. Place the piece, saw cuts face down, on the asbestos mat. Measure carefully where the ring is to join the setting and mark with a pencil.

Make sure that the ring is round by hammering it on the ring stick. Then file the outside of the seam flat. This ensures that there is enough silver in contact with the base of the claw setting to make a good join. The filed area need not be more than 3mm ($\frac{1}{8}$in) wide and makes it possible to stand the ring upright when soldering. Borax the area where the ring is to be joined on the base as well as the section of ring which has been filed flat. Balance the ring in position. Put a small bit of easy (otherwise called 'soft') silver solder on either side of the join and heat. Move the flame around, making sure that the setting is as hot as the ring. Easy solder runs more quickly than the medium so work carefully, taking the heat away as soon as the solder begins to run. Pickle and rinse. File away excess solder and rough edges around base.

Place the stone on the setting, making sure that the saw cuts at the bottom of the claws lie just outside the edge of the stone. File away any of the setting that protrudes beyond the stone. Remove the stone before filing to avoid damaging it. The stone should now fit flush on the setting. File the claws so that they are all the same size with no rough edges. With square edged pliers, bend the claws gently upwards so that they stand at right angles to the base.

Hold the stone in position and press the claws over the edge of the stone until it is held firmly. Press down opposite claws one at a time. This can be done with the flat edge of a pencil. Make sure there are no gaps between the claws and the stone, then polish with jeweller's rouge. It is best to polish each piece before soldering as this makes the final polish much easier.

Stones for jewelry

Lapidary (collecting and polishing your own pebbles and semi-precious stones) is an increasingly popular pastime which combines well with the skills covered in this metalwork course. You can mount your stones in silver-plated wire rings, tooled pewter settings or in soldered silver jewelry as here. But using ready-cut and polished stones is almost as satisfying and they are not expensive – a flat-backed oval cabochon is the ideal stone for a claw setting. You can combine stones with enamel work too.

Enamel

The art of enamelling

Enamelling is a craft dating back to the fifth century B.C., when the Greeks used enamelled gold inlays to decorate sculptures. In early examples of Greek and Celtic enamel work only opaque enamels were used, and it was not until the twelfth century that glowing, translucent colours appeared in the work of the Gothic enamellers. The craft gradually developed to a highly sophisticated level, and fine examples were produced by the jewellers of Elizabethan England and in France in the eighteenth century. By the early years of this century, the enamellers of the Russian firm of Carl Faberge were producing work of incredible precision and technique. Many enamel artifacts are now museum items, and museums and exhibitions are the best places to go if you wish to see what craftsmen have accomplished on this fascinating medium. Many beautiful works of art today are created in enamel, as these examples demonstrate. Moreover, with modern equipment it is one of the simplest techniques for making unique and practical decorative surfaces, as well as exquisite jewelry. Once the techniques of enamelling have been mastered, there are many ways in which you can use enamelled plaques to decorate your home – as splashbacks along small areas behind kitchen or bathroom sinks, on box lids, ashtrays, drinks trolleys, or purely as wall decorations. An enamelled surface is heat resistant, and splashes can easily be wiped off.

Metalwork and enamelling are skills which can be used together in many ways. You can cut your own metal blanks for jewelry (see page 23), progressing from copper to gold and silver as you become more experienced. And the more advanced techniques of cloisonné and champlevé require a certain skill to fashion the metal for a truly professional finish. When you first start working with enamels it is worth remembering that small pieces of enamel jewelry can be fired using a blowtorch or bunsen burner. But for anyone wishing to take up the craft seriously, a kiln will be required. The superb results that can be obtained will soon repay the initial outlay.

Enamel is fundamentally coloured glass, usually ground to a powder. This powder is spread on to a clean metal surface and

The enamelled pieces illustrated opposite and above show what can be achieved when more advanced techniques have been perfected. The scrolled wall plaque combines transparent and opaque enamels. The poppy design uses line technique, and the beautiful pendant is typical of the high standard achieved during the Art Nouveau period.

A wall mural made in four separate panels using very advanced techniques

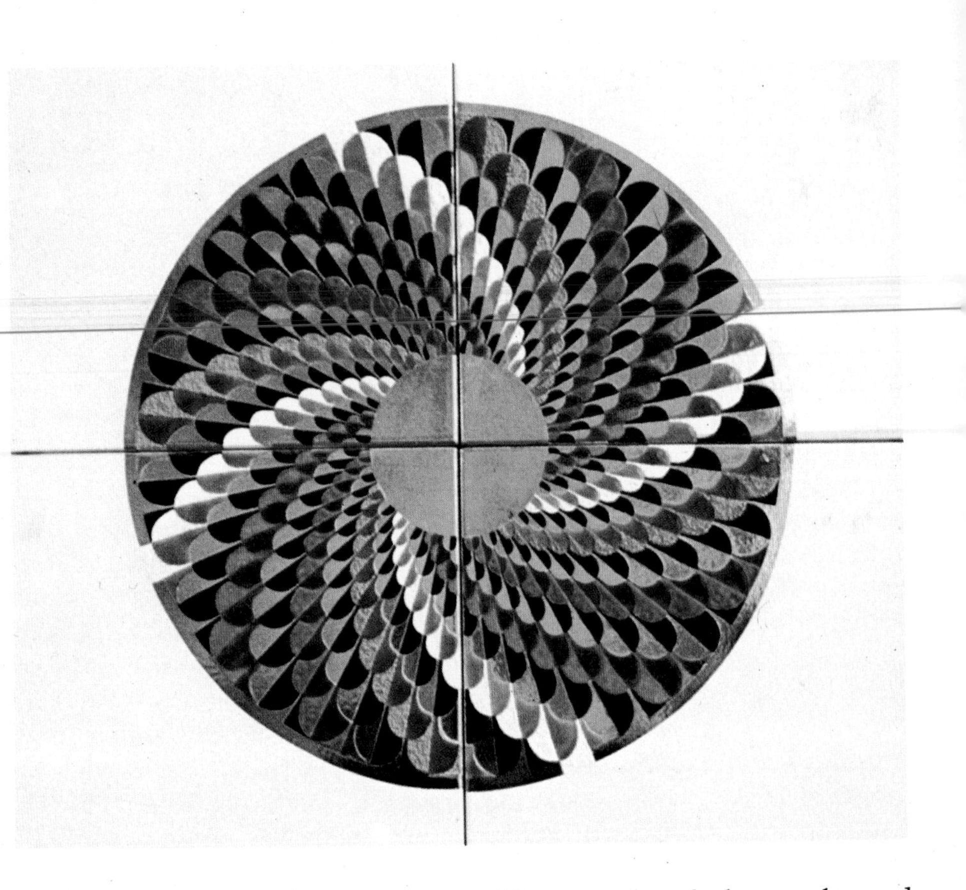

fired in a kiln at high temperature. The powdered glass melts and fuses on the metal creating a hard, smooth surface. The brilliance of the colours used in enamelling gives it its distinctive appearance. Ready-made enamel powders of good quality save much time and effort in tedious grinding and are available in many good colours and in varying quantities.

Once you have gained some experience you might want to grind your own enamel. Enamel in lump form – which has to be ground before use – has a longer shelf life than the powdered variety which tends to get mixed with dust and deteriorate. However, if you buy small quantities of powder and store them in containers with tight-fitting lids you should have no problem.

Generally enamel powders melt at a temperature of around 750°C (1380°F) but there are some with a lower melting point which are sold as soft enamels. If a higher temperature is required they are termed hard enamels.

Types of enamel

Opaque enamels normally have a high gloss, smoothly textured finish, but different textures – such as a coarser, low gloss surface –

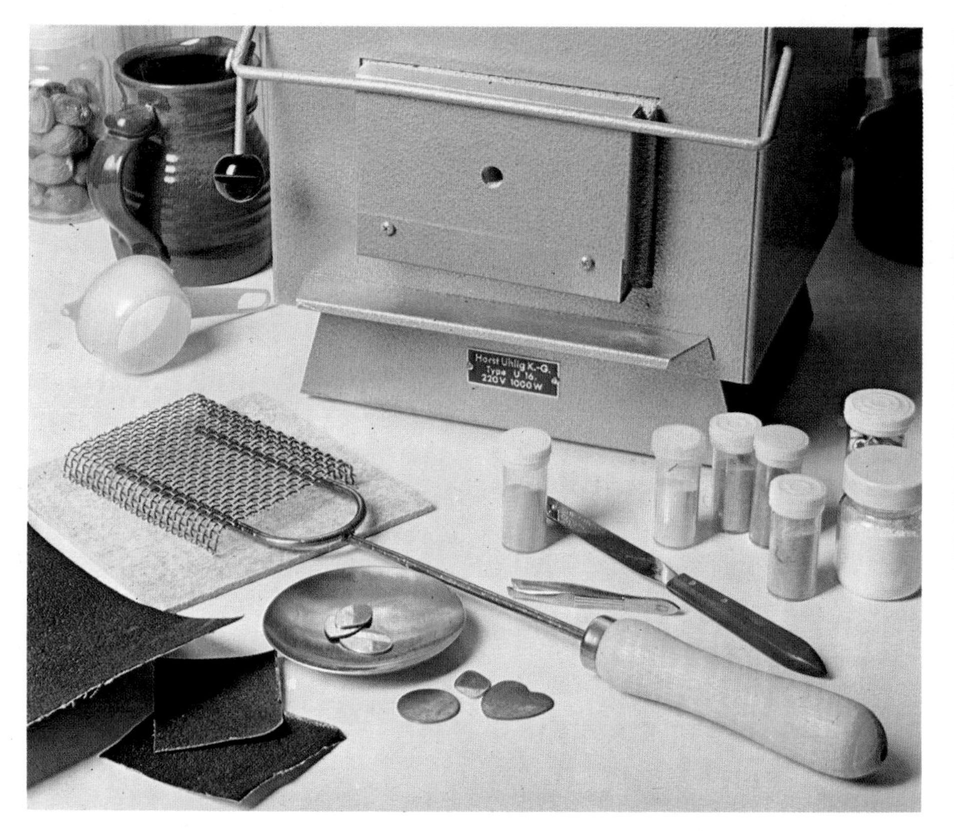

wrong

right

can be obtained by leaving the piece in the kiln over a shorter period of time. This process is called underfiring.

Transparent enamels melt at a slightly higher temperature than opaque enamels and are similar to coloured glass in appearance. **Transparent enamels** must be fired on a clear, colourless enamel undercoat – called a flux – otherwise their colour will be too concentrated. This is a slightly more involved technique than that of using opaque enamels. Transparent enamels cannot be underfired to vary their texture, though different types of flux can be used to give different effects to the transparent colours fired on them.

Self-cracking enamel This is a special variety of enamel powder that cracks or crazes like old porcelain. It is fired over an undercoat of whatever enamel you choose. Its textured surface looks old and interesting. For an example see the green dish on page 106.

Metal blanks

The metal blanks are the bases on which enamel powders are fired. Gold and silver provide the most reflective surface to use for transparent enamels, but as these are of course expensive you may prefer to use copper or gilding metal until you have acquired more

Above left: The materials you will need to do enamelling can be set up and used in a kitchen or on any suitable working surface. You will need a sieve, wire mesh stand, 2-pronged fork, palette knife, tweezers, enamel powders, metal blanks, emery paper and a kiln.
Above, top: Enamelling small test pieces. The copper blank covered with powdered enamel is placed on the wire mesh stand and then lifted into the kiln. The diagram shows the right and wrong way to powder a blank with enamel: the enamel should be slightly thicker along the edge of the metal than in the centre.

Basic enamelling equipment

Asbestos glove Although not always essential, particularly with modern kilns, an asbestos glove does protect your hands from the heat when opening and closing the kiln door or lid.
Long-handled, two-pronged fork This is useful for lifting pieces of work in and out of the kiln. You could improvise with an old toasting fork.
Sieve for sifting the enamel powders on to the metal blank. The sieve can be made of metal or plastic and should be very small. A coffee sieve would be quite adequate.
Emery or glasspaper [sandpaper] to remove the firescale which forms on the uncovered metal surface during firing. This cleaning is easier if you paint the metal surface with an anti-scale liquid before firing, although this is optional.
Steel wool pads or abrasive sponges are useful for cleaning and shining metal.
Brass tongs or long-handled tweezers These are used to grip the metal once it has been cleaned. You must not touch the metal with your fingers before firing as this will leave traces of grease on it which may spoil the finished result.
Stainless steel wire mesh stand These are specially manufactured for enamelling in a kiln. The pieces to be fired are placed on the stand which is then transferred into the firing chamber.
Palette knife for lifting the metal pieces on to the wire mesh stand—you could use a kitchen spatula for this.

expertise. Copper is good to begin with as it fuses well with the different powders and it is fairly soft so that you can cut and shape your own blanks. Stockists of enamelling powders usually carry a wide range of metal blanks that are suitable to begin with. They are available in various shapes so that you can easily fire small test pieces before you embark on more ambitious projects. Mica sheets can be used instead of metal blanks, but the techniques involved are difficult and best left until you are familiar with the materials.

Kilns

The largest single item will be the kiln. Think carefully about what sort of work you want to do and what size kiln you will require. The smallest kiln is suitable for pieces of jewelry, but if you want to do larger pieces such as bowls and panels you will need a kiln with a larger muffle (firing chamber). It is difficult to make more than one piece at a time, regardless of the size of the kiln, because of the amount of time and handling involved. A kiln takes a while to warm up to the required temperature – the larger the muffle the longer it will take to reach firing temperature – so bear this in mind too as running costs will be higher in the long term.
Some kilns are equipped with heat regulators. This is useful if you are firing gold and silver, which have relatively low melting points, but if you are using copper or gilding metal you will soon learn after a few test pieces and some practice to recognize the bright orange of the firing chamber which indicates the correct temperature. More details about the types of kiln available and their advantages and disadvantages are given on page 107.
Small pieces can be fired with a low torch or a bunsen burner so if you have either of these you can try this method. Remember that the enamels must only get the heat from the flame and not be touched by the flame itself. A piece of metal held over the flame, with the enamelled piece resting on it, will distribute the heat evenly. You will not be able to fire large pieces using this method, but it is a good way of learning how to handle enamels.

Working area

Your working area does not have to be large but it is essential to have the kiln on a heat resistant surface, preferably an asbestos mat, and the surrounding area should be fairly uncluttered. Place another asbestos mat next to the kiln on which to put the hot pieces. Cover the surface on which you work with sheets of paper. It makes it easier to collect any spilt powders and to clean up afterwards. Do not work too close to the kiln as the heat it generates can make you uncomfortable.

Simple enamelled shapes

Before you actually make anything it is useful to fire a few test pieces to become familiar with the equipment and the materials. Make a record of firing times and results. Even test pieces can be made into attractive brooches or earrings.

Plug in the kiln so that it can warm to the required temperature, about 750°C (1380°F). Prepare the piece of copper by rubbing it down with emery paper. The blank should be completely free of grease so do not touch it with your fingers once you have rubbed it down. Then, holding it with the tweezers, rinse and dry it without letting it touch your fingers. Apply the anti-scale liquid, if you have it, to the underside of the metal blank and leave to dry.

Using the tweezers, turn the blank face up and place it on a sheet of clean paper. Select a colour and sieve it on to the blank. The powder must be evenly spread with slightly more powder around the edge. Do not use too much powder – just enough to hide the metal from sight with no lumps or mounds anywhere on the surface.

Insert the palette knife or spatula carefully underneath the powdered blank. You might find it easier to do this if you keep the blank in position with the tweezers. Lift the blank on to the wire mesh stand. The powder left on the paper can be returned to its container if it is completely free from any dust or grit. Successful enamelling depends very much on the enamel powders and metal bases being scrupulously clean.

The kiln should now be a glowing orange colour which means it is ready for firing, Wear the asbestos glove if you feel safer with it and open the kiln door. Using the fork, lift the wire mesh stand and place it in the firing chamber as far back as possible, then close the door of the kiln.

The enamel powder on the blank will become darker and then begin to melt. Once it looks wet it is fired. Have the fork ready and as soon as the surface appears wet remove the wire mesh stand from the kiln. Put it on an asbestos mat and close the kiln door. The enamelled piece will change colour as it cools down. If the colour is not even once it is cool, you can sieve some more powder on to it – slightly thicker around the edges – and fire it a second time. First remove any firescale and apply more anti-scale liquid. Then apply the second layer of enamel.

Once the piece has cooled down completely, clean the back with a piece of emery paper. You can glue jewellers' findings [backings] to the back to make brooches and earrings from the pieces. It is a good idea to make a number of these small enamelled shapes to acquire some expertise in working with enamels before progressing to more ambitious techniques and projects.

Simple enamelled shapes

You will need:
Opaque enamel powders
Pre-shaped copper blanks, available from craft stores
Emery paper
Anti-scale liquid (optional)
Kiln and basic tools as described opposite.

Counter enamel and millefiore

Below: The piece, covered with enamel powder, is placed on the stilt for support. Contact between the counter-enamelled surface and the stilt must be kept to the smallest possible area to avoid spoiling the surface.

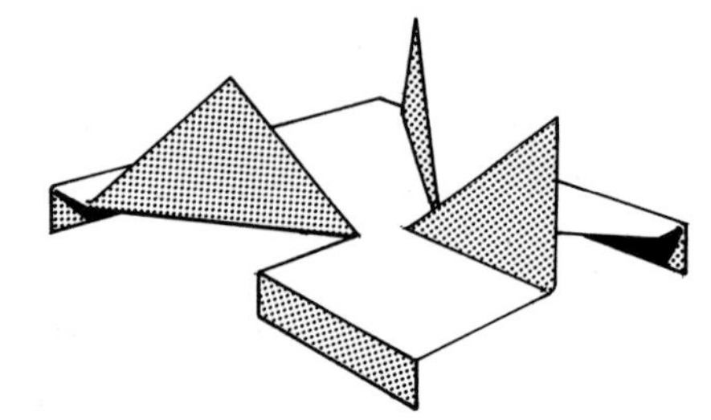

1 *A flat-bottomed stilt must be placed on a wire mesh stand. This stilt can be used without a stand.*

There are many techniques you can use in enamelling but, initially, there is enough fascination in the brilliantly coloured surfaces to inspire you to become familiar with handling and firing simple pieces. So don't rush to try out all the different techniques at once: it is wiser to perfect the basic skills first, which needs practice.

When a piece of metal is fired (subjected to extreme changes in temperature) the metal becomes permanently softer. So, once a piece has been fired, remember to handle it gently as it could bend more easily than you expect. The enamel, when fired, also effects the metal. When the metal is heated it expands and, once removed from the muffle, it contracts again. Because the enamel does not contract to the same degree, it causes the metal to curve slightly. This will not matter when you are firing small pieces but when you come to make larger pieces – especially if you want flat surfaces in order to combine the fired pieces to make up a larger piece – it does become important.

First of all, don't try to press it flat! To counteract this curving, the metal should be counter-enamelled – ie both sides of it must be enamelled. The enamel at the bottom will pull the metal to the same degree as the enamel on the upper surface and so keep the piece flat. It is also useful when a small piece is being fired a number of times to achieve a more intense colour effect, to prevent the metal curving to extremes thus causing the enamel to chip off.

Tools

Apart from the techniques and tools already described, all you will need to counter-enamel a piece is a tripod or stilt on which to rest the piece while it is being fired. You can buy a stilt specially made to form three points on which the piece can rest for the second firing (fig.1).

Counter enamelling

The counter-enamelled side of a piece is enamelled first and it forms the wrong side of the completed piece. Enamel powder made especially for counter-enamelling is available from craft stores but you can also use ordinary enamel powder. Counter-enamelling powder is similar to the enamels used for industrial purposes. It is extremely durable and because it is hard-firing – ie requires a high temperature to fire – the enamel will fire before the counter-enamelling becomes completely molten. Once you have counter-enamelled a blank you fire the second side with an enamel powder and the counter-enamelled side will remain intact.

Prepare one surface of a piece of metal by rubbing it down with a piece of emery or glasspaper [sandpaper]. The surface should be quite textured and criss-crossed with fine scratches. This will be the wrong side – ie the counter-enamelled side eventually. Do not handle it with the fingers.

Using tweezers turn it over and apply anti-scale liquid. This is not essential but it does make it easier to remove the firescale – in fact it should flake off quite easily once the piece has been fired. Leave the anti-scale liquid to dry. Switch on the kiln if you have not already done so. Turn the piece over with the tweezers and sieve on the counter-enamelling powder. Fire the piece until the enamel looks wet and has an even surface. Remove the piece from the muffle and let it cool. If your kiln does not have a temperature regulator then leave the door open so that the kiln will not over-heat. Once the piece has cooled, remove the firescale with a piece of emery paper and continue rubbing until you have a scratched surface. To enamel the right side sieve the enamel powder on to the piece as before. (You do not have to use the same colour as previously). Place the tripod or stilt on to the wire mesh and place the powdered piece on it. Once a piece has been enamelled and the other side is to be fired as well, the stilt must be used, as the first enamelled side on which the piece will be resting becomes molten. So, to keep the surface intact the supporting points from the tripod or stilt should be as small as possible. Fire the piece, remove it from the kiln and leave to cool. If the colour is not even, use more powder and fire again. There is no reason why both sides should not be equally attractive. Fire a few small pieces first, remembering to remove all the firescale after each firing. If you are careful about placing the counter-enamelled metal on the stilt and keeping it clean, both sides will be perfect.

Counter-enamelling is especially useful when making jewelry such as a bracelet which is worn next to the skin. The counter-enamelling will prevent the metal from discolouring the skin, which it would do otherwise.

Later on in the course you will see that it is possible to enamel both sides of a piece at once by using adhesives, but this can be a tricky operation for a beginner.

The pieces are enamelled and then millefiore or Venetian glass beads are placed on the upper surface and fired again to create flower designs. The pieces should be counter-enamelled to prevent the enamel from chipping.

2 The millefiore is placed on the enamel and fired. The longer it is left in the kiln the flatter it becomes.

Millefiore

Millefiore are enamel rods of different colours which have been joined and then cut into 'slices' to reveal a flower-like design on the flat ends of the slices. These can be bought in varying sizes and colours. Working with millefiore is fascinating, and you can achieve really beautiful results even from early efforts.

To apply the millefiore first fire the pieces until they are complete – ie both sides are fired. You should always counter-enamel a piece when you use the millefiore. Powder the piece again, place the millefiore in the required position and refire. Remember to rest it on the tripod otherwise you will spoil the underside. The firing time will take longer than previously as the millefiore and the enamel powder have to combine and run into each other.

The heat will cause the millefiore to melt and run flat on the existing surface. Fire it until it has the finish you want (fig.2). The powder will become molten and then fuse with the millefiore. The longer you leave it in the kiln the flatter the millefiore will become as it spreads. Once it is flat and smooth remove the piece from the muffle and leave to cool.

Transparent enamels

Transparent enamels show the surface of the metal on which they are fired or, if fired without a metal base – a very difficult process – they allow light to pass through them. The beautiful colours combined with the transparent qualities make these enamels very exciting. They can be used in the same manner as opaque enamels to make jewelry and napkin rings etc. With opaque enamels it is the solid brilliance of colour that is appealing but with transparent enamels the colours are less intense and they appear to be set below the surface, producing a quality similar to that of semi-precious stones.

Generally transparent enamels are hardfiring, ie they require a high temperature to fire, often higher than the temperature required for opaque enamels. So, despite their transparent qualities, these enamels will stand up to a certain amount of wear and tear, depending on what they are used for. However, enamels are basically glass so they can break when dropped. They should not be subjected to extreme changes in temperature as the expansion and contraction can cause the enamel to crack and chip.

Flux

Flux, when fired, forms a clear base on which to fire the transparent enamels. This gives the enamel more depth while still showing its colour. The flux is not essential – the transparent enamel can be fired directly on to the metal – but, as it improves the transparent quality of the enamels, it is usually used. Flux is powdered enamel and available in different grades which fire at different temperatures. Start with what is called a normal flux and do some test firings. It is a good idea to keep the tests you make so that if you change the type of flux you are using you will be able to compare the results. The different grades of flux affect the colour rendered by the transparent enamels. If you use a very hard flux, for example, it has an orange tint and will affect the final colour of the transparent enamel fired on it.

Soft flux requires a lower firing temperature than the enamel and is most useful for repair work. If a fired enamel is slightly burned

This enamel and silver brush set is a fine example of 1930s enamelling. The silver base has a patterned surface which accounts for the rayed effect.

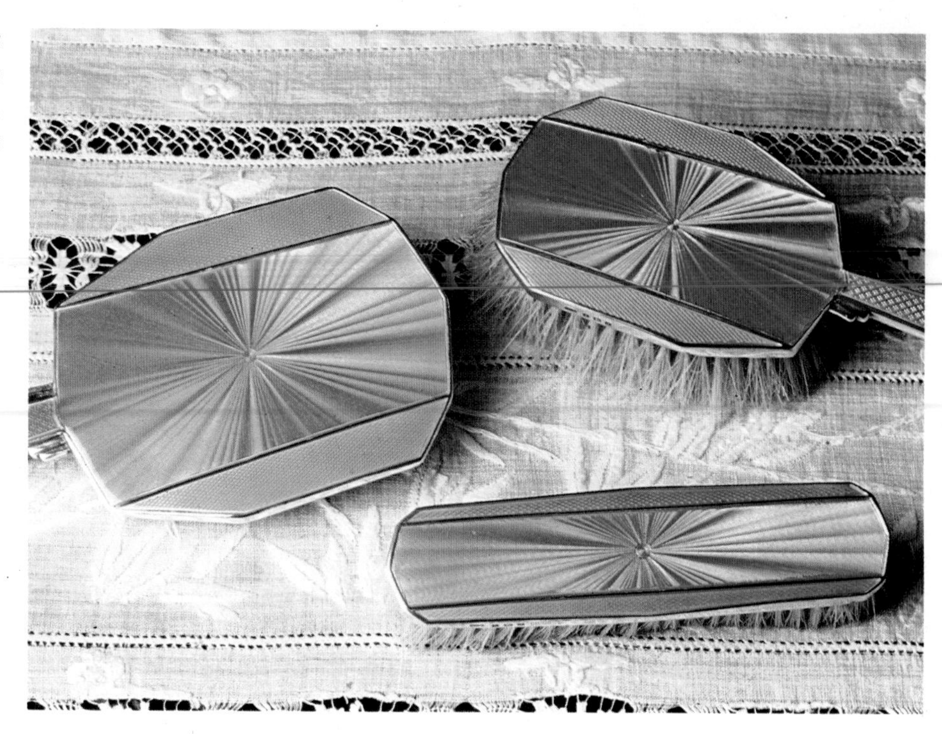

around the edges or has an uneven surface, fire it for a short period until it is red hot. Remove it from the kiln and immediately powder it with a soft flux. The enamel will be hot enough to melt the flux which will form a clear glaze over the enamel.

Test firing

Test firings are essential. The pieces you test fire do not have to be large and are therefore inexpensive. They simplify later work when you will want to know the colour values of a particular powder.

The colour of the metal blank will, of course, show through the transparent enamel. Some transparent enamels might look completely different when fired on copper, silver or gilding metal. Keep tests of the various results so that in future you can refer back to them when planning a design. You can attach the sample firings to the enamel powder containers for easy reference. It will save you making the same tests at a later stage and will also prevent any chance of not achieving the effect you want. Make two firings for each colour – do one firing with flux and the other without. The flux influences the depth of the reflective qualities of the enamel and for some colours the flux will not be necessary – this is largely a matter of taste. Also, as different brands of enamel will give varying results, always do test firings when using a new brand of powder to check how it handles and fires.

Counter-enamelling

Counter-enamelling is not essential for very small pieces but transparent enamels are seldom fired without it (see previous chapter). If the flux is fired thinly and evenly and if the transparent enamel is applied in the same way, counter-enamelling is not essential. But firings like this are seldom satisfactory as they tend to be flat and pale in colour. To get more depth and colour you need to apply the transparent enamel fairly thick or do two firings – the latter method being recommended. To achieve this successfully the metal must be counter-enamelled or else the enamel will flake off. The enamel flakes off because the contraction of the metal is different from that of the enamel. The combined thickness of flux and enamel when cooling stresses the copper, causing the enamel to crack. Counter-enamelling will prevent this.

The flaking or crackling of the enamel can also be prevented by using a thicker gauge for the metal blank. But, as this is not always possible and can be expensive, counter-enamelling is used to counteract the stress between the cooling metal and the enamel which can ruin a carefully fired item.

If an enamel should crack, throw it away or hit it with a hammer to remove all the enamel, then clean the metal and refire it. Do not run your finger over the cracked enamel. It is extremely sharp and will shave the skin from your finger tip.

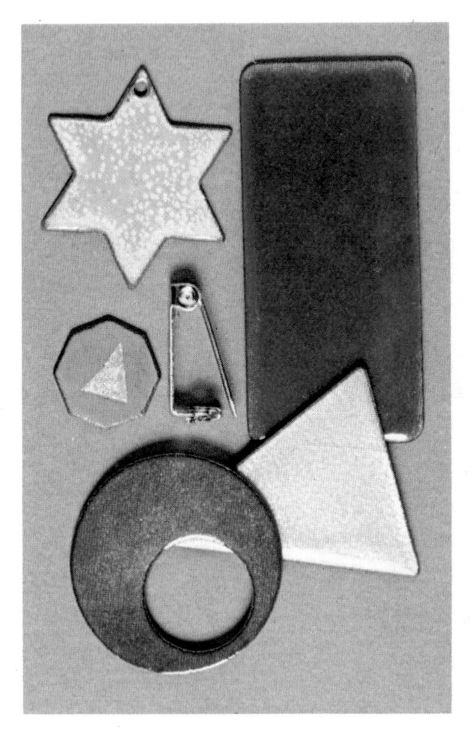

Transparent enamels fired on copper. The octagonal pendant was fired over a layer of silver foil.

A transparent enamel brooch

Switch the kiln on so that it will be hot enough, ie bright orange, by the time you have prepared the metal blank. Rub down one side of the blank with the emery paper until it is shiny and covered with small scratches. Paint the other side with anti-scale liquid and leave it to dry. Sieve the counter-enamelling powder on to the scratched metal surface and fire it. When it is cool remove the firescale. It should flake off if you used the anti-scale liquid. Clean, shine and scratch this surface with emery paper. Sieve the flux on to it. Using the spatula, lift the metal on to the stilt and place this on the wire mesh stand. It is better to use the flux fairly thick because, if it is too thin and fires unevenly, it cannot be corrected – you will have to start again. Fire the flux in the same way as for the enamel powder. Remove from the kiln and leave to cool.

Once it is cool lightly rub the sides of the metal with emery paper to clean it but be careful not to damage any of the fired surfaces. Sieve the transparent enamel powder on to the flux base evenly. Place this on the stilt and the wire mesh stand, and fire. If the colour is too pale you can add more powder and refire it or, if the colour is uneven, you can also add more powder and refire.

A transparent enamel brooch

You will need:
Transparent enamel powders in any colours you like. Keep them in airtight containers as moisture or damp will ruin them.
Metal blank
Brooch finding [backing] from a jeweller to attach to the back of the metal blank
Counter-enamelling powder—optional
Normal flux
Emery paper
Anti-scale liquid—optional
Equipment as already described

Leave it to cool completely and then glue the brooch finding [backing] to the counter-enamelled side. Always compare your results with the test samples you have made and keep a record of the results and the number of times the pieces was fired to achieve a particular colour. This will be very useful to refer back to when you want to combine different colours.

Firing on metal foil

To give transparent enamels more interest you can fire them on silver or gold foil. Silver and gold are ideal metals to use as a base for enamelling on but are very expensive. So, to obtain the effects of these precious metals, you can use silver or gold foil. They are available in sheet form, are extremely thin and fairly inexpensive. The results are most attractive as the foil has a more reflective surface than copper or gilding metal and therefore shows the transparent enamels at their best. Start with silver foil and experiment with small blanks.

Begin by counter-enamelling the blank. Clean the top surface with emery paper and apply a coat of enamel powder and fire it. This surface must be smooth without any imperfections. Fire it a second time if necessary until you are satisfied with the result.

Cut the foil the shape of the blank but make it slightly larger so that the edges of the foil overlap all round the blank. Wet the blank and smooth foil over it. Using a carborundum stone (available at hardware stores) remove the excess foil by working it downwards against the edge of the blank.

Leave the blank on top of the kiln for a few minutes to dry. Fire it to fuse the foil with the enamel underneath. Watch while it fires and should the foil lift around the edges use the spatula to push it down gently. Remove it from the kiln and let it cool before applying the transparent enamel. You do not need a flux when you are using foil.

If you are firing larger pieces you must avoid trapping air underneath the foil. To do this prick the foil once it is in position with a pin or needle. If you do this the bottom layer of enamel should be flux or transparent. And again, if the silver foil lifts during the firing, press it flat with a spatula. Instead of water special enamelling oil can be used with the foil. With oil you will not need to wait for the blank to dry before firing. Beginners may find oil easier to use.

If initial results are not satisfactory try again. With practice you will be able to achieve beautiful effects with metal foil.

This is an effective technique which is well worth perfecting. Try using it for small brooches, or make a set of shiny transparent enamel buttons for a special garment.

Sgraffito and stencils

After firing single-coloured enamels you will now be familiar with the actual firing technique and counter-enamelling. However, if you start to combine different colours and do a few experiments you will come up with varying results.

For example, if you powder a blank thinly with one colour, say blue, and then powder it with yellow, the end result is not green, as you might expect, but a speckled combination of the two colours. Alternatively, if you fire one colour first and then fire another colour on top of it, the top colour, depending on the firing time, would dominate it, showing specks of the initial colour used.

These colour combination can be very interesting, especially if you use different tones of the same colour eg a turquoise blue on a midnight blue. However, there are definite ways to control the colour combinations you want, rather than just haphazard results.

Sgraffito – The powder is sifted on to an enamelled surface and the design is then drawn in the powder with the end of a brush.

Position the stencil on an enamelling surface.

Sift powder evenly on to the cut-out area.

Remove the stencil with very great care.

Use a fine brush to remove any stray powder.

You can continue to make small enamel pieces for jewelry or buttons and if you want something larger, as a sampler, you can make a number of small square or oblong enamels and glue them together on a base, perhaps to make a large table mat.
To apply the techniques described you will need the equipment and tools described on page 80, and you will also need a stilt or tripod for counter-enamelling.

Sgraffito

Sgraffito is an Italian word meaning 'scratched'. In enamelling, the technique of sgraffito produces a pattern or design by scratching lines or shapes in the enamel powder, before the firing, to show another colour underneath.
You can experiment with opaque and transparent enamel powders but generally opaque enamels give more satisfactory results because they show stronger contrast and combination of colours. It is possible to do extremely fine and intricate designs with sgraffito, but you need a steady hand and patience for it.
If you have some enamels from previous firings that are not satisfactory, you can practice with these because it is possible to overcome imperfections on the surface by firing another layer off enamel on top of it. The first colour will eventually form the lines of the pattern and the second colour will be the surrounding area forming the bulk of the colour. So choose your enamels with this in mind. You can make abstract designs with simple lines or, if you are good at drawing, you can make almost any other pattern or outline you like.
To rectify an imperfectly fired piece decide on the colour you want to use and powder it, making it slightly thicker where the faults occur. Use a sieve for this, or attach a clean piece of muslin by means of an elastic band, over the top of an open jar and use the powder direct from the jar.
Take any pointed tool – the end of a water colour brush or a match-stick will do – and draw the design. The first coat of enamel must show so, to begin with, do not make these lines too fine. Make sure that the lines are of even width, if that is the effect you want, and also make sure that all the enamelling powder is cleaned off the lines of your design. Good strong lines and a fairly simple design are best until you are more familiar with the technique.
Fire the enamel. Watch it closely because you must remove it from the kiln as soon as it is fired and before the enamel begins to flow. If you over-fire the enamel it will flow, breaking the lines and making the enamel flat.
If you start with a new blank clean it by rubbing it with emery

paper. Counter-enamel the piece, then clean off the firescale. Apply the base coat of the design and fire. Apply the second colour to the base colour and draw your design. Fire the piece on the stilt, watching it carefully, and remove when fired.

Stencils

To create solid areas of colour rather than simple patterns created with lines you need to use a stencil.

Make a stencil from cardboard which is stiff but not thick. A scalpel or trimming knife is useful for cutting the stencil, especially for intricate work. Cut the desired shape from a piece of cardboard slightly larger than the blank you are using. Start with fairly simple shapes such as a square, circle or heart-shape, before trying anything more complicated.

Counter-enamel the piece, then fire the blank with the first colour. Place your stencil on the side fired top-side and powder the cut-out area evenly. You can use either the cut-out stencil or the cardboard surrounding the cut-out area. Whichever you use make sure the edges are neat and smooth. Remove the stencil carefully and, should there by any powder outside the area of the stencil, use a small soft brush to remove it.

Fire the piece again, and remove it as soon as it is ready. If you over-fire it the enamel will flow, breaking the lines and spoiling your design. So remove the piece as soon as the enamel becomes molten. By now you will recognize the point at which a piece is just fired.

Creating your own designs

There are many variations to the above techniques. You can combine stencilling with sgraffito, or you can use more than one stencil and a number of colours. Each different colour and stencil will need to be fired separately.

You need not be limited to cut-out stencils. Use any dried leaf or fern which lies flat and place it on the enamel to give you an outline. You will find the tweezers are useful for removing the delicate stencil motif and you can then use a pointed tool to tidy up any fine lines that are not clear enough.

Don't be afraid to experiment with all sorts of stencils, whether they are purpose-made or just odd pieces of metal or natural objects. The only limitations are those of your imagination. As you progress and learn different techniques you can combine them and adapt them as you please. You will already have experimented with different types of enamel and various colour effects. Some experiments will inevitably fail – but you will also create some original masterpieces.

General hints

Red enamels are slightly more difficult to perfect as they tend to burn when over-fired. Remove them from the kiln as soon as they are fired. Yellows are also fairly difficult to fire but most other colours are hard-firing and will withstand a number of firings.

Pitted surface and black patches
If these are formed it is because the enamel has been applied unevenly, and too thinly. This can be corrected by applying another layer of enamel, making it slightly thicker where the imperfections occur, and re-firing it.

Uneven enamel, without any signs of burns, means that the enamel is under-fired and should be returned to the kiln.

Store enamels in airtight containers. If the enamel is damp it will lift from the metal during firing. If the enamel contains any dust this will make holes in the fired enamel. Keep your working area as clean as possible and do jobs such as removing firescale well away from your powders.

Re-using blanks Enamels which are quite beyond repair are still not a complete waste. You can place the enamel between a fold of cloth and hit it with a hammer to shatter the enamel and salvage the metal blank for re-use.

Polishing the edges The edge of the enamels will be slightly black where the copper has formed firescale. You can remove this by carefully rubbing it down with a carborundum stone. It is a delicate procedure because you could damage or chip the enamel. However, it does improve the appearance of your work if you can show the copper edge of the blank rather than a black firescale edge, so it's worth the effort.

Coping with curved surfaces

Previous chapters in this course have covered various techniques which involve applying enamel powder on to a dry surface. It is also possible to dust the powder on to a surface which has been coated with an adhesive. This is essential when enamelling a curved surface, such as a bangle or bowl.

Bowls, ashtrays, napkin rings and bangles look most attractive when enamelled, but their curved surfaces present a problem if you are using enamel in powder form. Enamel powder sifted on to the dry, curved surface of a bowl will fall to the bottom, leaving the metal exposed on the higher surface. Similarly, it will fall off a bangle, leaving no powder on the metal at all. This problem is easily overcome by using an adhesive.

Adhesives

There are several water-based adhesives available, including tramil, gum tragacanth, gum arabic and cellulose wallpaper pastes which contain tragacanth.

Tramil can be bought ready to use from craft shops.

Gum tragacanth is available from craft shops ready-mixed or it can be bought in powder form from a chemist [drugstore] and made up at home but this is a lengthy process. Unfortunately, once mixed, gum tragacanth tends to ferment within a fairly short time. To counteract this add a little formaldehyde to the solution.

Gum arabic can be bought in liquid form from craft shops, and then mixed with distilled water (1 teaspoon of gum arabic to about 0.28 litre ($\frac{1}{2}$ pint) distilled water).

Cellulose wallpaper paste, which has a tragacanth base makes an excellent substitute, is readily available and is the choice of many experienced enamellers.

An adhesive can be applied direct on to the metal blank or on to a surface already enamelled (when applying a second coat of enamel). Apply adhesive smoothly to the surface to be enamelled (if you are applying it direct on to the copper this should first be prepared as usual). Use a sable or camel hair brush. Alternatively you might prefer to use a scrupulously clean finger tip as brushes

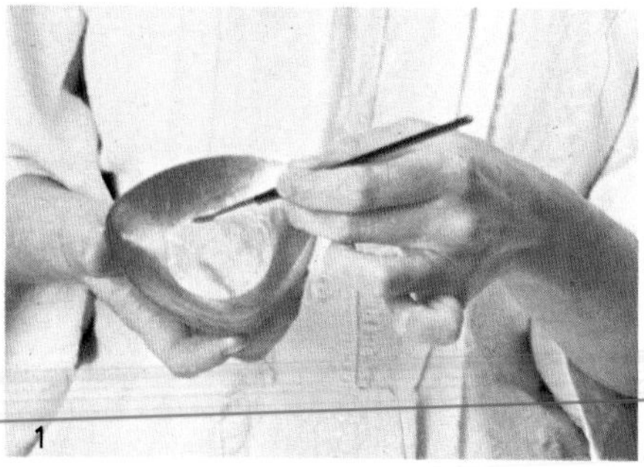

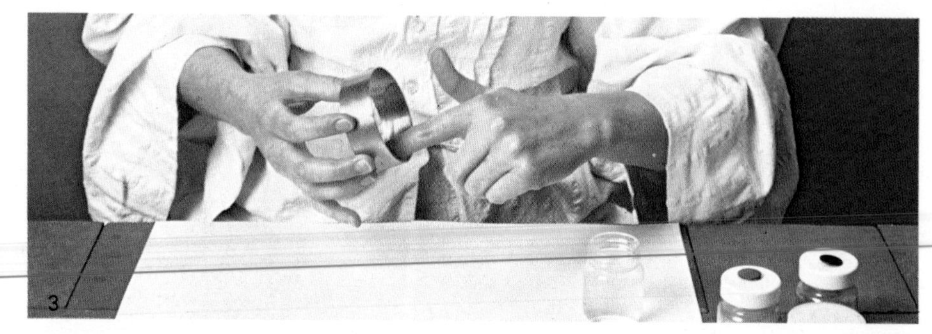

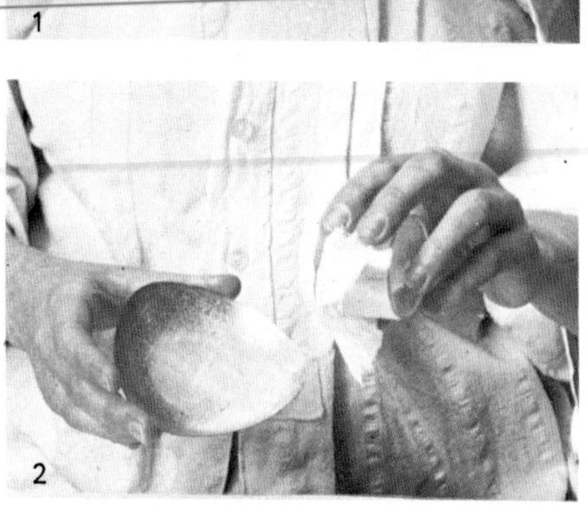

1 Applying adhesive to a curved dish using a sable water-colour brush
2 Sifting enamel powder on to the dish, tilting it to ensure even coverage
3 Applying adhesive to the inside of a bangle using a clean fingertip
4 Dusting the inside of the tilted bangle with enamel powder, working over clean paper

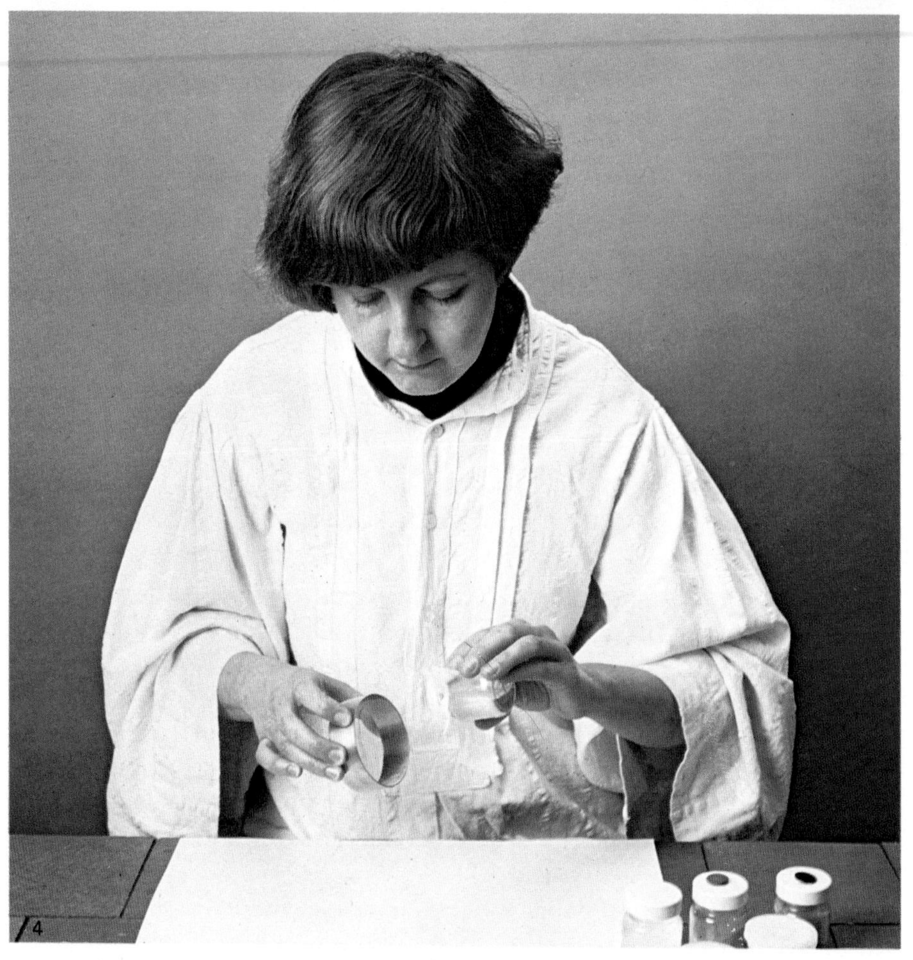

can creat air bubbles. Working over a sheet of clean paper, sift the enamel powder over the surface using a sieve. Or, attach a clean piece of muslin, by means of an elastic band, over the top of an open jar and use the powder direct from the jar. If the surface to be enamelled is curved, tilt the object so that the powder will adhere evenly to the whole surface.

Tip the powder which has fallen on the paper back into the jar. Place the object on top of the kiln for a while to dry out all the moisture before firing. If this is not done the moisture will lift the powder as it evaporates during firing.

Enamel bangle

This is not an easy technique and may take some time to master, so don't be discouraged by early results. When working on a bangle or napkin ring it is easier to counter-enamel the inside and then to enamel the outside.

Switch on the kiln so that it will be hot enough by the time you are ready to fire the bangle. Prepare the inside of the bangle for enamelling in the usual way. Paint the outside with some anti-scale liquid and leave it to dry.

Smooth adhesive on to the inside of the bangle and dust on the powder, turning the bangle as you dust so that the whole inner surface is evenly coated. Remember to work over clean paper. Place the bangle on top of the kiln until all the moisture has dried out, then place it on the mesh stand and fire.

When the bangle has cooled remove the firescale from the outside. It should flake off easily if anti-scale liquid has been used. Prepare the outside for enamelling and coat with adhesive, turning the bangle as before. Place the bangle on the stand and fire. Leave it to cool and then clean the edges with the carborundum stone.

Enamel bangle
You will need: Equipment as already described and a stilt Metal blank Enamel powder Adhesive Emery paper Carborundum stone Anti-scale liquid (optional)

Other uses for adhesives

If you wish to fire both sides of an enamelled article at the same time, prepare both surfaces, apply an adhesive to the underside and sift powder on to it. Place on the top of the kiln for the adhesive to dry out. Turn the object over and apply the adhesive (not essential unless the object is curved) and then sift on the chosen colour. Place it on the stilt with the underside down and fire.

It is not advisable to use red enamel on the top surface and another colour on the underside when working by this method as red, which is not hard-firing, tends to burn before many other colours are properly fired. (See general hints on page 92.) Another use for an adhesive is to hold powder in position when working with stencils or doing sgraffito (see page 89). When working with a stencil an adhesive makes it easier to obtain a neat edge. It also prevents small specks of powder from remaining on the surrounding area after the stencil has been removed because you can lightly blow these away without disturbing the unfired enamel powder on the design. As it gives you more control of outlines you are able to produce more intricate work.

Threads, lumps and embedding

This chapter explains some more of the decorating techniques which can be used when enamelling. These are the use of enamel threads, lumps or chips and simple embedding using copper wire, jump rings, odd pieces of enamel and even small nails.

Thread enamel

Threads of enamel, or thin rods, sometimes called enamel strings, come in a variety of colours. They are a fascinating medium for decoration. You can use enamel threads to make initials, geometric shapes, or purely haphazard patterns in rich colour combinations. The piece to be decorated is first counter-enamelled in the usual way. Next a thin layer of enamel powder is sifted on to the item and the threads arranged on this using tweezers to place threads exactly where they are wanted.

1 Bending an enamel thread over the low flame of a blowtorch.
2a Using a blowtorch to fuse threads together on a square of asbestos.

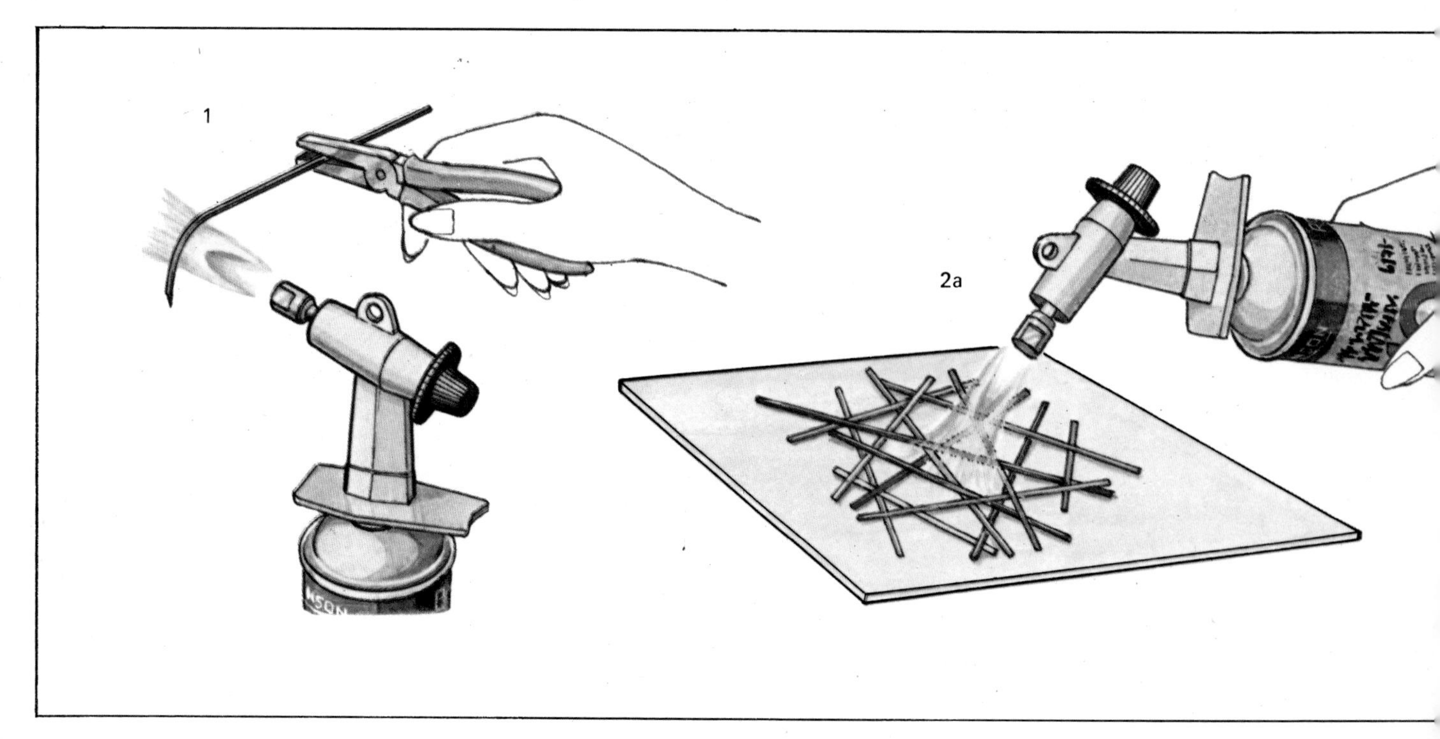

The simplest type of design is a linear one made from threads broken to the required length, but it is quite possible to bend threads. To do this pick up a thread of enamel with tweezers – or pliers – and hold it over the low flame of a blow torch. Alternatively you could use one of the burners on a gas stove. The enamel will soften and bend (fig.1) – sometimes rather suddenly. This process is not very easy to control at first, but it is fun to try and, as with so many things, practice makes improvement if not perfection.

You can, if you wish, join threads together in a pattern on a square of asbestos by fusing them with the very low flame of a butane blowtorch (fig.2a) and then when the threads have cooled, lift the entire pattern on to the prepared enamel piece, using tweezers or a small palette knife (fig.2b). The tree in the landscape was fused in this way. This is a useful process because sometimes carefully arranged, overlapping threads put straight on to a piece can slip out of place as it is being put in the kiln, which means starting all over again.

When the design is satisfactory the piece is placed in the kiln for firing. With a short firing the threads will remain in relief from the background. One possible snag, however, is that the threads may crack off in the cooling process. With a normal firing threads will flatten and with a longer firing they will spread. It is necessary to keep an eye on the piece in order to obtain the desired result.

2b Lifting the pattern of threads off the asbestos

3 and 4 Making a 'finding' from copper wire. Instructions are on page 99.

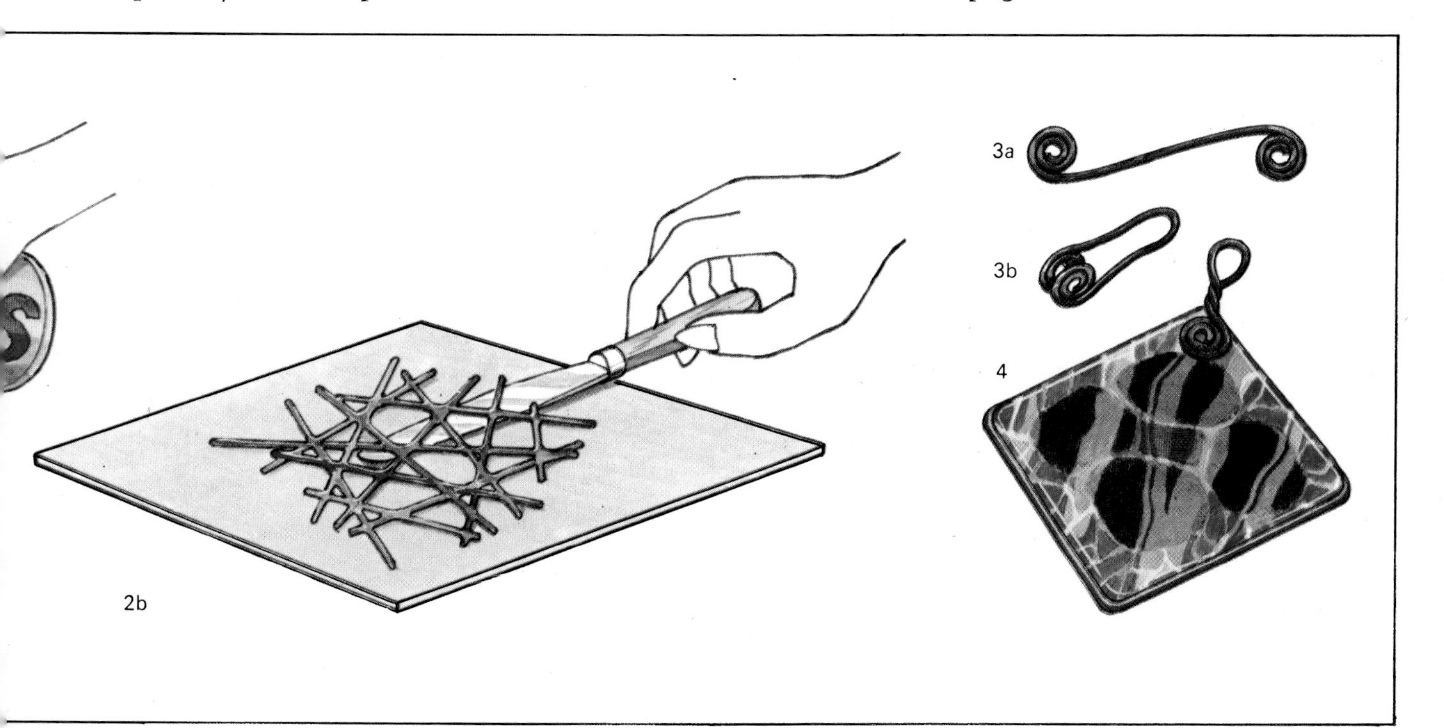

Pendants and brooches decorated by embedding small metal objects and copper wire, and by using threads and lumps of enamel.

Once you are used to working with thread you can achieve some interesting effects by firing one pattern and then placing more threads on top and refiring: you will have a partly flat and partly raised design.

Lump decoration

Lumps of enamel in mixed colours, and sizes (up to about 2cm ($\frac{3}{4}$in) across), are readily available from craft suppliers. Shot enamel is similar to lump, but the pieces are smaller in size. This is a simple way of obtaining a very satisfying result, giving the effect of pools of contrasting or harmonizing colours on a plain background. You can use as few – perhaps just one – or as many colours as you wish. Prepare the blank as for thread decoration. It is advisable to dust a thin second coat of enamel, the same colour as the first, on to your pre-fired piece before placing the lumps in position. This prevents them from sliding about on the shiny enamelled surface when the piece is being placed in the kiln.

It is best to choose similar sized pieces when planning a design as large lumps will take longer to fire than small ones. As with threads, by varying the length of firing time you can have either raised areas of colour or a completely flat surface. Two or three lumps placed close together will fuse as they melt, and by careful placing of the lumps pleasing flower designs can be made, which look deceptively difficult to produce and are in fact quite simple after a little practice.

Embedding

Yet another simple way of decorating an enamelled surface is by embedding small metal objects. Small circles and squares of copper can be bought, but for a truly original touch use bits and pieces, such as jump rings, links from a chain, small nails and copper wire. Wire can be coiled and bent easily to make initials or personal emblems. The wire should then be thoroughly cleaned with emery paper as there is a natural lacquer on the wire which may otherwise prevent it from fusing with the enamel.

If you have worked with wire to make silver or copper wire jewelry you can now experiment with embedding flat wire patterns to make unusual earrings. Or a ring or pendant bearing the owner's initial or zodiac sign makes a wonderful present.

Start off by counter-enamelling the piece, and then fire a coat of enamel on the top surface. Next dust on a second coat of the same colour. Place the pieces of metal in position with tweezers, return to the kiln and refire.

Some hints on making jewelry

If you are following both the metal and enamel courses in this book, you will find many ways in which the two skills can be combined. For instance, you can make your own chains and findings [backings] for enamel jewelry. Maybe there is one piece which you would like to make into a pendant but it has no hole for a chain or thong. To remedy this take a piece of copper wire about 10cm (4in) long and coil each end tightly with a pair of pliers – one end one way and one the other way – leaving a length of wire in between the two coils (fig.3a). Bend the wire in half so that the two coils are flat against each other (fig.3b). Slip the piece of enamel in between the coils, hold it tightly in place with one hand and twist the loop with the other (fig.4). The wire itself will indicate the direction of the twist. Glue into place with a quick-setting, strong adhesive, such as Araldite Rapid [Duco Cement is a suitable adhesive in US].

As you produce more and more enamelled pieces – and your early efforts will be quite small and therefore most suitable for jewelry making – you will need to know more about findings [backings]. Refer back to the Metal chapters which deal with jewelry-making for information on how to make jump rings and simple fastenings. A later chapter in this course, 'Facts, findings and transfers' on page 105, is designed to extend your knowledge of all facets of enamel work, including how to fit bell caps and how to soft solder brooch pins and ear clips to your finished jewelry (available at jeweller's suppliers and many craft stores).

Scrolling for decoration

Scrolling, or swirling as it is sometimes called, it quite one of the most exciting techniques for decorating pieces of enamel. In the decorative techniques covered so far the patterns have been created before firing. With scrolling, although the colours are positioned before firing, the effect is created by manipulating the molten enamel with a scrolling tool while it is in the kiln.

It is probably easiest for a beginner to use thread and lump enamel for scrolling. Many enamellers, however, scroll with dry enamel powder positioned carefully with a spatula. (You can make your own spatula by hammering flat one end of a piece of stout copper wire and inserting the other end into a cork). Other people prefer a 'wet pack', a paste made from enamel powder mixed with distilled water in which a little gum arabic (about 1 teaspoonful per 280ml ($\frac{1}{2}$ pint) has been dissolved. The wet pack is positioned using a small spoon-shaped tool (a salt or mustard spoon would do) to scoop up the paste and a spatula to push it into position.

A scrolling tool is a length of stout wire, about 30cm (1ft) long, with 1.5cm ($\frac{1}{2}$in) at one end bent at right angles and sharpened to a point. The other end is inserted into a wooden handle (fig.1). It is possible to make your own tool from a piece of wire coathanger and part of an old broom handle.

Technique

Position your chosen colours on a pre-enamelled surface. If thread and/or lump enamel is used it is advisable to sift a thin layer of the base colour on to the surface before these are positioned so that they do not slide around when the piece is placed in the kiln for firing.

Put the piece in the pre-heated kiln, remembering where you have placed your colours, as it is almost impossible to distinguish them when the piece is molten. When the enamel is red hot and in a molten state, open the kiln door, heat the point of the scrolling tool against the wall of the kiln, then gently stir round the colours,

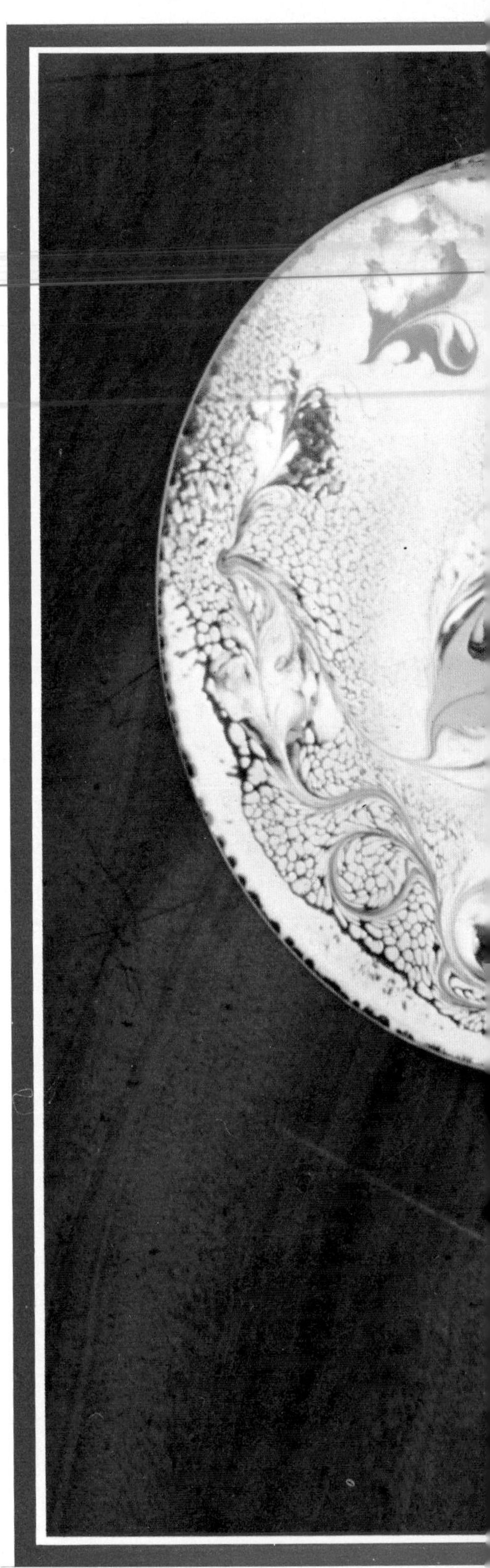

Scrolling for an ice bucket lid and bottle stoppers.

Both these pieces are the result of controlled scrolling.

Scrolled brooch
You will need: Equipment as already given plus a tripod or stilt and a scrolling tool A 5cm (2in) square metal blank Counter-enamelling powder Opaque white enamel powder Enamel threads Small transparent lumps Anti-scale liquid (optional) Emery paper Carborundum stone or fine file Brooch finding [backing] Strong adhesive

being careful not to press down too hard. It is only the top layer of the enamel you want to scroll so do not dig down to the metal itself. Some enamellers heat the scrolling tool in a gas flame such as a butane torch and thus avoid the possibility of the scrolling tool cooling and pulling the enamel off the work – the cause of many a disaster. High firing is required for scrolling so the operation should be done as quickly as possible because the kiln will start to cool with the door open. Close the door for a few seconds to allow the enamel to flatten out before removing it from the kiln. To begin with you will stir round your colours completely at random and perhaps achieve some fabulous results by sheer accident – and some disasters! Exciting though these happy accidents can be, the main purpose is to control the medium and not let it dictate results. It is very rewarding as you progress, to find that it is possible to exert a remarkable degree of control over your design both by the placing of the colours and by the direction in which you scroll. This, however, takes some time and practice to achieve.

Designing

Artist enamellers using the scrolling technique, decide beforehand on the effect they want. They compose the work before firing, placing the powdered, packed, lump or thread enamel into known position and known shapes before any work is done in the kiln. Thus the enameller knows where the colours are positioned when the surface of the work is in a molten state. Several firings may be necessary to obtain the required result and other layers of enamel may be added as the work proceeds.

The number of layers and firings possible is largely determined by the thickness of the metal and of the layers of enamel. It is essential that the piece is counter-enamelled if it is to be fired several times, using a hard-firing counter enamelling powder.

The two pieces in the photograph are both examples of controlled scrolling. The green piece was scrolled as shown in fig. 3, and full instructions for the white scrolled piece are given below. The same arrangements of threads could be scrolled in a random manner to produce a quite different effect.

Scrolled brooch

The instructions are for the white piece illustrated.

Switch on the kiln so that it will be hot enough, ie bright orange, by the time you have prepared the metal blank. Prepare and counter-enamel the piece in the usual way.

Remove the firescale using emery paper. Clean the edges of the piece with a carborundum stone or a fine file (fig.4), and finish off with emery paper. This ensures that when you place the piece on to the stilt to fire the other side just the very edge of the metal, and no enamel, is in contact with the stilt. Therefore the enamelled surface is not damaged. Using tweezers rinse and dry the piece and place it on a sheet of clean paper. Sieve the opaque white powder evenly over the copper, paying special attention to the edges. Place the piece on to the stilt (fig.5) and fire. When cool clean edges again. If you use anti-scale liquid wherever possible you will find that this tedious process of cleaning the metal after each firing is much reduced. You can also use a pickle solution as described below to remove firescale after counter enamelling.

Choose your coloured threads and lumps. Here mainly light and dark green threads have been used and transparent blue, amber and ruby lumps. Break the threads into more or less even lengths and have them ready with the chosen lumps on a clean piece of paper. Sieve a thin second coat of white enamel on to the piece, to prevent the lumps and threads from sliding about on the shiny enamelled surface when placing the piece in the kiln. Using tweezers, place the threads and lumps carefully in position (fig. 6 overleaf) with a light and a dark thread in each pair.

Using the spatula, lift the piece on to the stilt (placed on the mesh stand if the stilt has a flat bottom) and place it in the kiln. It is important to remember which way round you have placed the piece as it is not easy to see the pattern when the enamel is molten. When the piece is red hot, open the kiln door, heat the point of the scroller on the inside wall of the kiln and then gently scroll in the direction of the arrows (fig.7). Close the door for a second or two to allow the enamel to flatten, remove the piece from the kiln and leave to cool. When the piece is quite cold you can attach a brooch finding [backing] with a suitable adhesive. You cannot use solder if the piece has been counter enamelled.

Using a pickle bath Another way of cleaning metal blanks and freeing them from any grease is to immerse them in what is called a pickle bath. An effective and safe pickle bath can be made from a solution of approximately one rounded teaspoon of common salt to half a cupful of vinegar. Put the solution into a shallow bowl or dish and immerse your metal blank. Leave it there for three or four minutes, then lift it out with tweezers, rinse in clean water and dry. Enamel in the normal way. This solution is quite useful for getting rid of some of the firescale after one side has been enamelled. In constant use the solution should last for several days. After a time it will begin to go a greenish colour. This does not mean it is any less effective, but when you find that it takes a lot longer to clean the metal then it is time to throw it away and mix some more.

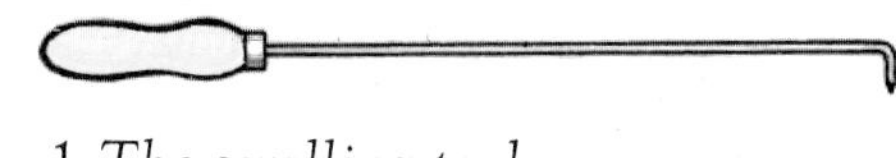

1 *The scrolling tool*

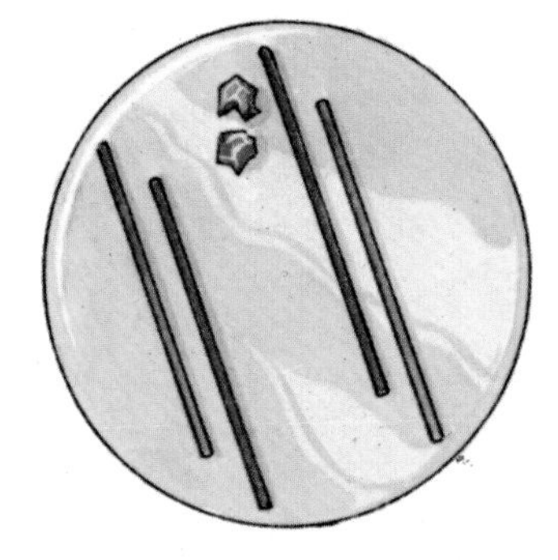

2 *Positioning the threads and lumps for the green brooch*

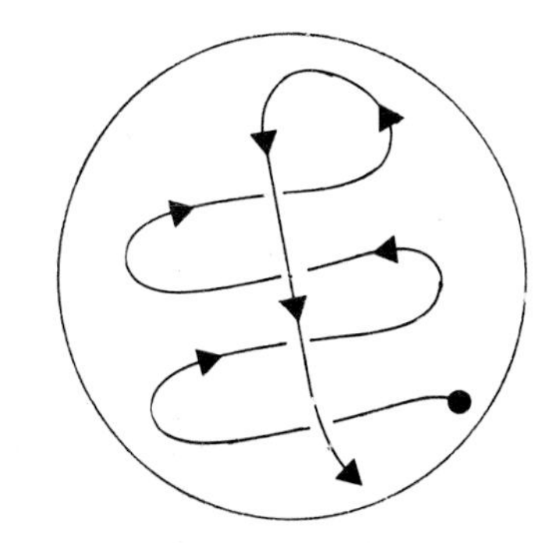

3 *Controlled scrolling*

4 *Cleaning the edges of the blank*

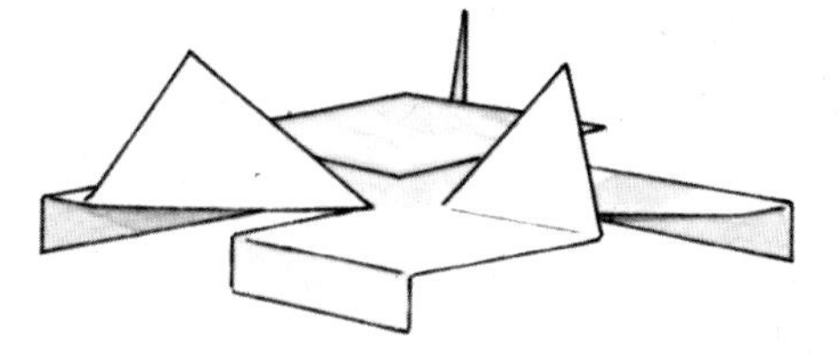

5 *The piece positioned on the stilt ready for firing and scrolling*

6 Positioning of threads and lumps for the white scrolled piece.

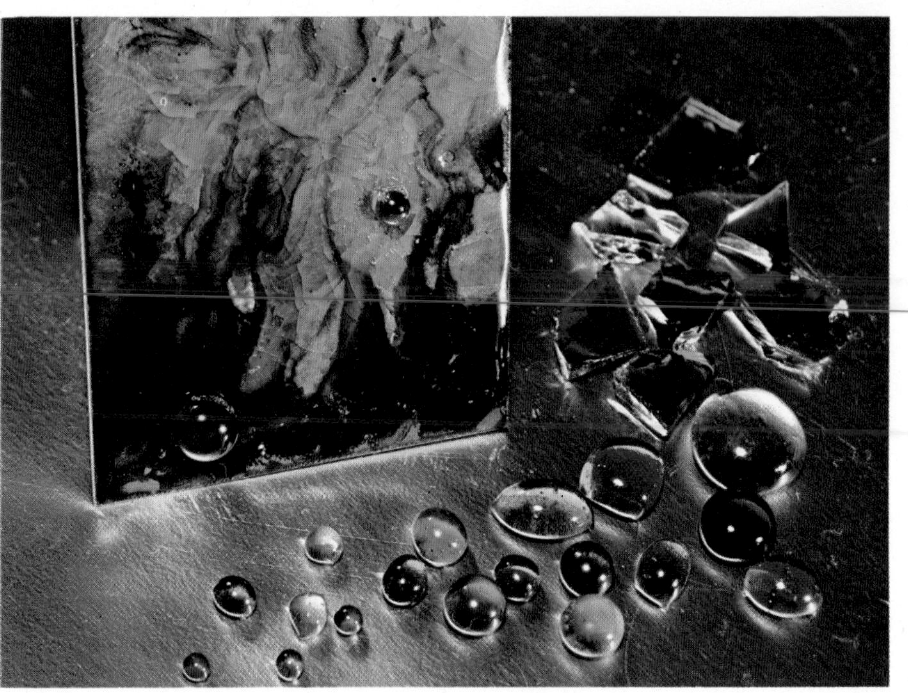

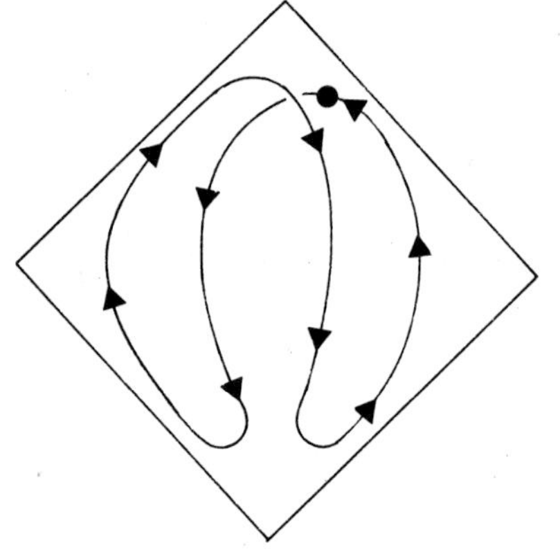

7 For a controlled effect, scroll in the direction and pattern indicated on the diagram.

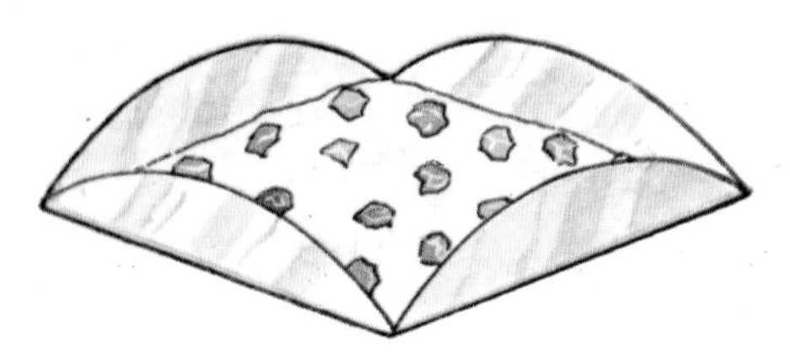

8 A 'tray' made from a circle of tin. Use it to make glass 'jewels' for decoration, like those illustrated here.

Making your own 'jewels'

Sometimes it is nice to have a raised jewel-like decoration on a brooch or pendant. Very effective jewels can be made by melting small pieces of glass. Old wine bottles are readily available and can be smashed by placing them between thick layers of newspaper and hitting them sharply with a hammer. Be careful not to touch the sharp fragments with your fingers.

It is not practical to fire pieces of coloured glass in the kiln with enamels, but it is possible to melt glass on its own and then glue it to the enamelled piece. To melt the glass you need a flat bed of plaster of paris – the glass will not stick to this. Make a 'tray' from a circular piece of tin by bending up the sides (fig.8). Spread an even layer of plaster in this, pressing it down flat. With tweezers pick up the bits of glass and place them gently on to the plaster. Choose pieces more or less the same size as larger ones will take longer to melt. Put the tray on to the wire mesh stand and place it in the kiln. You will have to open the door periodically to see how the glass is progressing. The pieces will contract slightly and take on a dome-like form. When this happens, turn off the kiln and leave the glass to cool slowly in the kiln. This is important, as it may crack if cooled too quickly. Any bits of powder which adhere to the glass 'jewels' can be easily brushed off. Wash the pieces, discarding any that are not satisfactory, and they are then ready for use.

Facts, findings and transfers

This chapter is designed to extend your knowledge of enamels: to give more details about how they are produced, and additional information on care of enamels, blanks and kilns. There is also more about findings [backings] and using bell caps for jewelry.

The manufacture of enamels

Traditional enamels contain lead so: do not put them into your mouth; do not smoke or eat while using them; handle them carefully, in small quantities, and store them in a closed container; and always wash your hands after using enamels.

The newer leadfree enamels, however, are recommended as much safer to use, especially for children, and give satisfactory results. Just take the same precautions as for lead with the reds, yellows and oranges which contain cadmium.

To produce a range of colours a basic enamel is evolved. This is a transparent, colourless enamel, usually described as 'flux'. Small additions of metallic oxides are made to the raw batch and melted in to produce different colours, such as cobalt oxide for blues, copper oxide for greens.

To give accurate control over the melting and the tint of the colour the enamels are melted in special fireclay crucibles, each containing only 1.8-2.3 kilograms (4-5 pounds), at a temperature of 1100°-1150°C (2012°-2102°F). When the materials are completely combined, and the colour has been checked against a standard sample, the enamel is poured on to a steel plate. When it has cooled, the cake of enamel is then crushed and graded to obtain the small lumps used for scrolling. Pieces which are 3mm-6mm–($\frac{1}{8}$in-$\frac{1}{4}$in) are described as 'cracked', 1.5mm-3mm ($\frac{1}{16}$in-$\frac{1}{8}$in) as 'crushed', and the particles smaller than 1.5mm ($\frac{1}{16}$in) which are used, by sprinkling on a base coat, as the simplest form of decoration for enamelled pieces are described as 'crystal'.

The main bulk of the enamel is crushed and ground on a porcelain ball mill to pass through a very fine mesh sieve and is supplied as a powder for application either through a sieve, dry, or by the wet pack technique (see page 100).

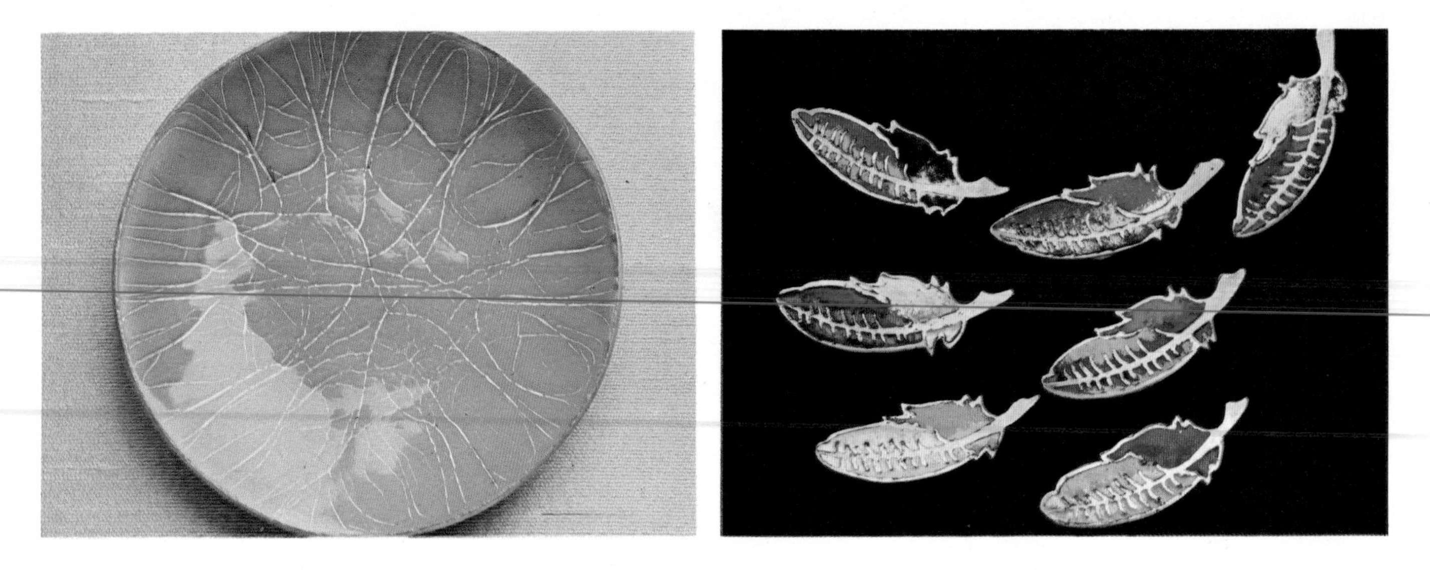

Above: Green self-cracking enamel is used over a white base coat on this dish.
Above right: These feathers show a modern, decorative use of the champlevé technique, explained in a later chapter.

Storage and care of enamels

When enamels are stored in powder form, they tend to be attacked by the moisture in the atmosphere and deteriorate. This deterioration of the enamel powder is the reason why early books on enamelling recommend the purchase of lump enamel to be ground with a pestle and mortar immediately before use. This is no longer essential as this fault is less liable to occur with modern enamel powders, as long as they are kept clean and dry.

The first stage of deterioration is a slight opacity showing in transparent enamels, the next is pinholing of the surface and finally complete devitrification, so that the enamel does not fuse or flow. If the enamel is 'washed', ie stirred in clean water, and the milky liquid is poured off, the very fine particles which have begun to deteriorate are removed and the enamel improved in quality. Continue this process until the water poured off is clear, without any trace of cloudiness. The powder will then be usable.

Thickness of copper

20 gauge (.9mm) thickness is suitable for general use and good quality blanks are of this thickness. For large panels or dishes a thicker gauge is advisable, 18 (1.25mm) or 16 (1.60mm) gauge. If the design is to be produced by deep etching the heavier gauge is required. Use of a thinner gauge than 20 brings a risk of the enamelled piece being fragile, particularly if it is not counter-enamelled and, if bent, the enamel will fracture.

When firing it is advisable to use a lower temperature for a thinner metal, or a longer time for a heavy metal. For example, if it takes three minutes to fire enamel on 20 gauge (.9mm) copper at 820°C

(1805°F) then, if using 18 gauge (1.25mm) copper, fire for four minutes at 820°C (1508°F). As with all aspects of enamel work, you will learn to judge from experience.

More about kilns

If you have not yet invested in a kiln but have been working with a blow torch think carefully before buying. If possible choose a kiln by independent recommendations as it is difficult to evaluate a kiln without using it. Consider, too, how much you can afford and how much you expect to use your kiln.

The cheapest kiln is the hotplate type. In this the element is embedded in the fireclay base and it usually has a light, lift-off aluminium lid. The size of piece which can be enamelled is 7.5cm-10cm (3in-4in) across and the maximum temperature is usually 800°-820°C (1472°-1508°F) but some models will go to a temperature of 900°C (1652°F).

Larger kilns are of two main types. The first consists of a fireclay muffle with the element as a continuous wire winding round it. This gives an even distribution of heat but not usually a fast rate of increase of temperature. Replacement muffles are available should it be necessary.

The second type is built of insulating bricks in a steel casing with separate elements in grooves in the floor and sides. This gives a fast rate of climb with a good reserve of heat at working temperature. Replacement elements are available.

Regulator If a kiln is not fitted with a heat input regulator, by adjustment of which the kiln temperature can be held steady, there is a risk of overheating. When using a kiln without a regulator it is advisable to switch off the current from time to time to prevent overheating and also economize on running costs.

Pyrometer The cost of a pyrometer, which is fitted to the kiln to indicate the temperature of the oven, may seem high in relation to the cost of the kiln, but many enamellers consider the cost well justified by the improved results of consistent firing.

Safety cut-out Some kilns have a safety cut-out which operates so that when the kiln door is opened the current is shut off.

Points to consider Is the size adequate? Is the rate of heating to 800°-850°C (1472°-1562°F) reasonable? Does it give even heating over the whole floor space? Is the insulation good enough to prevent the casing getting too hot for comfort?

Is it fitted with a regulator, pyrometer and safety cut-out? (It is better to purchase these fitted to the kiln rather than later.)

Your kiln represents your biggest financial outlay, so it is only wise to consider carefully all these points.

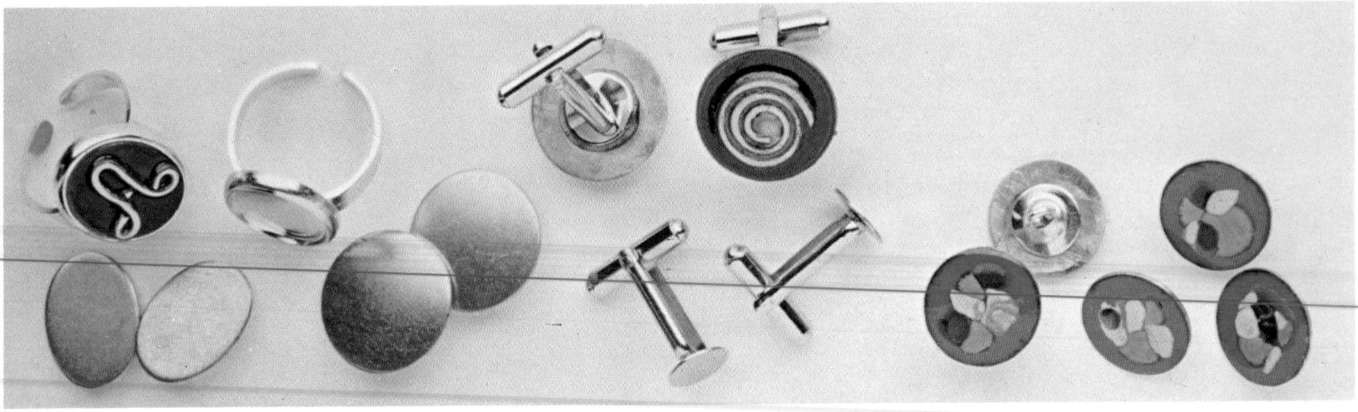

Just a few of the findings [backings] available for enamellers are ring sets, cuff link fittings and button shanks.

Findings

These are the accessories that are used to hold jewelry in position, for example brooch pins in various sizes, ear clips, hooks and screws, tie clips, jump rings for attaching chains, bolt rings for necklace or bracelet fastenings and bell caps. Cuff links, adjustable rings and key rings also come under this heading.

Bell caps Sometimes it is a good idea instead of making a pendant with a hole for a jump ring and chain, to attach a bell cap to the top, add a jump ring and thread the chain through the loop on this. Bell caps can be obtained in various sizes and designs and add to the decoration of the pendant.

Before gluing into place with a strong adhesive, it is necessary to flatten the bell cap to fit the enamelled piece (figs 1a and b). You can probably do this by squeezing the bell cap into place with your fingers but if necessary use pliers.

Bell caps are also very useful in making a bracelet. Having enamelled your blanks on both sides, attach one or two bell caps to opposite sides of each piece (figs 2a and b) and join the pieces together with jump rings. Add a bolt ring at one end and a jump ring at the other and the bracelet is complete. You can also use bell caps to make interesting and unusual necklaces.

Attaching brooch pins and ear clips If the piece has been counter-enamelled, the finding must be glued in place with strong adhesive. If the piece is a small one which has been enamelled on one side only the pin can be glued or attached by soft soldering. Soft soldering can be done with a blowtorch or soldering iron (see page 42), but sometimes this weakens the spring of an ear clip and it can also discolour the finding.

Soft soldering, however, can be done very satisfactorily on top of the kiln while you are enamelling other pieces. On many kilns the heat is sufficient to melt the solder without damaging the enamelled

surface. You will need soldering flux – paste sold in a tin – and electricians' solder, in the form of wire, which is sold on a reel or in a small container.
Clean the back of the piece thoroughly. If the metal is not clean and shiny the solder will not fuse. Using a match stick put a little of the paste flux on to the metal where the finding is to be placed, together with a small piece of solder wire. The piece of wire should be about half the length of the brooch pin (fig.3a). Place the piece with enamel side down on a tin lid and put the lid on top of the kiln.
Put a little paste flux on to the finding and position this on the enamelled piece. Obviously it will not sit flat because of the piece of solder but, in a few minutes, the solder will melt. At this stage it is possible to move the finding if it is not quite in the correct position. Using a palette knife lift the tin lid and the piece off the kiln and place it on an asbestos sheet to cool.
This method can be used to attach any finding to the back of a piece which has not been counter-enamelled.
Some brooch pins have one, or two, small holes in the bar. In this case you can lay the pin flat on the piece to which it is to be attached and put a curled piece of solder over each hole (fig.3b). The heat will draw the solder through the hole and the pin will be soldered to the piece. Tidy the solder with a fine file if necessary and clean the pin if it has become at all discoloured.
You can also buy ring mounts, chains for pendants and numerous other accessories from specialist suppliers and jewellers.

Transfers

Transfers for firing on to enamelled jewelry, boxes, matchbox covers and plaques come in a range of sizes and types, eg there are birds, flowers, animals and pictorial scenes. The transfer consists of paper backing, printed picture and a protective coat of lacquer.
Apply the transfer to a *smoothly-fired* surface to which any additional decoration, eg a scrolled border, has already been added. Choose a hard-firing colour for the base coat for best results. The colours on the transfer are very thin so the base colour of your piece will show through. Thus a yellow rose fired on a blue base will appear green, while a yellow rose on a paler yellow base will appear more yellow.
Soak transfer in water until it will slide on the paper – this should take about half a minute. Slide transfer on to enamelled surface and with a soft cloth smooth out the transfer from the centre to remove excess water and air bubbles. Leave the piece overnight to dry naturally.
This process *cannot* be speeded up. For best results place the piece

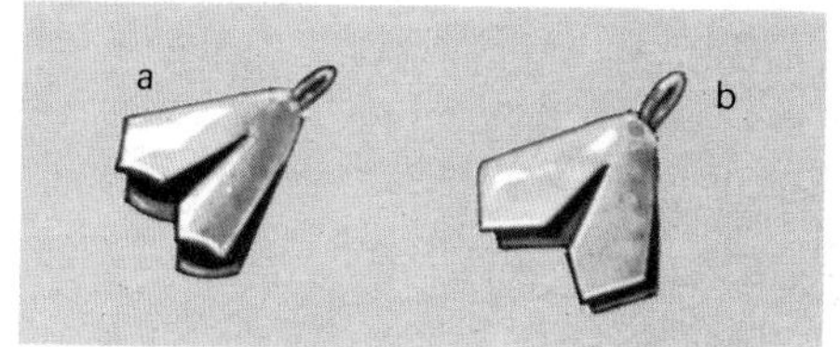

1a.*An unflattened bell cap*
1b *A bell cap after flattening with fingers or pliers*

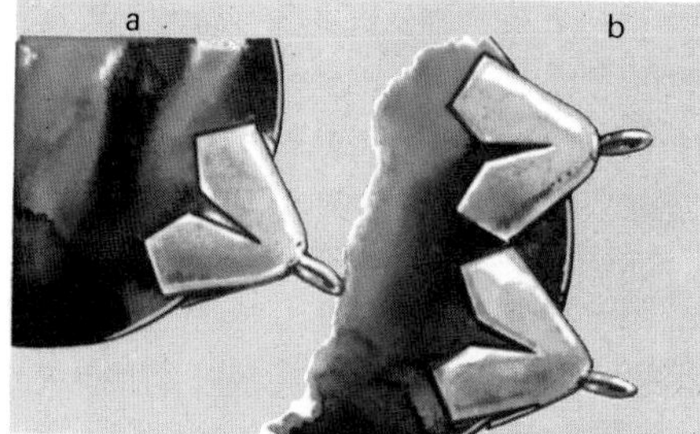

2a *and* 2b *For a bracelet attach one or two bell caps to opposite sides of blank.*

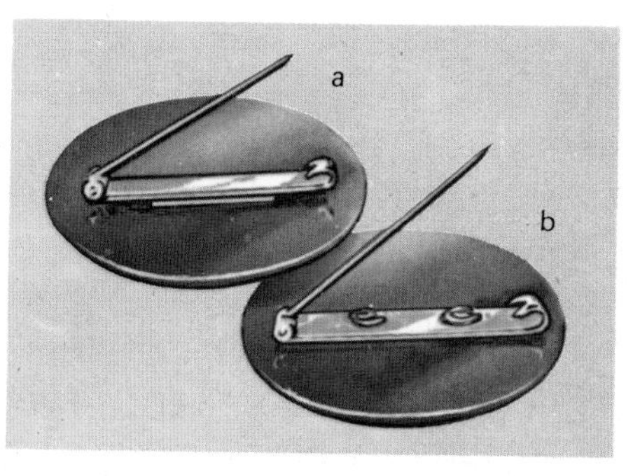

3 *Two ways to solder a brooch*
a *Brooch pin on top of solder wire.*
b *Solder wire on top of holes in brooch pin.*

There are many transfers available for enamel work, and they can be used to produce instant and professional-looking designs.

in a cold kiln, then switch it on. This ensures that the lacquer coating on the picture deteriorates gradually and does not burst into flame and spoil the picture. If your kiln has a hole in the door you can work with it shut but if not open door from time to time to let out the fumes from the lacquer. Remove the piece carefully from the kiln from time to time. (If the colours are clear and glossy when the piece is held in the light the transfer is fired.)

Gold transfers As the gold looks glossy all the time you cannot apply the same test to see if the piece is ready. Therefore, start with coloured transfers, note the time they take to fire, then try a gold one and fire it for the same length of time.

When you have had some fun with these 'instant pictures' you can go on to create your own designs with painting enamels.

Painting with enamels

Painting with enamels can be used to produce extremely fine enamel work. Specially finely-ground enamel powder is mixed with a medium and actually painted on to a pre-enamelled base and then fired in the kiln in the usual manner.

Painting with enamels first began in Limoges in the 16th century. At the time it was a complete innovation as, up until then, *champlevé* and *cloisonné* were the only techniques practised by craftsmen in the art of enamelling.

The painting technique is not easy to master and considerable patience and practice are necessary. This chapter covers the basic technique for painting and also explains a method, known as 'line technique', which can be used to create sharply-defined pictures or designs by containing dry or wet-packed jewelry enamel powder within a drawn line.

The technique of enamel painting is not easy to master but beautiful results like the examples shown here are possible. Above is 'Woman in a turban' by Valerie Bexley.

Technique

Painting enamels are applied to a pre-enamelled (and counter-enamelled) base. Use ordinary jewelry enamel powder for the base coat but make sure it is spotlessly clean. If necessary wash the powder, rinsing it until the water is completely clear. Also, ensure that the base is smooth – a slightly under-fired surface will detract from the finish.

A white base is generally used, though pale colours are also possible. As painting enamels are mainly pigment and have hardly any substance of their own they are greatly affected by the base on which they are applied. Therefore, blues and black keep their colour when painted over a pale blue base but reds lose their brilliance and take on the shade of the base coat.

Painting enamels are specially designed for their purpose. They are finely ground, and unlike jewelry enamels, soluble, which means they can be mixed with a medium for painting. Tramil or a specially-produced painting medium (an oil), obtainable from suppliers of enamel powders, can be used for this purpose.

Alternatively, some enamellers prefer to use two or three drops of liquid gum arabic mixed into about half a cup of distilled water (or

'Badgers playing', a beautiful painted enamel by Diana Harding

distilled water on its own) to mix the painting enamels. If you use gum arabic granules or crystals available from some hardware stores instead of liquid gum arabic, dissolve two or three granules in about half a cup of hot water. Allow to cool before mixing with the enamels and then add a little at a time until you have a suitable consistency. Don't try to rush this process.

Use a small palette knife to mix the powder and medium as this crushes any small lumps and ensures a smooth consistency. Do not make the mixture too thick – if you want a rich, dark colour build it up in separate coats between firings. A sable hair brush is excellent for applying the paints, as it helps to give a smooth, even finish. With practice, it is feasible to cover an area evenly with colour or to produce gradual changes of shade, merging one colour with another without unsightly splodges.

A big advantage that painting enamels have over jewelry powders is that they can be mixed together to make new colours. Adding white to produce pale shades and the simple mixing of two colours presents no difficulties. But if you mix several colours the strongest colour will always predominate when fired.

Firing It is usually the firing of painting enamels which causes the most problems. The drying out of the enamels before firing is very important. Dry them slowly near the kiln – not on top of it. Sudden and extreme heat changes will cause the wet colours to move just enough to change some fine shading into a muddy pool, or produce hairline cracks running through the colours. If you are using the oily painting medium, it will not be possible to get rid of all the oil by drying the work near the kiln. Therefore, having dried the enamels as much as possible, bring the kiln up to normal temperature – about 750°C (1382°F) – and place the piece in it for a few seconds. When you withdraw the piece you will see vapour rising as the medium evaporates further. Continue drying in this manner until the vapour ceases to rise. Do not close the kiln door during this drying out process or the piece will catch fire and be spoilt. This additional drying out in the kiln is not necessary when using other media. Bring the kiln back to normal temperature if it has cooled slightly and you are ready to fire. Painting enamels usually fire very quickly (in a little under half a minute), though some of the colours take slightly longer to fire than others and the softer ones, such as reds and yellows, fade.

Colours which have faded can be touched up easily afterwards and the piece re-fired. It is possible to fire a piece several times between applications of colour. When the piece is finished every colour should shine so that a clear overglaze of soft flux is not necessary, though you may apply this if you wish.

Problems and their causes

Bubbling, pitting and matt finish These are caused by using the wrong medium to mix the enamels, mixing colours too thickly or by firing before the enamels have dried out completely.

Matt finish alone This can simply be a case of under-firing, sometimes at too low a temperature as well as for too short a time. Re-firing at the right temperature should correct this.

Hairline cracks running through base coat can result from firing the painting colours at too low a temperature.

Fading and speckling are caused by over-firing – the speckling is the base coat pushing up through the colours.

Line technique

Jewelry enamel, whether applied dry or wet-packed on metal, tends to flood or spread when fired, unless it can be held in position. In the past, this positioning of enamel was made exact by the use of cloisonné and champlevé methods. It is now possible, however, to hold enamel in position, dry or wet-packed, by the use of an oil-based felt-tipped pen or a ceramic wax pencil. (Do not use a water-based felt-tipped pen).

The design is usually drawn on a pre-enamelled base and dry or wet-packed enamel placed inside the design. It is necessary to take considerable pains with the drawings so that the enamel will stay where it is put and not flood. This means that the drawing must in itself form 'cloisons' or pockets in which the enamel is placed. There should be no gap in the line through which the enamel can escape into another colour. When the drawing is completed to your satisfaction, fill the pockets with enamel (wet-packed is the easiest method), then dry and fire the piece in the usual manner. You can give the piece a professional look by outlining the design with a painting enamel powder mixed with distilled water and applied with an ordinary mapping pen or any pen with a fine steel nib.

Line technique is useful for outlining figure shapes, drawing in features, and signing work.

Building up a panel

Using pre-shaped copper blanks for enamelling limits designs to the shapes available. It is not difficult, however, to cut shapes from metal, and by making your own blanks the scope for designing is much greater. Cutting metal decoratively is part of the metalwork course, so refer back to page 23 for full instructions. You can use tin snips [metal cutters], but for fine work a jeweller's piercing saw is better as it is much more accurate on curves and angles.

This chapter gives instructions for making a rectangular panel by mounting together a series of small cut shapes. This method makes it possible for those with small kilns to build up large panels. The 'King of Hearts' panel is made following this 'stained glass window' principle, with the component parts of the design mounted on to a piece of wood representing the border of a playing card. The wood can be white as on a real playing card, gold as shown here, or any other colour. The finished card is illustrated on page 118.

King of Hearts

This project involves using several of the techniques covered in previous chapters: using an adhesive, stencils, working with threads and lumps of enamel, and painting with enamel.

The method of transferring the design to the copper with carbon paper, used for this project, is an alternative to gluing the tracing of the design straight on to the metal. It is a good method to use for simple shapes but not suitable for very intricate cutting.

Switch on the kiln so that it will be hot enough, ie bright orange, by the time you are ready to fire. Take a tracing of the trace pattern given overleaf, including the numbers.

Make sure your piece of copper is completely flat. In its new state the metal is quite shiny and will therefore take neither pencil marks nor carbon marks. To give a matt finish to the copper, brush it over with salt and vinegar pickle and leave to dry. When pickle is quite dry, wipe over the copper gently with a clean rag. The surface will now take pencil or carbon marks.

Lay your piece of copper on a board, place the tracing over it, and keep the tracing in place with drawing pins at the two top corners

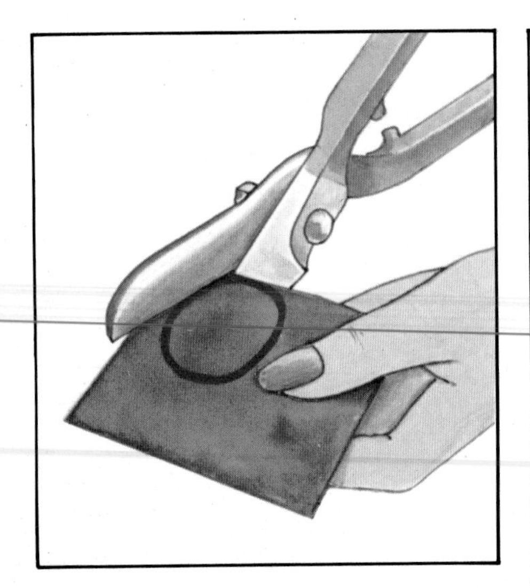

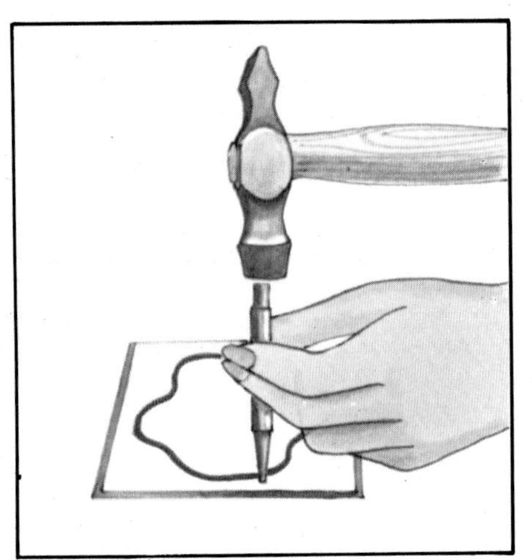

Two methods of cutting metal blanks from copper—Top: Cutting simple shapes using tin snips [metal cutters]. Above: Before using a piercing saw, make a mark just off the cutting line of the work with a nail punch and hammer, and drill a hole for the saw blade.
Right: Trace pattern for the King of Hearts panel

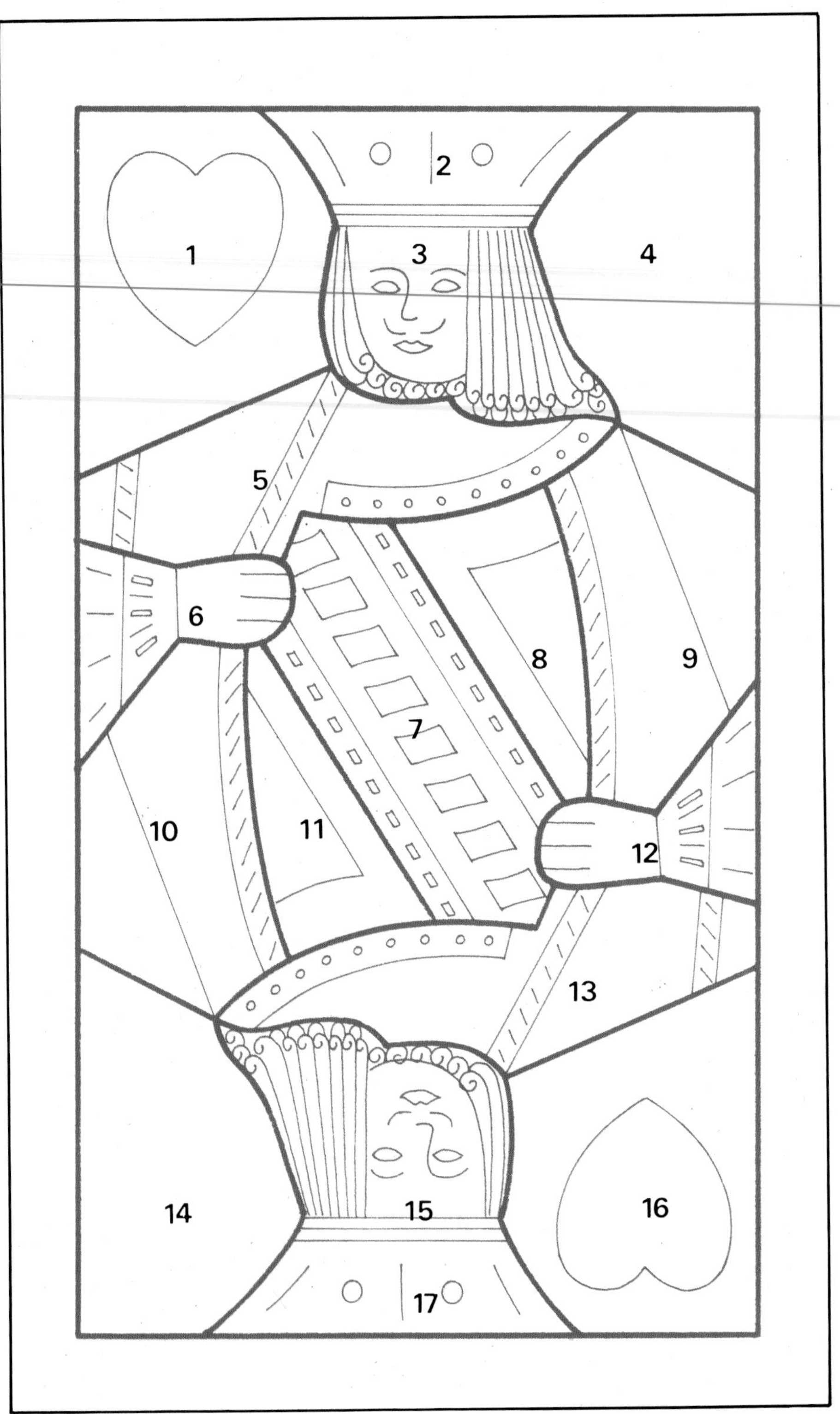

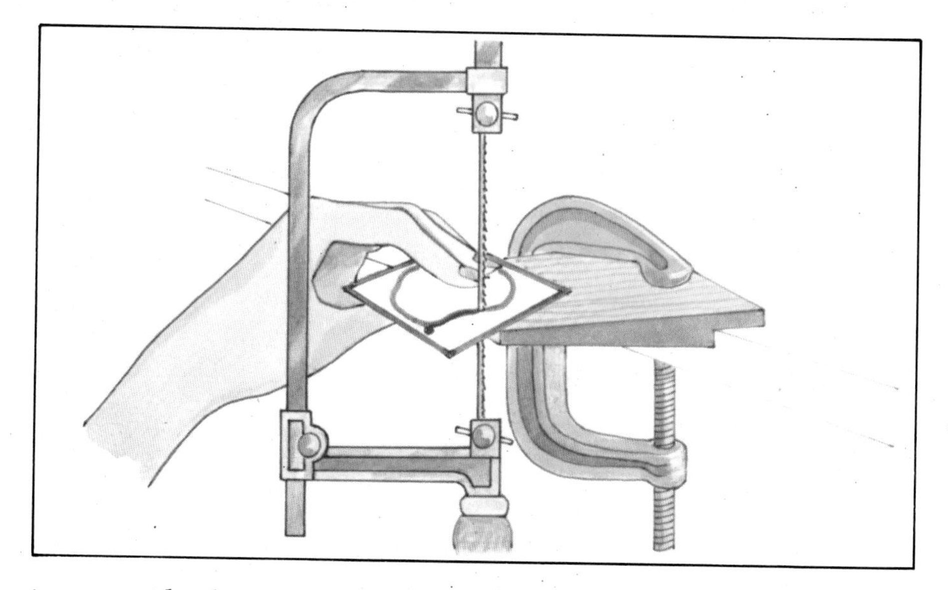

Cut along the edge of the design, sawing gently with an up-and-down motion.

just outside the copper. When the pins are pressed down tightly the heads will keep the copper secure. Slide a piece of carbon paper under the tracing and draw carefully over the lines to transfer the design on to the metal. Mark in the numbers for each section as well as the outlines to avoid any confusion as you work.

Cut out the shapes indicated by the thicker lines on the trace panel. There is, of course, no need to stick the tracing on to the copper as you have traced out the design with carbon paper. As each piece is cut, lay it in its proper place on the tracing. You will notice that as the card looks the same either way up, the shapes are in pairs, except for the middle one, number 7. This means you can enamel pieces in pairs, saving time and electricity.

Prepare pieces 1, 4, 14 and 16 and then give them a coat of white enamel powder. If you wish, give them a thin coat of adhesive before sifting on the powder. Many enamellers prefer to use this method for all their work (see page 100). Place pieces singly or in pairs (according to size of kiln) on to the wire mesh stand. Dry out if using adhesive, and fire. Remove loose firescale (this must be done after each firing).

Trace off a heart shape from diagram, and with a scalpel, cut a stencil from card (see page 92). Place the heart-shaped stencil in position on pieces 1 and 16 in turn. Paint a thin coat of cellulose wallpaper paste on the heart area and sift on a coat of red enamel powder. Remove stencil carefully (tweezers are useful for this job). Tidy up the edges of the heart shape with a slightly damp, fine brush. Dry piece on top of kiln and then fire.

Next, prepare and give the two crowns, numbers 2 and 17, a coat of yellow enamel. Put threads, broken to length, and chips of

King of Hearts

You will need:
Basic enamelling equipment
Wire cutters
Piercing saw and other cutting tools
A piece of wood or hardboard 22.5cm × 15cm (9in × 6in), painted in the chosen colour, for mounting the work
1.25mm (16-18 gauge) or .9mm (19-20 gauge) copper 17.5cm × 10cm (7in × 4in)
2 small bought heart-shaped copper blanks (or cut your own)
About 15cm (6in) copper wire
Opaque enamel powders in white, bright yellow, royal blue, scarlet and black
Transparent deep blue enamel powder (use an opaque enamel if you prefer)
Black painting enamel powder
Salt and vinegar pickle solution for cleaning the metal (see page 103)
Anti-scale liquid (optional)
Enamel threads in assorted colours
Small lumps (or chips) of enamel
Fine paint brushes
Tracing paper, fairly hard pencil, carbon paper and drawing pins
Thin, stiff cardboard
Scalpel
Fine file or carborundum stone
Blow torch (optional)
Cellulose wallpaper paste with tragacanth base, gum tragacanth or tramil
Brushes for applying pickle and anti-scale liquid
Strong adhesive

The King of Hearts panel is made up in sections, enamelled separately and then mounted on to a wooden base.

enamel in place for the decorations, having first sifted on another thin coat of yellow powder to stop them slipping about. Prepare and give the faces, numbers 3 and 15, a coat of white enamel. Then paint in the features and hair carefully with black painting enamel (see page 114) and fire.

Prepare and give numbers 5 and 13 a coat of red enamel, being careful not to over-fire as red enamels are not hard-firing.

Break the threads for decoration. Cut a shape for each piece in thin cardboard to mask all but the sections to be black and yellow. Shake on these two colours and carefully remove the mask. Drop threads in place with tweezers, and fire.

Paint in the finger lines and fire in the same way as you did the features. Alternatively, these could be put in with very thin black threads at the same time as the sleeve decoration.

Prepare and give numbers 9 and 10 a coat of red enamel. Break threads for decoration. In order to achieve the necessary curve you will have to break the threads into short lengths or bend a thread to shape using a blowtorch (see page 97).

Mask off the area on each piece to remain red and shake blue enamel powder over the rest. Remove mask and shake a little red enamel along the edge where the threads are to be placed. Position the threads carefully with tweezers and fire.

Prepare and give numbers 8 and 11 a coat of royal blue enamel. Then mask off the area on each to remain royal blue, give the remaining area a coat of deep blue as before, and fire.

The centre piece, number 7, is the last to be enamelled. Have ready the white threads – thin ones for the long stripes and short pieces of thicker thread for the dots. Prepare and give the piece a coat of black enamel. Sift on another very thin layer of black powder, position the threads with tweezers and fire.

Cut a stencil for the shapes down the middle of the piece, position it very carefully and shake on red enamel powder. Remove stencil and fire the piece, being careful not to over-fire.

Clean the backs of all the pieces with emery paper. Clean off the edges of all the pieces with a fine file or carborundum stone and then with emery paper in order to ensure that they are really bright and smooth. Care in cleaning the edges makes a tremendous difference to the final result.

Mark the rectangular shape of the enamel 'card' on to the piece of wood with a fine pencil line. Scratch the surface within this area with a sharp point (a nail will do) and scratch the backs of the copper pieces. This gives a key for the glue. Stick the pieces in place on the wood with a strong adhesive. Cover the work with several layers of newspaper and put a heavy weight on top until the glue has set – preferably overnight.

The two Ks are made from flattened copper wire. Anneal the wire first. Annealing copper consists of heating the copper to red heat in the kiln and then plunging it straight into cold water.

This serves two purposes: it burns off any grease on the metal and, at the same time, softens it so that it can be easily bent into any shape. It is a useful technique when working with wire.

Place the wire on a very hard surface (a steel block if you have one) and hammer it to flatten it slightly. Cut with wire cutters into the required lengths. Clean it in the salt and vinegar solution with the two little heart-shaped blanks. Enamel red and fire. Glue into place on the finished card.

A striking champlevé necklace of cut shapes enamelled separately.

The art of cloisonné

Modern silver peacock feather pendant made using traditional methods. This is an example of concave cloisonné.

The term '*cloisonné*' comes from the French word *cloison*, meaning partitioned area. The term is used in enamelling to refer to partitions or fences of rectangular flat wires, placed along the lines of the design and then fixed by various methods to the metal base. The cells formed between these wires are filled with enamels of the desired colours, each *cloison* being quite separate from its neighbour so the colours cannot run together when fired.

Greek goldsmiths produced the first *cloisonné* enamels in the 4th and 5th centuries BC, but the most beautiful *cloisonné* works were Byzantine gold enamels of the 9th to 11th centuries. One of the most outstanding examples of Byzantine workmanship is the Pala d'Oro now in St Mark's Cathedral, Venice.

Cloisonné is a technique still practised today by many enamellers: it is especially suited to making jewelry.

Metal for cloisonne

Fine (ie pure) silver is an excellent metal for *cloisonné* as it has no added metals to cause oxidation and discoloration of the enamels when firing. Sterling silver (an alloy of 925 parts silver and 75 parts copper) and copper go black with oxidation. This necessitates cleaning off the metal after each firing. Beautiful results, however, are possible with thorough cleaning. Fine silver or copper rectangular *cloisonne* wires, in various sizes, can be bought ready made.

Technique

Traditional cloisonne In traditional *cloisonné* work the wires are annealed (see page 65) to make them pliable and then shaped along the lines of the design by manipulating them with the fingers, pliers or tweezers. Wires are then soldered to the base and the ends of the wires soldered together (soldering silver wire is covered in the Metal course on page 65). The silver peacock feather pendant (left) was made in this way.

The enamels are washed and laid wet into the *cloisons*. As enamel shrinks on firing, the cloisons are refilled and re-fired until the level of the enamel reaches the level of the *cloisonné* wires, or just

This exotic cloisonné dish and stand was made in China at the beginning of the twentieth century.

Pear-shaped pendant
You will need: Basic enamelling equipment, plus a stilt Pear-shaped copper blank 7.5cm (3in) overall length with a hole for a jump ring Fine silver rectangular *cloisonné* wire 1.5mm × .31mm (.06in × .012in) (If rectangular wire is not available use a round fine silver wire) Counter enamelling powder Gum tragacanth Round and square copper studs—various sizes Paintbrush Silver enamel flux Transparent enamel in chosen colours. Copper chain and jump ring Flat carborundum stone pencil Salt and vinegar pickle Water of Ayr stone (stick or pencil filed to a fine point) Tripoli and rouge (optional) Silver cleaning brush (soft nailbrush) Wire cutters Tweezers, jewelry pliers (optional) Polishing grade emery paper, grades 2/0, 3/0, and 4/0 Non-abrasive silver polish, which will not damage the enamel (optional).

above. The whole surface is then rubbed with a flat carborundum stone under running water until it is quite level. Refiring then gives a glossy surface.

Concave cloisonne The peacock feather pendant was finished by leaving the cells semi-filled. This method, known as concave *cloisonné*, gives added reflections which look most attractive on jewelry. It also means that fewer firings are necessary.

Pear-shaped pendant

A method of making *cloisonné* which does not involve soldering is used for making the pear-shaped pendant. Shaped silver wires are embedded into the first coat of enamel – silver grade transparent flux – thus avoiding soldering them in position. The pendant is made from a bought copper blank with silver wires forming the *cloisonné* work.

Switch on the kiln so that it is hot enough to use, ie bright orange, by the time you are ready to fire. Clean copper blank in salt and vinegar pickle (1 teaspoonful of common salt to half a cup of vinegar) for a few minutes and/or then rub both sides with emery paper. Swill under the tap and then dry with a clean cloth. Do not get finger grease on the surface.

Paint back of blank with gum and sift on counter enamelling powder. Pick up pendant carefully without disturbing the counter enamelling powder: slip a palette knife under the piece and pick up by the edges with your fingers. Turn over the piece very carefully and place it on the stilt. Dust silver flux on to the upper surface. With a brush, clear any enamel powder which may have fallen on to the stilt.

Place the stilt on the wire mesh stand on top of kiln to dry out the gum. When dry, fire the piece (on stilt and mesh), until the enamel is glossy. It is worth persevering with this method of firing counter enamel and surface base coat together as it saves work, time and electricity. Many enamellers use adhesives for all their work.

When cool, clean the firescale from the edges of the piece with a fine file or carborundum stone. Draw around the pendant on paper and create your design. Alternatively trace the design from the photograph. The lines of the design should be curved and flowing and never cross. Avoid sharp angles and narrow straight cells – they tend to crack the enamel.

Switch off, or unplug, the kiln and allow to cool until the chamber has a dull red glow. If kiln is too hot silver wire will melt when placed in it. Switch on kiln again and anneal the *cloisonné* wire to remove 'springiness' and make it pliable. Place wire on stilt and mesh before putting it in kiln. Do not overheat wire or it could melt, ie remove when it is the same dull red as the firing chamber. Shape the wire over your drawing using fingers, tweezers or fine jewelry pliers. The pieces of wire can be of any length but must have at least one curve before they will 'stand up': a straight piece of rectangular wire would fall flat. Cut wire with wire cutters. Transfer the shaped wires to the pendant and glue them carefully in place with the gum. Place pendant on stilt and mesh and then dry out gum by placing it on top of kiln.

Cool the kiln to dull red glow and then, when adhesive is dry, fire the piece until the wires just begin to sink into the coat of flux. Take great care not to overfire or the *cloisonné* wires will melt and appear to sink right down into the enamel, eventually firing out altogether, leaving a horrible mess. Keep the kiln at this low temperature, by switching off from time to time, for the rest of the firings. Clean the edges with the file when the piece is cool, and between subsequent firings.

Wash enamels carefully, especially the transparent ones. A good

Silver wire cloisonné on a copper blank. The wires are embedded in transparent flux enamel before enamelling the pendant.

method is to place some dry enamel powder in a deep spoon and run tap water very slowly on to the powder, stirring gently. The water will immediately become cloudy. Allow all the cloudy water to run away, leaving larger clean fragments of enamel behind. Keep washing until no more cloudiness appears. Examine the residue for any opaque white bits and remove them.

Lay in the colours wet, packing in as much enamel as possible. Use a small spatula or palette knife to pick up enamel and position it in the appropriate cell with the end of a paintbrush. If the enamel is too wet it will run where you don't want it to – if it is too dry, the enamel already in the cell will not 'accept' the enamel on the spatula. You will soon learn how wet the mixture should be.

This cloisonné and gilt box, in the form of a mythological beast, is about 15cm (6in) high and was made in China in the eighteenth century.

Place piece on stilt and mesh and dry out thoroughly on top of the kiln. Do not try to rush this drying process.
Fire carefully until the enamel has a shiny surface, withdrawing the pieces from the kiln at the first possible moment. When cool, clean edges of copper and then repack the cells with more enamel as the first lot of enamel will have shrunk considerably. Dry and re-fire. Repeat this process until the *cloisons* are full.
Place copper studs and opaque shot colours where required, sticking them in place with a little gum. Dry and then fire until they have just fused into the enamel surface.
When cool, clean the edges of the copper with a fine file. Then polish with successively fine polishing grades of emery paper if necessary. Descale and polish the copper studs using wet water of Ayr stone, which has been filed to a fine point. Clean the silver *cloisonné* wires in the same way. Clean each stud and wire separately taking great care not to scratch the enamel. Wash the piece with washing-up liquid and water to remove all traces of water of Ayr stone, using a soft silver cleaning brush. Once again take care that enamel is not scratched. Dry on a soft cloth. Polish with silver polish, then wash and dry as before. Alternatively, polish with tripoli and rouge. Attach the chain with a jump ring.

Polishing with tripoli and rouge

These two very fine abrasive materials are used to give a professional finish to silver or copper. They can be used for either hand polishing, as described here, or on an electric jewelry polishing machine. (An electric polisher will obviously speed up the process, but excellent results are possible with hand polishing). Wrap a piece of chamois round your finger or a stick and rub the chamois on a block of tripoli. Polish the piece of work by rubbing the chamois with tripoli on it over the piece. Then wash off the tripoli with washing-up liquid and water, using a silver cleaning brush. Dry with a soft cloth. Repeat the process with a block of rouge.
Always use tripoli first and then the finer abrasive, rouge – not the other way around. The blocks of abrasive tend to dry out in time. If they do put a little pure oil – any sort will do – on the block of tripoli before rubbing the chamois on it. Rouge can be moistened with paraffin in a similar way to tripoli if the block dries out.
After polishing you should always wash the item well with washing up [dishwashing] liquid and a soft silver cleaning brush. Metal parts can be polished with a non-abrasive metal polish. As you progress to using more expensive materials, perhaps to enamelling on gold and silver, careful cleaning at each stage and painstaking polishing and finishing will become more important.

The art of champlevé

Champlevé, or inlaid enamelling, is the technique in which depressions are hollowed out of the metal surface and these are then filled with enamel. For many hundreds of years *champlevé* was one

This champlevé hair ornament shows a good combination of exposed metal and enamel.

of the main techniques of the enameller's craft. It is not widely practised today, however, which is a pity because it is comparatively easy. Moreover, it is a technique which cannot be mistaken for anything other than what it is – enamelling on metal, ie both enamel and metal can be easily seen in the finished work. In this respect it is similar to its sister technique, *cloisonné*. *Champlevé* is, however, quicker to execute and is more suitable for large panels and plaques. It is also more durable than *cloisonné*. After the decline of the Byzantine *cloisonné* workshops the centre of enamelling moved to Western Europe where, for the next three hundred years (1200-1500AD), *champlevé* was almost the only type of enamelling produced. It was made in such large quantities that thousands of *champlevé* artifacts still exist and can be seen in all the world's prominent museums.

Metal for champleve

Copper is suitable for *champlevé* work. The copper used should be no thinner than 1.25mm (16-18 gauge). The copper must be thoroughly cleaned on both sides with steel wool and scouring powder. Once the metal has been cleaned take care not to get finger grease on it as this will stop the resist used on the areas of the metal not to be etched from adhering properly to the copper.

This double headed sea serpent pendant is another example of the beautiful results that can be obtained with champlevé.

Technique

In medieval times hammer, chisel and gouge were the tools used for hollowing out the metal. Today this hard work has been replaced by an acid bath. Nitric acid is most commonly used as the etching medium. It can be bought as a liquid concentrate. For very fine work however, use ferric chloride.

Nitric acid To make an acid bath using nitric acid, add one part nitric acid to three parts water. In an acid-proof container (plastic or glass) large enough to accommodate easily the piece to be etched, make up enough of the acid solution to cover the piece by about 1.5cm ($\frac{1}{2}$in). It takes about five hours for this nitric acid solution to eat down to half the thickness of 1.25mm (16-18 gauge) copper and correspondingly longer for thicker gauges.

WARNING Great care must be taken when dealing with nitric acid – if it touches the skin it will burn and it gives off poisonous fumes. When preparing an acid bath with nitric acid *always* pour the acid into the water. Never pour the water into the acid as this will cause dangerous splashing.

Always wear rubber gloves and, to avoid inhaling the poisonous fumes, *never* stoop over the acid bath longer than is strictly necessary. For the same reason, use a container with a lid for the acid

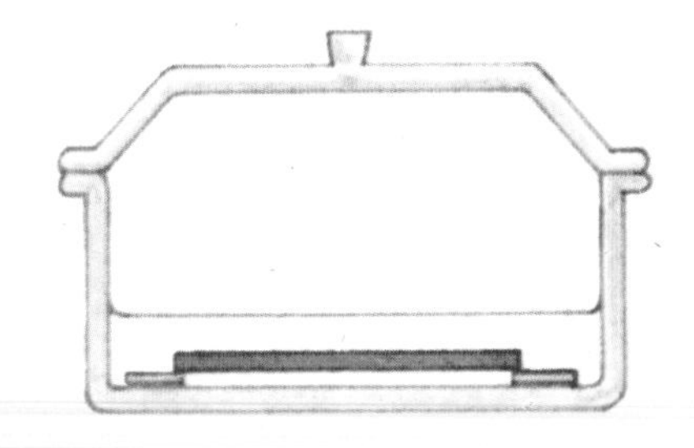

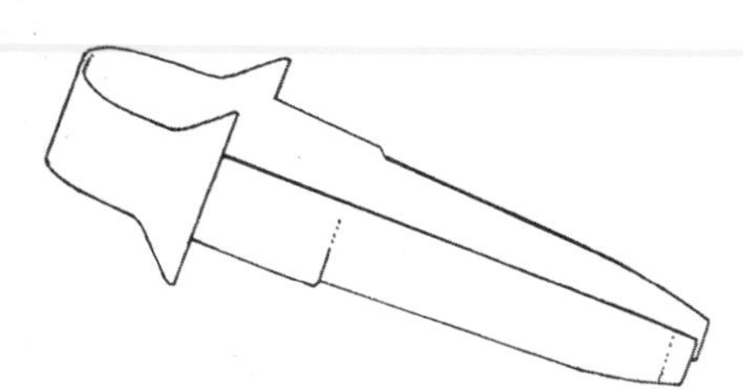

1 Headless matchsticks are positioned under the metal to allow the acid to flow underneath the piece which is to be etched in the acid bath.

2 Plastic forceps like these can be used to put the work into the acid bath and to lift it out.

bath, such as an old oven-proof glass casserole or any other suitable glass container. Keep well away from children and animals.
Work in a well-ventilated room or workshop.
Dispose of the acid bath, preferably when it has become weakened or 'tired' from etching copper. (Tired nitric acid becomes very blue in colour.) Turn on the cold tap and leave it running while you pour the acid bath slowly and carefully down the sink. Leave the tap to run for about ten minutes after pouring away acid. (This method is not suitable for disposing of large quantities of acid.)
Ferric chloride does not require such great care in handling as does nitric acid – if ferric chloride touches the skin it will merely stain, it does not give off poisonous fumes and does not splash. Ferric chloride is obtainable in lump form. To make an acid bath, the lumps should be crushed to a rough powder with a knife or spoon and mixed with an equal quantity of water. An old jam jar is useful for this. To be etched with ferric chloride, a piece of copper must be turned face down in the acid bath with matchsticks placed under it to keep the metal off the floor of the container (fig.1). It takes about seven or eight hours for ferric chloride to etch 1.25mm (16-18 gauge) copper.
The resist The areas of the metal which are not to be etched must be protected with a layer of acid-resistant paint or varnish – called a resist. The stopping-out varnish sold by craft suppliers for etching is a suitable resist. Alternatively the bitumen paint used to paint outdoor ironwork is quite satisfactory. As the acid will attack any exposed metal, no matter how small the area, you must be very careful to cover with the resist all parts of the metal you have planned to leave exposed in the completed work. You must also protect the edges and the back of the piece. This can be done by painting on melted candle wax.

The blue dish

Although the instructions given here are for making the dish in the photograph the method can be applied to such items as copper panels and pieces of jewelry.
Cut the tracing paper and carbon paper to fit the dish. Remove paper from dish and clean copper thoroughly with steel wool and scouring powder. Then rinse and dry with soft cloth.
Trace off design from the photograph. Place carbon paper and tracing back on to dish and, with a hard pencil, transfer the design on to the metal.
Paint the resist on to all areas which are to be left exposed in the completed work. Take care not to get any finger grease on the metal while transferring the design or painting on the resist or the resist

The finished dish, shown full size. Use this as a trace pattern for the design.

will not adhere satisfactorily to the copper. Leave the resist to dry completely if using stopping-out varnish or until almost dry if using bitumen paint (about 2 hours). Have a final look at the design and tidy it up if necessary with a craft knife.

Protect the edge and back of the dish by painting on melted wax. Allow wax and resist to dry. Put on rubber gloves and prepare the acid bath, remembering to add the acid to the water and not vice versa to avoid the risk of dangerous splashing.

Place dish gently in the acid bath, right way up, using plastic forceps. Leave the acid to work, checking occasionally to see how the etching is progressing and to give the acid a gentle stir with a feather or spoon. After about three hours the acid solution will have lost some of its strength. Make up a new acid bath in the same way as before and then, using plastic forceps, transfer dish to it. When the acid has eaten down to half the thickness of the copper (in about five hours), remove dish from the bath with plastic forceps. Retain the acid solution. Wash the dish under hot running water. This will also remove the wax. Remove the resist by scraping it off with an old knife, while still under the

Blue dish

You will need:
3 acid-proof containers with lids (plastic or glass)
Plastic tongs, tweezers or the plastic print forceps sold by photographic suppliers (fig.2)
Craft knife
Rubber gloves
Feather or wooden spoon
Old knife
Old nail brush
Enamelling equipment plus a stilt
1.25mm (16-18 gauge) almost flat copper dish, 10cm (4in) in diameter
Nitric acid
Resist, as previously described
Melted candle wax
1 teaspoonful gum arabic mixed with 0. 28 litres ($\frac{1}{2}$ pint) distilled water
Brushes for applying resist, candle wax and gum arabic solution when counter enamelling
Turpentine
Opaque enamels in chosen colours
Counter enamelling powder
Carborundum stone
Steel wool and scouring powder
Fine steel wool and washing-up [dishwashing] liquid
Tracing paper, hard pencil and carbon paper
Metal polish or the clear lacquer sold by art suppliers for lacquering oil paintings
Soft cloths

hot, running water. The resist will have been softened by the acid and should come off fairly easily. Remove last traces of resist with turpentine and a soft cloth. Clean the piece again under hot, running water, using scouring powder and steel wool. Rinse throughly and dry with a cloth. The piece can now be enamelled. The design originally drawn on the metal with the resist will appear in relief. Switch on the kiln so that it will be hot enough, ie bright orange, by the time you are ready to enamel.

Paint the back of the dish with gum arabic solution and sift on counter enamelling powder. Pick up the piece by the edges, without disturbing the counter enamelling powder. Turn the dish right way up and place it carefully on the stilt. Fill the cells on the right side of the dish with enamel, leaving the relief parts to serve as walls to contain the seas of colour. Use the wet pack method, ie mix enamel to a paste with gum arabic solution. Be sure to pack plenty of enamel inside the cells for this layer of enamel will sink to about half its depth when fired, necessitating another layer of enamel of the same colour.

Place stilt on the wire mesh stand and dry out the enamel thoroughly on top of the kiln before firing. Fire the piece. Refill the cells, dry as before and re-fire. After this second firing, you will probably find that the enamel projects a little above the metal design. The work must be stoned down with carborundum until enamel and metal are level.

Wash the work under the tap, using a nail brush to remove any traces of grit left from the stoning process. Re-fire to bring back the sheen to the enamels. The exposed copper parts of the design will now be black with fire-scale.

In another acid-proof container make up a solution by adding one part of the acid bath left over from the etching process to ten parts of water. (Tired acid becomes very blue in colour.) To loosen the firescale from the exposed copper on the dish, place the work in this weak acid bath. (A weak acid bath is used so that the enamel does not get attacked by too strong acid.) Leave dish in the weak acid bath for between half an hour and an hour (checking after half an hour to see if firescale has loosened sufficiently). Then remove dish with plastic forceps and wash carefully under a running tap, using very fine steel wool and washing-up liquid. Do not use scouring powder or a rough grade of steel wool which would damage the surface of the enamel. Polish the piece all over with metal polish and a soft cloth. Alternatively, lacquer over the metal parts only to prevent the copper from tarnishing.

This is the basic method for all champlevé work. To economize on time you could etch several items at once.

Enamelling on silver

Enamelling on silver combines the highest skills of both metal and enamel work to create beautiful items which are works of craftsmanship and art.

Fine (pure) silver, Brittania silver (an alloy of 958 parts silver and 42 parts copper) or sterling silver (an alloy of 925 parts silver and 75 parts copper) can be used for enamelling, but sterling silver, because it is not pure silver, develops dark grey patches, called fire stain, when it is heated in the kiln. These patches have to be rubbed off with wet and dry paper – a tedious, time-consuming task. Brittania silver is less liable to fire stain than sterling silver.

Enamelling on silver is different from enamelling on copper as silver is a softer metal. Because of this, the metal is sometimes inclined to warp in the kiln, causing the enamels to crack. Also it is a shame to completely cover a beautiful metal such as silver with enamel, so designs usually feature exposed silver.

A major rule concerning enamelling on silver is that the edge of the enamel must be protected by a border or rim of metal soldered into place. This border prevents the enamel from chipping away at the edge and helps to prevent the silver warping. If the border also works as part of the design, so much the better. The jewelry in the photographs illustrates the use of a practical and decorative border. The most practical border is one which is fairly even in width. Larger pieces will also require a border on the back for extra support.

Tools and materials

The solder When soldering silver which is to be enamelled later and then exposed to a high temperature in the kiln, use hallmarking quality silver, enamelling solder or hard silver solder. These have a high melting point so they should not run when the enamel is fired. Enamelling solder is the best to use as it has the higher melting point of the two. (The melting point of enamelling silver solder is 730°C-800°C (1346°F-1472°F) and the melting point of hard silver solder is 745°C-778°C (1373°F-1432°F).

Soldering flux This is a substance that prepares the metal surfaces for joining. It also helps the molten solder to flow smoothly. Borax is usually used as the flux for silver enamelling solder and hard silver solder. It is a white powder which is mixed with water to the consistency of cream and painted on to both surfaces to be joined by the solder. Avoid getting borax near the eyes. As an alternative, you can use one of the special-purpose commercial products.

Pickle When the piece has been soldered it must be placed in a pickle of one part concentrated sulphuric acid to eight parts water in order to remove all traces of borax before the work can be enamelled. Concentrated sulphuric acid can be obtained from chemists. The chemist may do the diluting for you but if you have to do it yourself *always* add acid to water and not the other way round – this is most important when dealing with sulphuric acid. Treat this pickle with great respect and take the same precautions when handling it as described for handling nitric acid (see page 127). When putting the silver into the pickle, or when taking it out, hold it in plastic, copper or brass tweezers, not iron or steel tongs. Iron and steel will ruin the pickle and any silver subsequently

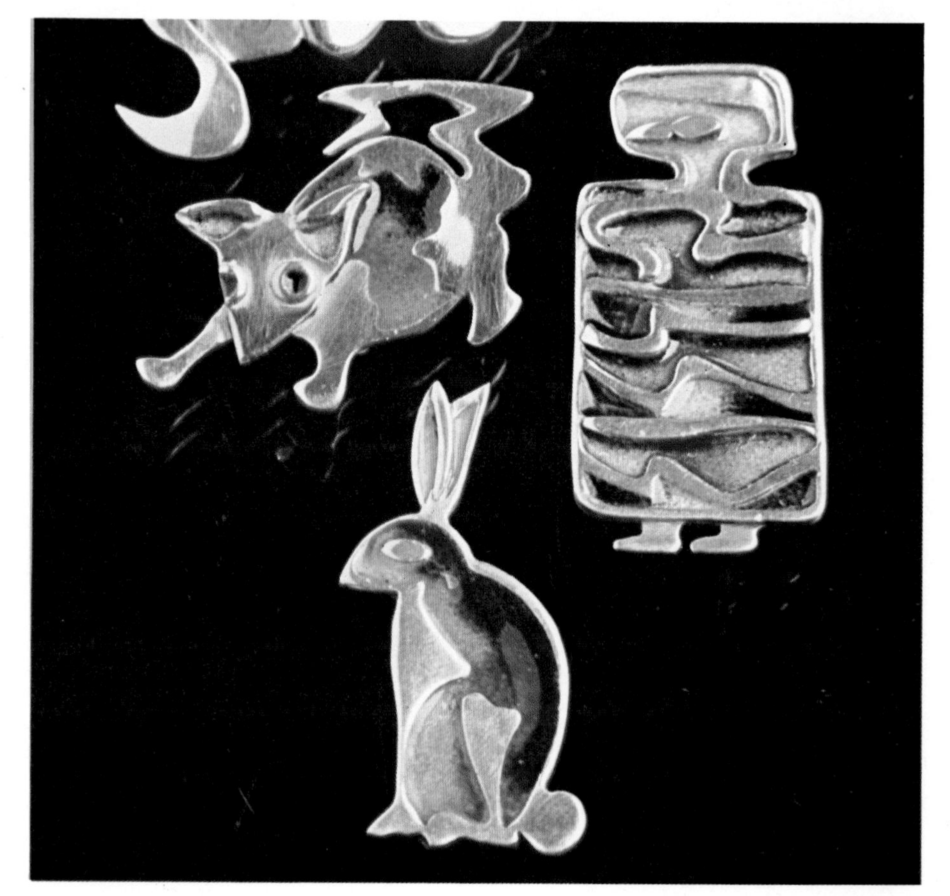

These delightful brooches were all made up and enamelled in the way described here.

pickled will come out dull grey and have a most unpleasant texture. Rinse the silver under running water.

The enamels used on silver are usually the transparent ones which allow the metal to be seen through them (see page 85). The enamels are applied in thin layers, each layer being fired separately, starting with a silver grade transparent flux. (By applying the enamels in thin layers you can obtain subtle shadings in the colouring.) After the transparent flux, it is best to put on the light colours first and then the darker colours in successive layers.

If you prefer, you can use just one colour on a piece. In this case first apply and fire a thin layer of transparent flux and then fire successive thin layers of one colour. Alternatively, after the flux, fill the whole area within the border with one colour, topping up as necessary in the way described for *cloisonné*. If desired, the enamel can be left below the level of the border.

A small brooch

The instructions here are for the rabbit brooch in the photograph but you can adapt them to any small design.

Draw out the shape of the border and rabbit's eye on paper and stick the paper to the silver. Cut out the shapes with the piercing saw and then wash off the paper template.

Place the eye face downwards on the base of the stilt. Place the stilt on a fire brick. Paint eye with flux and place a snippet of solder on top of it. (It is best to flatten the solder with a hammer first so that fine snippets can be cut off easily). Then, using the blowtorch with a low flame, melt the solder on to the eye. Any wide areas of the border, in this case the rabbit's thigh, should also be treated in this way.

The next step is to solder the border to the remaining piece of silver which will form the base. Place the border on the remaining piece of silver and, with a pencil, draw round the inside and outside edges of the border. Paint the underside of the border and the area within the pencil lines on the base with flux. Do not use too much flux and do not take flux right up to the inside line to avoid solder running on to exposed area of base. Lay the border right side up, inside the drawn outline, on the piece of silver which will form the base. Cut snippets of solder as before and lay these along the *outside* edge of the border (fig.2). Lay in any extra details, in this case the eye, right side up, having first painted a spot of flux on to the back.

Make up pickle in glass bowl, by adding one part acid to eight parts water. (Do not add water to acid as this will cause dangerous splashings). Pick up the base piece of silver with the border on top.

1 Glass brush 2 Solder laid outside border 3 Silver heated from underneath.

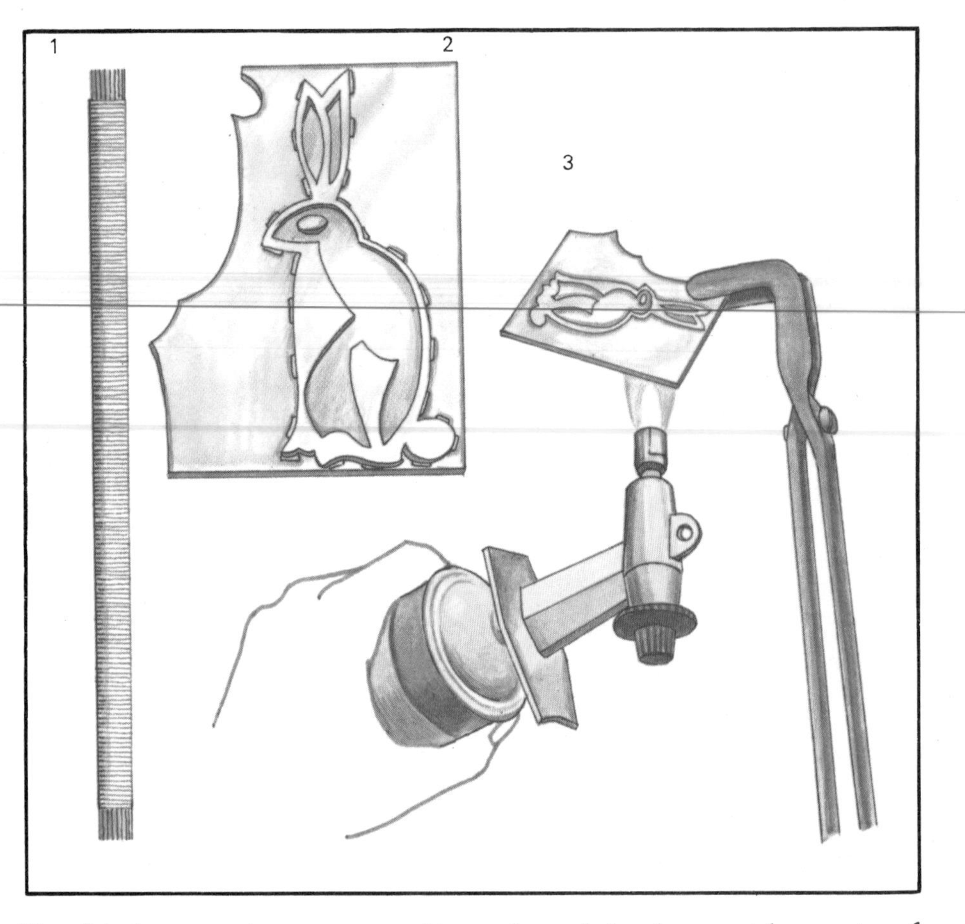

Small brooch

You will need:
Equipment as for cutting copper with a piercing saw (see page 23)
Enamelling equipment plus a stilt
Piece of .71mm (21-22 gauge) fine silver, more than twice the size of the finished brooch
Acid-resistant vessel with a lid for pickle, such as old oven-proof glass casserole
Metal tongs, glue, paper and pencil
Plastic, copper or brass tweezers, or plastic forceps
Blowtorch with adjustable flame
Fire brick
Sulphuric acid (handle with great care)
Silver brooch fastening
Borax or other suitable flux
Soap-impregnated steel wool
Silver grade, transparent flux enamel powder
Transparent enamel powder in chosen colours
Wet and dry paper—from finest grade (800) up to 600 grade
Polishing grade emery paper 2/0, 3/0 and 4/0 (if available)
Non-abrasive silver polish (optional)
Tripoli and rouge (optional)
Silver cleaning brush or soft nail brush
Glass brush (fig.1) sold by jewellers' suppliers (if available)
Electric jewelry polishing machine (optional)
Paintbrush
Absorbent kitchen paper
Two pieces of chamois (optional)
Soft cloths for polishing

Do this by grasping a protruding edge of the base with a pair of tongs. Then gently heat the whole area of silver from underneath with the low flame of the blowtorch (fig.3). Gradually increase the heat until all the silver begins to glow pink and the solder runs. You will find that the solder underneath the eye and the thigh will melt and stick it to the base (this process is called 'tacking').
When the solder has run, allow the brooch to cool down naturally until it is quite cold. Then holding it with the tweezers or forceps (not the tongs) put it into the pickle to dissolve any remaining flux. Leave for about 10 to 15 minutes. By cooling the brooch before putting it in the pickle you avoid splashing. Rinse the brooch thoroughly under running water. Dry with a soft cloth.
Using the piercing saw in the usual way, cut away all the silver surrounding the border. Finish the edges with the needle files. You now have a silver shape with a relief border and detail. Solder a silver brooch fastening on to the back of the brooch, minus the pin. (Pin is fitted after enamelling).
Switch on the kiln and bring it up to normal firing temperature, ie

the firing chamber should be bright orange. It is a good idea at this stage to put the silver shape in the hot kiln and give it a 'dry run' firing, ie without any enamel. This is to make sure that no flux remains in the cracks. If there is a residue of flux undissolved by the pickle it will run out of the cracks and turn black during the firing. If such a residue is visible after the brooch is taken out of the kiln, cool the brooch, put it back into the pickle, rinse and dry it and do another dry run. Continue this process until no more flux appears.

You are now ready to enamel. The colours are applied in several thin layers starting with colourless, silver grade transparent flux. Wash the flux enamel powder carefully. Then, with a small paintbrush, place a thin layer of it inside the border on the front of the brooch. The water will help to spread the enamel in an even layer. The layer should be as thin as possible without letting the silver show through. If preferred, the enamel can be applied with a pen nib or the quill end of a feather cut off diagonally.

Draw off the excess water with a piece of absorbent paper. Place the brooch on a clean stilt, so that it rests on its fastening, and dry out on top of the kiln. When the powder has dried, check to see that there are no grains of enamel on exposed silver border. If there are any grains on the silver, brush them off. Fire the piece.

Reverse the brooch and fire a coat of enamel on the back around the fastening, having first washed the powder.

Turn the brooch right side up once more and lay in the lightest colour if several are being used. Continue adding successive layers of enamel (in successively darker colours if desired) until the enamel has reached the required depth. By now the silver will be greyish with black solder showing around the edges. Clean it first under a running tap with soap-impregnated steel wool. If the enamel has got on the silver border, stone it off carefully under running water with a carborundum stone. Wash off all traces of carborundum under running water using a soft silver cleaning brush or soft nail brush.

Dry the brooch on a soft cloth. Then polish the silver with successively fine polishing grades of emery paper, if available. Rinse away any traces of emery from the brooch, using a soft brush as before, and dry it between each grade used. If there are any deep scratches on the silver use wet water of Ayr stone to smooth the metal. Rinse, using a soft brush as before, and dry. You will probably find that the enamel has been scratched during the cleaning of the silver. If so, give the piece one more quick firing to bring back the gloss to the enamel. Do not over-fire. Then polish with tripoli and rouge by hand or with a polisher.

Index

Photographers:

Steve Bicknell-6; 10; 8TL; 84; 114; 118; 111;
Camera Press-14; 17; 19L;
Stuart Dalby-56;
Alan Duns-53T; 108; 110;
Louis Fordaan-29;
Geoffrey Frosh-50/1;
Horatio Goni-106; 126; 127;
Melvin Grey-5; 31L; 60; 64; 79; 100/1; 129;
Peter Heinz-13T; 38; 46; 104; 40;
Paul Kemp-71; 123;
David Levin-72;
Chris Lewis-11; 55; 119;
Sandra Lousada-68T; 120
Dick Miller-cover; 22B; 26BL; R; 27L, TR, BR; 28; 31R, 33; 34; 32; 48; 89; 75; 90; 91;
Keith Morris-87;
Coral Mula-128;
Alasdair Ogilvie-2; 12; 35; 36; 98; 102;
Jill Paul-68B;
Roger Philips-86;
Ruth Rutter-78;
Kim Sayer-112; 132;
Rodney Todd-White and Son-76; 77;
Jerry Tubby-21; 58/9;
Mike van der Vord-53B;
Victoria and Albert Crown Copyright-121; 124;

Artwork:

Victoria Drew-8TR, B; 9;
Barbara Firth-116; 117;
Trevor Lawrence-16; 18; 62; 82; 83;
Paul Williams-20-; 22; 24; 25; 26; 37; 41; 42; 43; 47; 51; 52; 57; 63; 66; 67; 73; 96/7 103; 104;